ALLE ZEIT WACH 1842

C.L. Campbell D.M. Benson (Eds.)

Epidemiology and Management of Root Diseases

With 26 Figures and 18 Tables

Springer-Verlag
Berlin Heidelberg New York
London Paris Tokyo
Hong Kong Barcelona
Budapest

C. Lee Campbell, Professor, BS, MS, PhD
D. Michael Benson, Professor, BS, MS, PhD

Department of Plant Pathology
Box 7616
North Carolina State University
Raleigh, NC 27695-7616
USA

ISBN-13: 978-3-642-85065-3 e-ISBN-13: 978-3-642-85063-9

DOI: 10.1007/978-3-642-85063-9

Library of Congress Cataloging-in-Publication Data. Epidemiology and management of root diseases / C.L. Campbell, D.M. Benson (eds.). p. cm. Includes bibliographical references and index. ISBN 0-387-57579-0. – ISBN 3-540-57579-0. 1. Roots (Botany) – Diseases and pests. 2. Soilborne plant, diseases. 3. Soilborne plant pathogens – Control. I. Campbell. C. L. (C. Lee), 1953– . II. Benson, D.M. (David Michael), 1945– . SB732.87.E65 1994 632′.3–dc20 94-14991

Softcover reprint of the hardcover 1st edition 1994

Typesetting: Best-set Typesetter Ltd., Hong Kong

SPIN 10021707 39/3130/SPS – 5 4 3 2 1 0 – Printed on acid-free paper

We dedicate this book to
Dr. Ralph Baker
Pioneer, original thinker, mentor, innovator,
truth seeker.
Exemplum tuum nos moderetur.

Preface

Root diseases continue to pose special challenges to growers and to plant pathologists. Whereas the symptoms of most foliar diseases are apparent almost immediately after infection has occurred, those of root diseases remain hidden beneath the soil line. Propagules of soilborne pathogens reside in the soil and lie in wait for the passing of an unsuspecting root or, in some cases, may actively seek it out. Only when the "deed is done" do the above ground symptoms appear. To the grower the challenge is to anticipate when and where the root disease will strike, because to be effective the management practices must usually be in place before the disease occurs. To the plant pathologist, and the epidemiologist in particular, the challenge is to quantify the propagules of the causal pathogens, which are bound in the soil matrix, and to characterize the various components of the epidemics of these hidden diseases without destroying the crop of interest in the process.

Perhaps it is because of these challenges that less has been written about epidemics of root diseases than about epidemics of foliar diseases. Nevertheless, we felt that the science and practice of root disease epidemiology and management had progressed sufficiently to warrant a comprehensive examination. Thus, we sought to assemble as authors a group of researchers who had experienced first hand the dilemmas and opportunities of working with root disease epidemics and their management. We challenged the authors to explore not only what was known in their area but also what needed to be known, to tell us not just about problems but also about potential solutions.

The exploration begins with a critical examination of the components that make up the epidemic. In the first four chapters the authors consider the interacting factors that give rise to and participate in the epidemic – inoculum, host roots, mycorrhizae, and the soil environment. In the middle three chapters the authors deal with the integrated aspects of the actual epidemics – assessment of root disease intensity and the temporal and spatial aspects of epidemic development. In the final four chapters the authors address management issues associated with host resistance, chemical management, biological management, and cultural management of root diseases.

We want to thank the authors of this book who took time from their already busy schedules and answered our challenge. Through their responsiveness and diligence they made our task as editors an enjoyable one. We also want to thank Joyce Denmark for her assistance with the index. A special note of thanks is due our wives, Karen and Pat, and our families for their patience and tolerance as we spent time at home on editorial matters. We want to thank our editor Jürgen Kranz for his guidance and encouragement throughout the project as well as the editors and staff at Springer-Verlag for their patience as we moved beyond our initial deadline to the subsequent completion of the book. Their excellent assistance in making this book a reality is greatly appreciated.

Raleigh, North Carolina
Spring 1994

C. Lee Campbell
D. Michael Benson

Contents

List of Contributors

ALTEN, H. von, Inst. für Pflanzenkrankheiten und Pflanzenschutz, Universität Hannover, Herrenhäuser Str. 2, 30419 Hannover 21, Germany

BENSON, D.M., Department of Plant Pathology, North Carolina State University, Raleigh, North Carolina 27695-7616, USA

CAMPBELL, C.L., Department of Plant Pathology, North Carolina State University, Raleigh, North Carolina 27695-7616, USA

ENGELKES, C.A., Biocontrol Plant Disease Lab, Bldg. 011A, Room 275 BARC-W, Beltsville, Maryland 20705-2350, USA

ENGLISH, J.T., Department of Plant Pathology, University of Missouri, Columbia, Missouri 65211, USA

FRAVEL, D.R., Biocontrol Plant Disease Lab, Bldg. 011A, Room 275 BARC-W, Beltsville, Maryland 20705-2350, USA

GILLIGAN, C.A., Department of Plant Sciences, University of Cambridge, Downing Street, Cambridge, CB2 3EA, UK

GRUNEWALDT-STÖCKER, G., Inst. für Pflanzenkrankheiten und Pflanzenschutz, Universität Hannover, Herrenhäuser Str. 2, 30419 Hannover, Germany

MAC DONALD, J.D., Department of Plant Pathology, University of California, Davis, California 95616, USA

MITCHELL, D.J., Department of Plant Pathology, University of Florida, Gainesville, Florida 32611, USA

MORTON, H.V., CIBA-GEIGY Corporation, P.O. Box 18300, Greensboro, North Carolina 27419, USA

NEHER, D.A., Department of Plant Pathology North Carolina State University, Raleigh, North Carolina 27695-7616, USA

SCHÖNBECK, F., Inst. für Pflanzenkrankheiten und Pflanzenschutz, Universität Hannover, Herrenhäuser Str. 2, 30419 Hannover, Germany

Shew, B.B., Department of Plant Pathology, North Carolina State University, Raleigh, North Carolina 27695-7616, USA

Shew, H.D., Department of Plant Pathology, North Carolina State University Raleigh, North Carolina 27695-7616, USA

Sumner, D.R., Department of Plant Pathology, University of Georgia, P.O. Box 748, Tifton, Georgia 31793-0748, USA

1 Inoculum

D.M. Benson

1.1 Role of Inoculum in Root Disease Epidemics

In root disease epidemics, inoculum is but one part of the disease triangle with host and environment. Although inoculum is a critical factor in the development of root disease epidemics, the role of a favorable environment and presence of a host also contribute to root disease epidemics. Often the inoculum and host are present but the environment is unfavorable and hence no epidemic develops. Other times, a host is present in a favorable environment, but inoculum is not present in the infection court. In addition to host, and inoculum, the role of soil physical factors such as pH, soil texture, and bulk density can have a great influence on the extent and severity of root disease epidemics.

1.1.1 Concepts of Inoculum

Several concepts involving inoculum including soil fungistasis, inoculum potential, inoculum efficiency, and inoculum density need characterization prior to an analysis of inoculum-root disease relationships. Inoculum of soilborne pathogens should be viewed in terms of a dynamically changing state within the soil environment. Inoculum surviving in the form of pathogen propagules in soil is dynamic in the sense that the energy status of an individual propagule changes over time. For many propagules the lack of exogenous nutrients prevents germination. In combination with inhibitory substances, such as ammonia, a natural inhibition of propagule germination known as soil fungistasis provides a biological mechanism that ensures successful infection of soilborne propagules. Inoculum potential has been suggested as a way to conceptualize changes in efficiency (energy) of propagules to cause disease. Over time, the inoculum efficiency of propagules subject to soil fungistasis may be increasing, decreasing, or unchanged depending on environmental fluctuations and nutrient levels in soil but these changes in inoculum efficiency are not manifested until a host is present. Inoculum efficiency is a measure of the success of a propagule to incite an infection; over a population of propagules, inoculum efficiency is the per-

C.L. Campbell/D.M. Benson (Eds.)
Epidemiology and Management of Root Diseases

centage of propagules that infect. When energy levels in propagules fall below that needed to maintain vital metabolism and respiration, propagule death may follow. Inoculum density, that is, the number of propagules per unit volume of soil is a practical means of tracking changes in propagule numbers over time. The rate of decrease or the survival of inoculum is dependent on soil physical, chemical, and biological factors, nutrients levels, and environmental effects. Each of these factors as related to inoculum will be discussed below.

1.1.2 Soil Fungistasis

Propagules of soilborne plant pathogens surviving in soil are normally incapable of germination and growth unless the soil fungistatic factor is overcome. Dobbs and Hinson (1953) first described this widespread soil fungistatic factor extant in a variety of soils including forest soils, grassland soils, cultivated garden soils, waterlogged soils, and subsoils. The extent of germination inhibition of fungal spores decreased with increased depth of the soil sample in a profile. Although leaching soils with water did not overcome the inhibitory effect, autoclaving soils did which suggested a biological origin for fungistasis. In addition to fungal spores, Ho and Ko (1982) suggest that bacteria and actinomycetes are also sensitive to a soil "stasis" and coined the term "microbiostasis". Even though the term "soil fungistasis" may not accommodate all microorganisms affected by the factor, the term is so completely entrenched in present-day literature as to make it the term of choice. Soil fungistasis serves as a mechanism whereby inoculum of soilborne plant pathogens is able to survive in soil in the absence of a host until such time as conditions become favorable for host infection.

Understanding the concept of soil fungistasis is critical to understanding the survival, and ecology, of soilborne pathogens and the epidemiology of root diseases. Once the significance of the soil fungistatic factor was appreciated in the scientific community, a great deal of research was begun to describe the factors affecting fungistasis. The ultimate aim of the research was to find ways of altering the soil environment in order to induce germination and subsequent death of propagules in the absence of host tissue.

Research by J.L. Lockwood and colleagues at Michigan State University in the USA advanced a "nutrient sink" hypothesis to explain soil fungistasis. This hypothesis was based in part on the recognition of two types of fungal propagules in soil, i.e., carbon-dependent fungi such as conidia of *Thielaviopsis basicola* which require an exogenous source of nutrients for germination, and carbon-independent fungi such as of *Cochliobolus sativus* which do not require an exogenous source for germination of conidia (Lingappa and Lockwood 1964; Ko and Lockwood 1967). In a "nutrient sink" environment endogenous nutrients are constantly leached from carbon-independent spores by the activity of surrounding microorganisms. Indeed, through a series of experiments in which a leaching system with Millipore filters over a sand base was used, Ko and Lockwood (1967) and Hsu and Lockwood (1973) demonstrated that germination of carbon-independent spores was inhibited by the simulated nutrient sink. Lockwood (1975) found that loss of glucose from filter disks placed on natural or autoclaved soil was faster from natural soil than from autoclaved soil, suggesting the importance of carbon utilization by soil microorganisms in the natural soil.

Lockwood's nutrient sink hypothesis does not differentiate among soil microorganisms on, adjacent to, or some distance (microscopic) away from the fungal spore, but rather, indicates only that a gradient of nutrient diffusion away from the spore is established. Fradkin and Patrick (1985) devised a series of experiments in which conidia of *Cochliobolus sativus* were separated from surrounding soil by polycarbonate membranes of either 10- or 0.2-μm pore size. The 0.2-μm pore size excluded contact between soil bacteria and the conidia. With the 10-μm membranes, bacteria colonized the surface of conidia inside the membrane, but the bacteria population outside the membrane was not increased over controls. With conidia in 0.2-μm membranes, no bacteria were observed on the conidia and bacteria populations outside the membrane were not enhanced. When glass fibers containing 0.09 g dextrose were placed in the 0.2-μm membranes bacterial growth occurred on the exterior surface of the membrane and bacteria populations in the soil outside the membrane also increased. This was not the case with conidia in 0.2-μm membranes.

The experimental results of Fradkin and Patrick (1985) suggest that the concentration of fungal spore exudates is sufficient for bacterial growth only at the fungal spore surface but not beyond. Bacteria populations in soil separated from the conidia by only the 0.2-μm membrane did not increase. Thus, the diffusion gradient, i.e., the nutrient sink, does not extend beyond the spore surface. However, nutrient deprivation does not account totally for soil fungistasis, because conidia colonized by bacteria and buried between membranes in soil did germinate when moved to a nutrient-free medium. In other experiments, Fradkin and Patrick (1985) were able to demonstrate that the bacteria populations in soil produce inhibitory substances that restricted conidial germination and hyphal elongation from those conidia that did germinate.

Lockwood and his colleagues developed the nutrient sink hypothesis only after unsuccessful attempts to isolate inhibitory substances from soils known to be fungistatic. In other labs, investigators continued to look for inhibitory substances responsible for soil fungistasis. One such group was led by Ralph Baker at Colorado State University. In the alkaline soils typical of Colorado and other western states of the USA, a volatile fungistatic factor was characterized (Hora and Baker 1970, 1972a,b, 1974; Ko and Hora 1972; Romine and Baker 1972). Properties of the fungistatic factor included volatility at ambient temperature, water extractable from soils, emanation over extended periods, absorption by activated charcoal, decreased activity with soil depth, increased concentration of the factor with increasing soil alkalinity, and production of the inhibitor by soil microorganisms, but nullification with nutrients (Pavlica et al. 1978). The fungistatic factor was also produced abiotically by liming alkaline soil (Hora and Baker 1974). Acidic soils typical of Michigan, on the other hand, do not produce inhibitory volatile factors (Bristow and Lockwood 1975).

Several workers (Ko et al. 1974; Pavlica et al. 1978; Schippers and Palm 1973) have independently demonstrated that ammonia is one volatile responsible for the fungistatic effect. Ammonia may be generated in soil when the concentration of the ammonium ions in soil solution increases and exceeds the partial pressure of ammonium ion, thus liberating ammonia into the soil atmosphere. Ammonia, however, is not inhibitory to all fungi, notably *Fusarium solani* f. sp. *phaseoli*, in fact, ammonia at 1 to 100 μg/g actually stimulated germ tube elongation from germinating conidia (Pavlica et al. 1978). Other volatile compounds in addition to ammonia with the notable exception of ethylene may also be involved in soil fungistasis (Pavlica et al. 1978).

Watson and Ford (1972) proposed that soil fungistasis was caused by a complex balance of inhibitors and stimulators in soil and to investigate only inhibitors or stimulators (primarily nutrients) failed to adequately characterize the phenomenon. Changes in concentrations of inhibitors and stimulators affected the fungistatic balance in soil and resulted in induction, maintenance, or release from soil fungistasis. Indeed, Hora et al. (1977) evaluated the nutrient sink hypothesis and the volatile fungistatic factor in experiments with *Gonatobotrys simplex*, *Penicillium chrysogenum*, *Verticillium albo-atrum*, and *Fusarium solani*. The selection of these fungi provided the investigators with spores that were nutrient-independent (*G. simplex*), nutrient-dependent (*V. albo-atrum*), or intermediate in response to nutrients. Based on stimulation or inhibition of conidial germination, the investigators concluded that both inhibitory substances and nutritional factors in soil best explained soil fungistasis.

As the effect of inoculum in root disease epidemics is described further, the reader should keep in mind the role that soil fungistasis plays in limiting the germination and growth of fungal spores and other microbes in the absence of suitable substrates.

1.1.3 Inoculum Potential

Garrett (1956) defined inoculum potential as the energy of growth of a pathogen available for infection of a host at the surface of the host to be infected. However, Horsfall (1932) had already defined inoculum potential as the number of infective particles present in the environment of the uninfected host. Thus, subsequent authors have used inoculum potential in the abstract or as a measure of inoculum density. This led to confusion until Horsfall and Dimond (1963) redefined inoculum potential as an abstract concept that involves the interaction between the amount of inoculum and its virulence, the amount of available host tissue and its susceptibility plus the effect of environment on each. Baker (1965) attempted to improve the definition of inoculum potential by using Garrett's idea of inoculum potential to mean the interaction of inoculum density with capacity factors, i.e., the environment, and Grainger's (1956) idea of disease potential to describe the interaction of host susceptibility and disease proneness, i.e., the environment. Thus, Baker defined disease incidence as the interaction of inoculum potential and disease potential. Vanderplank (1975) argued that inoculum potential cannot be quantified because inoculum potential determines disease and disease measures inoculum potential. Baker (1978) later expressed inoculum potential in mathematical terms as

$$\log s = m(\log x + \log v + \log n + \log f),$$

where s is the number of successful infections, m is the slope of the inoculum density-infection curve, x is the inoculum density, v is pathogen virulence, n is the nutritional status of the propagules, and f is the effect of environmental influences on germination and penetration efficiency. From an experimental standpoint, however, absolute measurement of degree of virulence, nutritional status, and environmental influences on germination and penetration are difficult if not impossible given our current abilities to study soilborne plant pathogens. Bouhot (1979) and Mitchell (1979) defined inoculum potential in terms of the number of successful infections obtained under optimum environmental conditions on a standard susceptible host. Such a definition illustrates why Vanderplank (1975) argued against the usefulness of the concept.

Inoculum potential can be useful for conceptualizing ideas about the behavior of inoculum in soil prior to disease development. For a specific pathogen and host, inoculum potential should be thought of in hindsight rather than foresight. That is, differences in disease development for a specific pathogen can be expressed as changes in inoculum potential, and rather than quantitatively measuring inoculum potential, one should think in terms of relative increases or decreases in inoculum potential to explain resulting disease development. In the absolute sense, inoculum potential can never be assessed because once the attempt is made the assessment is no longer measuring "potential", but now is estimating inoculum "efficiency". Inoculum efficiency is the action form of inoculum potential and varies with nutrient status of the propagule, environment, genetic factors such as aggressiveness, and host.

1.1.4 Inoculum Efficiency

In terms of a population of propagules, inoculum efficiency is the percentage of propagules that are successful in initiating an infection. On the basis of an individual propagule, inoculum efficiency is the probability that a single propagule will cause an infection. Form of inoculum, its nutritional status, distance from infection court, and environmental conditions affect inoculum efficiency. Inoculum efficiency has been estimated for several soilborne plant pathogens with several different methods with regard to successful infections. Methods that assess individual infections may give the most meaningful estimates of inoculum efficiency, but other methods that estimate successful infections from the multiple infection correction of Gregory (1948) provide a means of estimating inoculum efficiency as well. Although inoculum efficiency is defined as the proportion of propagules that initiate an infection, a better definition could be based on the number of host infection sites available for infection versus the number actually infected and the proportion of propagules participating in the infection.

Bowers and Mitchell (1991) used oospore inoculum of the heterothallic fungus, *Phytophthora capsici* to infest a microwaved field soil which was planted subsequently to pepper in growth chambers. When oospores from 16 different pairings of isolates of *P. capsici* were used as inoculum, mortality ranged from 5 to 35% for one oospore per gram to 100% mortality at 100 oospores per gram. Inoculum efficiency of oospore inoculum was estimated from percent mortality data after transformation for multiple infections (Gregory 1948) and calculated by regression of the estimated number of infections plotted as $\ln[1/(1 - y)]$ where y is disease incidence (Vanderplank 1975) versus the number of oospores per gram of soil. Thus, the slope of the regression line for the estimated number of successful infections on the average per plant versus the number of oospores per gram soil is the estimated inoculum efficiency. For *P. capsici*, inoculum efficiency was 0.011 or about 91 oospores per gram of soil were required for a successful infection (Bowers and Mitchell 1991). In the *P. capsici*-pepper pathosystem oospores are thought to germinate indirectly to form sporangia that in turn release zoospores that infect plant tissues. The low inoculum efficiency for oospores is not unexpected, because oospores do not infect the host directly but rather through an indirect, multiple-step process. Likewise, English and Mitchell (1988) calculated inoculum efficiency of chlamydospores of *P. parasitica* var. *nicotianae* based on the observed number of infected tobacco roots and total chlamydospores in the volume of soil holding a tobacco seedling. On tobacco, inoculum efficiency ranged from 0.005 to 0.003 (200 to 333 chlamydospores/g soil per infection) for *P. p. nicotianae*. Because tobacco seedlings were only grown for 2 weeks, estimates of inoculum efficiency may have been higher if a longer period of time was allowed for roots to encounter more inoculum in the container.

In the two previous examples, successful infections were estimated with the multiple infection correction based on plant mortality counts or number of infected roots. When multiple spore stages are involved in the infection process, the multiple infection correction may not adequately correct for successful infections and this may result in low estimates of inoculum efficiency.

As an estimate of individual infections, Tomimatsu and Griffin (1982) cultured peanut roots from soil infested with microsclerotia of *Cylindrocladium crotalariae* and counted the number of colonies developing per unit length of root. Instead of using the number of microsclerotia per gram to calculate inoculum efficiency, the percentage of microsclerotia germinating within 1 mm of the peanut root was used to estimate the actual number of microsclerotia initiating an infection. On this basis, Tomimatsu and Griffin (1982) calculated an inoculum efficiency of 100% for microsclerotia with the cultivar VA-72-R. When the actual number of microsclerotia per gram of soil and observed infections per meter of root were used in the calculations rather than the estimated number of microsclerotia germinating in the 1-mm zone near the root, the estimate of inoculum efficiency was 2.3 (0.4 microsclerotia/g soil per infection) on the cultivar Florigiant. Estimates of inoculum efficiency for *C. crotalariae* based on colonies growing from infected peanut roots may severely overestimate efficiency compared to estimates based on actual root necrosis. Tomimatsu and Griffin (1982) also calculated inoculum efficiency with $\ln[1/(1 - y)]$ based on root necrosis as a measure of infection. In this case, inoculum efficiency was 0.0085 (118 microsclerotia/g soil per infection) for VA-72-R. Apparently, peanut roots can tolerate a large number of infections before root cells become necrotic.

Soil saturation period also can influence inoculum efficiency. Reynolds et al. (1985) grew seedlings of Fraser fir (*Abies fraseri*) at −100 millibars (mb) in soil infested with chlamydospores of *P. cinnamomi* that was subsequently flooded for periods of 2, 4, 8, or 16 h before resetting soil matric potential to −100 mb. As the period of flooding increased, inoculum efficiency increased from 0.10 (10 chlamydospores/cm^3 of soil per seedling) at 2 h, to 0.12 (8.3 chlamydospores/cm^3 of soil per seedling) at 4 h, and finally to 0.29 (3.5 chlamydospores/cm^3 of soil per seedling) after a 16-h flooding period. Inoculum efficiency increased linearly between 4 and 16 h of flooding. Presumably, the increase in inoculum efficiency for chlamydospores of *P. cinnamomi* after longer saturation periods was due to enhanced release of zoospores of sporangia formed at −100 mb that were subsequently able to find and infect seedling roots. Alternatively, the longer saturation periods may have predisposed fir seedlings increasing host susceptibility.

Gilligan and Simons (1987) evaluated inoculum efficiency based on the probability of infection given a certain radial distance of inoculum from the infection court. Gilligan (1985) had introduced the term pathozone to describe this zone about the root within which the center of a propagule must lie for infection to occur. Rather than assess the number of infections, a tedious and time-consuming exercise, Gilligan and Simons (1987) used presence or absence of infection of the host expecting that the probability of infection at a given distance from the infection court is expressed with increasing trials. Experiments with propagules of *Rhizoctonia solani* in the radish pathozone resulted in inoculum efficiency estimates of 0.65 (1.5 propagules/g soil per infection) for a propagule at the root surface to almost nil at a distance of 16 mm from the infection court. Increasing the number of propagules in a clump of inoculum of *R. solani* to eight propagules per clump increased inoculum efficiency to 1.0 (one propagule/g soil per infection) at the root surface, and 0.13 (7.7 propagules/g soil per infection) at a distance of 16 mm from the infection court. Apparently, clumped inoculum of *R. solani* may in some way pool the energy available for infection of host tissues.

1.1.5 Inoculum Density

As indicated above, estimates of inoculum efficiency require information on the number of pathogen propagules. In most of the examples used above, pathogen number was expressed on a per gram of soil basis. Inoculum density is a measure of the number of propagules per unit weight or volume of soil. The term is convenient for expressing the amount of inoculum in soil. For soilborne plant pathogens that may exist in multiple or unknown stages in soil, inoculum density may be expressed as colony-forming units (cfu) per gram of soil rather than as propagules per gram. Examples of these pathogens are bacteria as well as fungi that survive as both hyphae and sclerotia in soil. A few specific methods for determination of inoculum density of soilborne pathogens will be given later in the chapter.

1.1.6 Spheres of Influence on Inoculum

In the soil environment, inoculum may be stimulated to germinate and infect host tissue from zones of influence around both mycorrhizal and nonmycorrhizal roots, and around germinating seeds. The term "rhizosphere" was first coined by Hiltner in 1904 to describe the zone of soil around the root that influenced microbes. In the rhizosphere, microbial populations and relationships are different than in nonrhizosphere soil. Greater activity of microbes including soilborne plant pathogens is due to the stimulating effects of nutrient exudates in the rhizosphere. Many pathogens depend on nutrients in the rhizosphere to provide the energy for germination and infection of host roots. Other zones of influence on soilborne inoculum have also been described such as the spermosphere (i.e., the zone around a germinating seed), the hyphosphere (i.e., the zone around a hypha), and the mycorrhizosphere (i.e., the zone around a mycorrhizal root).

Activity of inoculum and other microbial populations is intense in the rhizosphere. The rhizosphere to nonrhizosphere soil ratio for counts of microorganisms is typically 5 to 20:1 but may be as high as 100:1 for bacteria (Rovira and Davey 1974). The nature and composition of root exudates have been studied extensively. As outlined by Katznelson (1965), almost any compound synthesized by plants can be found in root exudates including, simple sugars, amino acids, organic acids, nucleotides, flavonones, enzymes, auxins, glycosides, etc. Exudates may influence spore germination, mycelial growth, zoospore chemotaxis, and other pathogen processes involved in the early recognition and infection of host tissues. Exudate patterns are host-specific and even cultivar-specific. Patterns of exudation may be altered by changes in the environment and other physical factors affecting host growth and development. Altered exudation patterns may influence subsequent spore germination and infection in the rhizosphere.

Bean root rot caused by *Fusarium solani* f. sp. *phaseoli* illustrates the role of exudates in the spermosphere and in the rhizosphere around the developing hypocotyl. Chlamydospores of *F. s. phaseoli* germinated best in soil when amino acids such a L-asparagine are present (Cook and Schroth 1965). No germination occurred in the presence of ammonium sulfate or potassium nitrate alone. Chlamydospores germinated in the presence of glucose and available nitrogen in soil but not as extensively as with the amino acid. In some cases, however, extensive lysis of germ tubes was noted in soil amended with asparagine but not in soil amended with glucose (Cook and Snyder 1965).

Germination of chlamydospores of *F. s. phaseoli* in the spermosphere reached nearly 60% within 20 h but only about 20% germination was observed in the hypocotyl rhizosphere (Cook and Snyder 1965). Exudate patterns were different in the spermosphere of bean than in the rhizosphere around hypocotyls. In the spermosphere, more glucose and amino acids were found than in the hypocotyl rhizosphere where glucose but only trace amounts of amino acids were found (Schroth and Snyder 1961). Differences in quantity and composition of exudates in the spermosphere and rhizosphere probably accounted for differences in the observed germination of chlamydospores and subsequent lysis in the spermosphere. Enhanced activity of bacteria and other microorganisms resulting from exudates in the spermosphere was suggested as the cause of germ tube lysis.

Germination of propagules in the rhizosphere as a result of nutrients that overcome soil fungistasis is but one step in the infection of host tissue. Nutrients that favor germination may not have the same effect on penetration and lesion development. As summarized by Toussoun (1970), penetration of bean hypocotyls by *F. s. phaseoli* was most abundant with amendments of potassium nitrate, a compound that did not stimulate chlamydospore germination. Penetration was also abundant, however, with amino acids that favored germination. Glucose, on the other hand, resulted in sparse penetration of the hypocotyl by the pathogen. Subsequent lesion development was greatest with either nitrate or amino acid amendments and least with glucose amendment. The effect of nutrients on pathogenesis in the rhizosphere should be viewed from the standpoint that carbon and nitrogen are required for spore germination, but indirect effects involving saprophytic soil microorganisms as mediated by the quality and quantity of exudates in the spermosphere and rhizosphere may affect subsequent penetration and disease development.

Several investigators have sought to define the extent of the rhizosphere effect on microbial populations and inoculum of soilborne plant pathogens. Papavizas and Davey (1961) used a multiple-tube sampling device to estimate rhizosphere populations of bacteria, streptomycetes, and fungi at various distances from 18-day-old blue lupine seedlings. Populations were highest on the rhizoplane, i.e., the root surface, dropped drastically at 0–3 mm from the root surface, then gradually decreased at distances up to 18 mm from the root surface. Authors disagree on the extent of the rhizosphere effect. Foster (1985) cites research to indicate that most rhizosphere bacteria occur in the first 10–50 μm from the rhizoplane. The rhizosphere effect may be greater for fungi. Chlamydospore germination of *Fusarium oxysporum* was greater than 50% within 0.5 mm of the root surface at a point 13 mm from a peanut root tip, but dropped to zero germination at 1.8 mm from the rhizoplane (Griffin 1969).

The rhizosphere effect can vary depending on environmental conditions. For example, Reynolds et al. (1985) used the formula of Ferriss (1981) to calculate the rhizosphere width for infection of Fraser fir seedlings by chlamydospores of *Phytophthora cinnamomi* as a function of soil matric potential. Rhizosphere width was 4.01 mm at 0 millibars (mb) then fell precipitously to 0.46 mm at −10 mb (Fig. 1.1). Rhizosphere width was independent of inoculum density. The relatively wide rhizosphere at 0 mb probably reflects the optimum matric potential for zoospore release from

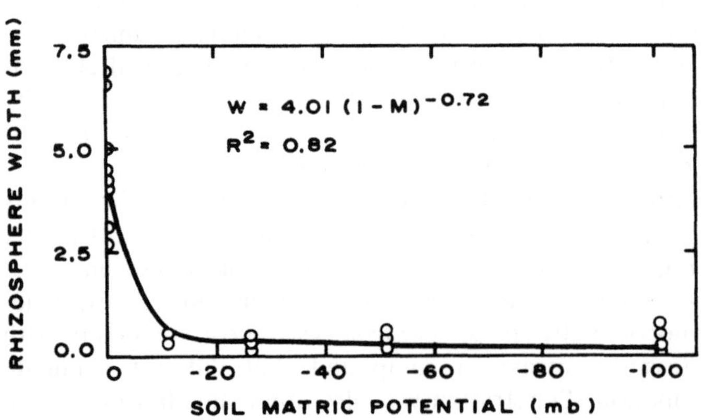

Fig. 1.1. Effect of soil matric potential on rhizosphere width of 2-week-old Fraser fir seedling radicles. Data are based on infection of seedlings in soil infested with chlamydospores of *Phytophthora cinnamomi*, determined by plating onto a selective agar medium. Soil moisture was maintained at a constant level in Buchner funnel tensionmeters. The data values indicated were calculated by using Ferriss' (1981) model for rhizosphere width. (Reprinted with permission of K.M. Reynolds and Phytopathology)

sporangia formed from germinating chlamydospores of *P. cinnamomi* and the high mobility of zoospores at soil saturation.

Exudation pattern, spore germination, and host infection can vary greatly in the spermosphere depending on environmental conditions and host genotype. Short and Lacy (1976a) found that yellow seeds of the pea cultivar Miragreen exuded severalfold more carbohydrate than green seeds of the same cultivar. Yellow-green seeds of the same cultivar were intermediate in exudation between yellow and green seeds. Yellow seeds develop as the seed matures and lose the normal green color, a maturation process called bleaching. Although temperature influenced exudation rate, total carbohydrate exudated by yellow seeds of the wrinkled-seeded Miragreen was greater after 96 h than exudate from green seeds of the smooth-seeded cultivar Alaska. Most of the difference in the amount of exudate from Miragreen was due to a higher exudation rate during the first 18 h in an artificial leaching system. When seeds were soaked in water for 24 or 48 h, yellow seeds exuded much more carbohydrate than green seeds, but subsequent exudation in the leaching system was minimal. Yellow seeds are thought to lose membrane integrity as seed matures. When seed is sown, exudation is more rapid until the membranes assume a normal function in the germinating seed.

The large differences in exudation patterns between yellow and green seed have a tremendous effect on spore germination in the spermosphere and disease development. Short and Lacy (1974) used a chlamydospore-infested soil column containing pea seeds to estimate chlamydospore germination of *F. solani* f. sp. *pisi* at various locations and distances from the germinating seed. Chlamydospore germination was greatest near the emerging pea radicle, but less, although similar, at other points in the pea spermosphere. Germination was also greatest in soil at 50% soil moisture (−8 mb) and less at 20% soil moisture (−33 mb). Chlamydospores in the spermosphere of green seeds of Alaska always had less germination than chlamydospores in the spermosphere of yellow seeds of Miragreen. In Miragreen, the spermosphere effect on germination extended to 5.5 mm near the radicle at 50% soil moisture, but to only 3.5 mm in Alaska. Soaking Miragreen seeds 48 h prior to planting removed exudates and resulted in sharply lower rates of chlamydospore germination at the seed surface and in the spermosphere. In effect, the spermosphere was shrunk from 5.5 mm for nonsoaked Miragreen seeds to 0.5 mm for soaked seed.

Not only do differences in seed exduation patterns related to seed color have an effect on chlamydospore germination, but exudation associated with seed color also affects subsequent disease development. In growth chamber experiments with a soil naturally infested with *Pythium ultimum* and artificially infested with *F. s. pisi*, seed and seedling rot was greater for yellow seeds of Miragreen than for green seeds of Alaska (Short and Lacy 1976b). As with exudation patterns, green seed of Miragreen also had less seed rot than yellow seed of the same cultivar. Seed soaked at 22 °C for 48 h had less seed rot than nonsoaked seed but those soaked at 10 or 30 °C had as much rot as nonsoaked seed. Loria and Lacy (1979) collected exudate from yellow seed (bleached) and incubated the exudate with green seeds (unbleached) of the same cultivar infested with chlamydospores of *F. s. pisi*. Seed rot was more severe on unbleached seeds incubated with exudate from bleached seed than on seed incubated with unbleached exudate. This effect was bypassed when inoculum was injected directly into the cotyledons where exudate effects on germination of chlamydospores and subsequent infection were inoperative. Exudate from bleached seed contained threefold more glucose and fructose and almost ninefold more sucrose than that from unbleached seed. However, the concentration of 14 amino acids was similar in exudates of both bleached and unbleached seed. The unfractionated exudate from bleached seeds containing carbohydrates stimulated a fourfold greater germination of chlamydospores than the unbleached fraction. Unlike the results with high germination and subsequent lysis of chlamydospores of *F. s. phaseoli* in the presence of amino acids, chlamydospores of *F. s. pisi* did not germinate in the amino acid or nonamino acid fractions of pea exudate.

The spermosphere effect in beans and peas can be correlated with exudation pattern, chlamydospore germination, and subsequent disease development. No doubt these types of relationships exist for other soilborne plant pathogens and the hosts they infect.

Two other concepts for describing inoculum and infection of host tissues, namely, competence distance or volume and pathozone, warrant discussion before leaving this section. Grogan et al. (1980) proposed the idea of competence distance and competence volume to describe the relationship of soilborne inoculum in root diseases. Competence distance is the maximum distance which a propagule can lie from the root surface and still cause infection. Competence volume is the volume of soil around host tissue from which a propagule can infect. The shape of this volume of soil is dependent on the growth dynamics of host tissues. For example, the competence volume of a carrot root can be calculated on the basis of the volume of a frustum of a cone minus the volume of the root itself (Punja 1986). Competence distance and volume were envisioned as terms to describe inoculum-root disease relations for inoculum that could germinate outside the rhizosphere. Sclerotia of *Sclerotium rolfsii*, for instance, can germinate on the soil surface at distances up to 3 cm from carrot roots and infect host tissue (Punja 1986). Although the epidemiology of diseases caused by pathogens with large propagules such as sclerotia can be described by computing competence distance and volume, the application of the concepts for smaller propagules of soilborne plant pathogens has not been attempted.

The pathozone concept of Gilligan (1985) is not limited to the rhizosphere, spermosphere, etc. or to specific host morphology and it does not imply a distinct zone of host exudates. In addition, pathozone can also be used to describe hyperparasitism of fungal propagules by antagonists. Thus, pathozone may be a more functional concept for studying the dynamics of inoculum in the infection court than some of the previously used terms.

1.2 Forms of Inoculum

Soilborne plant pathogens exist in soil in specific forms related to the morphological development characteristics of a given pathogen. Inoculum may be single cells a few micrometers in size for soilborne bacteria to multicellular structures approaching 10 mm for sclerotial fungi. In root disease epidemics, an understanding of the nature and form of inoculum surviving in soil and its ability to give rise to primary infections is critical to the development of root disease management strategies. Fungi may exist in soil as sporangiospores, oospores, chlamydospores, sporangia, zoospores, conidia, sclerotia, microsclerotia, or as hyphae associated with plant debris. Because the pathogenic bacteria are nonspore formers, single cells and clumps of cells associated with plant debris are the primary means of existence in soil.

Many soilborne plant pathogens cause root diseases that are monocylic, i.e., a disease within a given crop develops from infections resulting from primary inoculum. A few pathogens such as *Aphanomyces* sp., *Phytophthora* spp., *R. solani*, and *S. rolfsii* may induce polycyclic diseases, whereby secondary inoculum produced during disease development results in additional infections or new infections on other plants.

1.2.1 Primary Inoculum

The form of inoculum existing in soil that initiates infection of host tissue directly or indirectly is termed primary inoculum. Formation of primary inoculum may occur on host tissues during pathogenesis or as a result of saprophytic colonization of pathogen-killed host tissues. Microsclerotia of *V. dahliae*, for instance, are an example of primary inoculum formed saprophytically on host tissues following pathogenesis. In other cases, primary inoculum may be formed as a result of the conversion of propagules in soil. Macroconidia of *F. solani* formed on sporodochia on host tissues are converted to chlamydospores once macroconidia are introduced into soil. Primary inoculum is subject to the vicissitudes of survival in soil between crops. Environmental factors can influence the nutritional status of primary inoculum during survival and thus affect inoculum potential and efficiency.

Propagules of primary inoculum germinate directly or indirectly in soil to infect host tissues once soil fungistasis is overcome. In the case of *Fusarium* spp., germ tubes develop directly from chlamydospores, at the surface of the infection court and infect susceptible tissues. The extent of germination of chlamydospores of *F. s. phaseoli* can vary depending on host genotype and environmental factors. Schroth and Cook (1964) found differences in percent germination of chlamydospores in the spermosphere of three pinto bean cultivars as long as the seed coat was intact. Resultant preemergence damping off was related directly to germination rate in the spermosphere of each cultivar.

With *P. cinnamomi*, however, chlamydospores germinate indirectly at favorable matric potentials in soil to form sporangia, a process that may take several hours or days depending on temperature and moisture conditions. Sporangia then germinate indirectly under saturated conditions to release motile zoospores in the rhizosphere of susceptible crops. Zoospores, subject to chemotactic responses, move across carbon nutrient gradients in water-filled pores to encyst on host tissue. As encystment occurs zoospores form a cell wall from protein reserves in the now nonfunctional flagella. Finally, the zoospore cyst produces a germ tube that infects the host. Thus, what is a fairly simple process in the case of primary inoculum of *Fusarium* spp. involves multiple steps which are subject to an exacting soil environment in *P. cinnamomi*.

In homothallic species of *Phytophthora* such as *P. cactorum*, oospores formed as the result of gamete fertilization can take on the same role as chlamydospores through formation of sporangia. Oospores as primary inoculum, however, are subject to constitutive dormancy and do not germinate readily or uniformly in soil even though nutrients sufficient to overcome soil fungistasis are adequate. Germination triggers in oospores are poorly understood although certain wavelengths of light can be stimulatory. In general, many factors such as light, temperature, nutrition, enzymes, and maturity are involved in oospore germination but the exact combination of these factors is poorly understood (Ribeiro 1983).

Sclerotia and microsclerotia are two other examples of primary inoculum that persist for extended periods in soil. Sclerotia of *Sclerotinia* spp. and *Sclerotium* sp. develop from hyphae on the exterior surfaces of infected plant tissues. Cultivation can introduce sclerotia to soil and as crop debris decomposes sclerotia survive as primary inoculum. Sclerotia may be organized into a distinct rind and medulla tissues such as in *S. rolfsii* or form as a tightly woven mass of intertwined hyphae in the case of *R. solani*. Volatiles produced from decomposing crop debris may stimulate sclerotia and microsclerotia to germinate and infect host tissues (King and Coley-Smith 1968; Linderman and Gilbert 1973; Beute and Rodriquez-Kabana 1979; Punja et al. 1984). In the case of *S. rolfsii*, methanol produced from remoistened peanut hay was a primary compound that stimulated germination of sclerotia (Beute and Rodriquez-Kabana 1979).

Several types of sclerotial germination have been reported including myceliogenic, sporogenic, carpogenic (reviewed in Coley-Smith and Cooke 1971), and eruptive. Myceliogenic and eruptive germination are most common for soilborne plant pathogens such as *S. rolfsii*, *C. crotalariae*, *R.*

solani, and *V. dahliae*. Sporogenic germination is typical of *Botrytis* spp. where conidiophores develop directly on the sclerotial surface. Apothecia develop from sclerotia of *Sclerotinia minor* under certain environmental conditions during carpogenic germination. In the latter two cases, primary inoculum produces conidia and ascospores that normally infect above-ground plant parts after successful dissemination of spores by splashing rain and/or wind.

Sclerotia of *S. rolfsii* often germinate eruptively in soil to form a mass of hyphae in a strand that grows toward the infection court (Punja and Grogan 1981). Eruptive germination increases the distance from which sclerotia can lie from the rhizosphere and still infect host tissues. Eruptive germination increases the probability that the pathogen will have enough energy available for infection of the host but provides no means of survival should the mycelium fail to infect.

Microsclerotia are effective forms of primary inoculum for soilborne plant pathogens such as *Cylindrocladium* sp., *Macrophomina* sp., and *Verticillium* sp. Microsclerotia are usually formed in host cortical tissues by saprophytic growth following parasitic activity of the pathogen. Thus, microsclerotia are bound in host tissues until the tissue decomposes on death of the plant or after harvest. Pigmented cells in the microsclerotium of *V. dahliae* contain a single nucleus, mitochondria, ribosome lipid droplets, and other typical organelles (Brown and Wyllie 1970). The pigmentation of cells plus a mucilaginous pigmented matrix surrounding the cells gives the microsclerotium a rigid structure that protects the microsclerotium in the adverse soil environment. Pigmented cells, but not hyaline cells devoid of cytoplasm, gave rise to germ tube hyphae during germination (Brown and Wyllie 1970). Only mycelial germination has been reported for microsclerotia. Microsclerotia may germinate repeatedly as in the case of *Verticillium* sp. (Farly et al. 1971) if suitable host tissue is not present, a feature that can promote the effectiveness of primary inoculum.

Primary inoculum of bacterial pathogens includes single cells and clumps of bacterial cells in soil or in crop debris or roots of susceptible or immune plants such as weeds. Because bacteria are single-celled organisms, infection of host tissues takes place directly from inoculum residing in the soil through wounds such as those associated with the emergence point of lateral roots.

1.2.2 Secondary Inoculum

In root disease epidemics, secondary inoculum may be produced in or on infected plants. This secondary inoculum can induce additional infections during the cropping cycle and results in logarithmic increases in disease. For perennial crops like trees, it is not hard to visualize the importance of secondary inoculum in secondary infection of the root system of a large tree as roots intertwine over time. In some cases, secondary inoculum may also play a role in root disease epidemics for annuals. For example, Campbell and Powell (1980) and Campbell et al. (1984) demonstrated the importance of secondary inoculum in black shank of tobacco caused by *Phytophthora parasitica* var. *nicotianae*. Presumably, sporangia and/or zoospores formed on tobacco roots as a result of infection from primary inoculum and were disseminated by irrigation or rain down the furrows resulting in new infections on previously uninfected plants. Pfender and Hagedorn (1983) found that *Aphanomyces euteiches* on pea could spread from plant to plant in the row up to 18 cm away or to plants in adjoining rows during the cropping cycle. The form of secondary inoculum, mycelium, or zoospores was not determined. Koch et al. (1987) found that plant-to-plant spread of *Phymatotrichum omnivorum* was important in development of Phymatotrichum root rot of cotton. Plant spacing within rows of 20 to 69 cm inhibited plant-to-plant spread with the effective distance dependent on environment, cultivar, and edaphic factors. In the case of Phymatotrichum root rot, secondary inoculum consists of mycelial strands called runner hyphae that can grow on living roots or debris as the pathogen spreads in the row.

1.3 Surivival of Inoculum

The success of soilborne plant pathogens in inducing root disease depends primarily on dissemination and survival of inoculum. Survival is the most critical aspect of the ecology of soilborne plant pathogens. It is the survival phase that determines the amount of inoculum available for infection of host tissues when susceptible crops are planted. Soilborne plant pathogens exhibit a wide variety of mechanisms for adapting to surivival in soil. Space does not permit a discussion of the unique survival mechanisms associated with many soilborne plant pathogens. Reviews by Park (1965), Schuster and Coyne (1974), and books by Bruehl (1987) and Garrett (1970) illustrate the many aspects of survival of plant pathogens in soil.

1.3.1 Idealized Survival Curve

Persistence of inoculum in soil over time after dissemination can be visualized as a survival curve with three possible shapes (Fig. 1.2). Survival of most soilborne pathogens in soil progresses through three phases including an initial survival phase, a logarithmic death phase, and a residual survival phase over time. During the initial survival phase, inoculum may (1) increase saprophytically on plant debris, or commensally in the rhizosphere of non-hosts (Fig. 1.2, curve C); (2) persist at the initial population level (Fig. 1.2, curve B); or (3) enter the logarithmic death phase (Fig. 1.2, curve A). Inoculum in curve A (Fig. 1.2) enters the logarithmic death phase im-

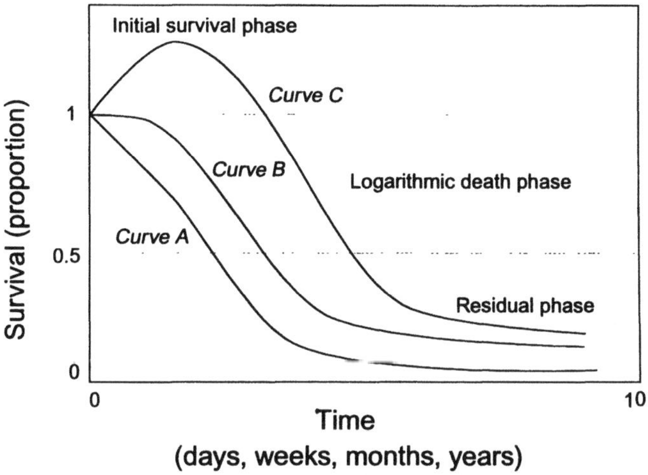

Fig. 1.2. Idealized survival curves for soilborne plant pathogens. *Curve A* represents inoculum that enters the logarithmic death phase immediately after dissemination. *Curve B* represents multicellular inoculum such as sclerotia that has a lag phase before entering the logarithmic death phase, and *curve C* represents saprophytic growth of inoculum upon dissemination to soil. Note the varied initial survival phase and the residual survival phase in which a small proportion of original inoculum persists for an extended period

mediately after dissemination and incorporation into soil. This pattern is typical of natural inoculum with single-celled propagules and artificially produced inoculum added to soil. Curve B illustrates an example of multicellular inoculum such as a microsclerotium in which several cells must die before the microsclerotium loses viability, thus there is an initial lag phase prior to the logarithmic death phase. Curve C represents inoculum that increases saprophytically on plant debris for a period of time, such as *Rhizoctonia solani*, before entering the logarithmic death phase. Consequently, the length of the initial survival phase is quite variable, depending on the soilborne pathogen. The second phase of survival is the logarithmic death phase in which a large percentage of the original population is lost over a relatively short time period. Baker (1991) has described several factors including environmental influence on inoculum at formation, intrinsic influences, and germination-lysis that may account for inoculum entering the logarithmic phase shortly after introduction into soil. The logarithmic death phase may last for a few days to several weeks for the more resistant forms of inoculum. Finally, inoculum transcends into a residual phase of survival in which a small proportion of the original population survives for a long time. Although three distinct survival curves can be described, the same pathogen may exhibit any of the three depending on nutritional status, and/or existing physical and environmental factors. When inoculum of *R. solani* was cultured on chopped potato soil then added to field soil on a 1:1

basis, the initial saprophytic growth phase of curve C was found (Benson and Baker 1974c), apparently because the mixing process exposed additional potato substrate that was available for colonization by *R. solani*. If lower rates of colonized substrate were mixed with soil, survival typical of curve A was found.

Organic amendments can affect the shape of the survival curve. Hakeem and Ghaffar (1977) amended soil with various crop residues to study survival of sclerotia of *Macrophomina phaseolina*. Saprophytic increase in sclerotial survival was found in nonamended and cotton-residue amended soils typical of curve C, whereas alfalfa, clover, and wheat residues all resulted in survival of sclerotia typical of curve A.

1.3.2 Transformation of Survival Curves

Survival curves are transformed so that comparison of the effect of various factors such as environmental variables on survival can be made. Several transformations for quantifying survival have been proposed. The half-life concept of Yarwood and Sylvester (1959) predicts the time required for death of 50% of the initial inoculum with the assumption that death follows a logarithmic rate. Critics of the half-life concept question the appropriateness of the equation to account for survival curves such as curve B when multicellular propagules exhibit a lag phase before entering the logarithmic death phase. Dimond and Horsfall (1965) proposed the semilogarithmic transformation to account for the lag phase wherein the logarithm of the ratio of viable propagules at time t_1 to propagules at time zero is plotted against time. Baker (1971a) suggested the log-probit transformation to describe survival of soilborne plant pathogens based on the approach that time could be considered a toxin, and that susceptibility of inoculum to the toxin (i.e., death) followed a normal distribution. Benson and Baker (1974c) tested the semilogarithmic and log-probit transformation for survival of *R. solani* under conditions of varying initial inoculum, temperature, moisture, pH, and composition of substrate for culturing inoculum. Even though different assumptions were made in developing the transformations, close agreement in terms of predicting the time required for 50% survival and the rate of propagule death (i.e., slope value between the two transformations) was found for the various factors tested.

The linear nature of proposed transformations does not account adequately for the residual phase of survival and thus results in underestimation of inoculum surviving during the residual phase. On the other hand, if data points from the residual phase are included, the transformations underestimate the logarithmic death phase and overestimate survival rate during the residual phase. Residual inoculum can have a profound effect on subsequent root disease epidemics and needs to be accounted for as new models are developed for survival of soilborne pathogens.

Nonlinear equations to describe both the logarithmic death phase and the residual phase along with differences in initial survival might be more useful for understanding the dynamics of inoculum survival over time. The pathozone models described by Gilligan and Simons (1987) for probability of infection by inoculum at various distances from the host might be adaptable to survival data but this has yet to be determined.

1.4 Quantifying Inoculum in Soil

Quantification of inoculum in soil is essential for studying and managing root disease epidemics. A significant amount of research effort has been expended in the development of methods to quantify inoculum of soilborne plant pathogens. A recent text on methods gives numerous procedures for quantifying soilborne pathogens (Singleton et al. 1992). The process of quantification is based on knowledge of the form of inoculum in soil for a particular pathogen. Inoculum may be quantified by direct count, bioassays, soil assays, serological assays, or by chemical assays. Often a large proportion of a research budget may be expended for materials and time to assay inoculum in soil. Thus, improvement in methods that can reduce the expense but increase the precision of assays is constantly sought. For instance, Campbell and Nelson (1986) used the variance of the general mean procedure to estimate efficiency of a soil-plating procedure for enumeration of microsclerotia of *Macrophomina phaseolina* in soil. Although precision of the method improved with the number of subsamples assayed per soil sample, cost considerations were used to set the level of precision acceptable for large-scale sampling. The following sections will describe general methods for assessing inoculum in soil samples. Sampling strategies for collecting representative soil samples will be discussed under "spatial aspects" in Chapter 7, Section 7.2.

1.4.1 Direct Counts

Pathogens with macroscopic propagules such as *Sclerotium rolfsii* can be assayed by direct counts. Leach and Davey (1938) collected soil samples from sugar beet fields, wet-sieved the samples through a series of nested sieves, then back-washed the residue on the sieve into a white-colored dish for enumeration. Viability of individual sclerotia was assessed from germination counts made after transfer of sclerotia to a moistened peat:soil mix. Rodriquez-Kabana et al. (1980) developed a modified, direct-count procedure for enumerating viable sclerotia of *Sclerotium rolfsii* in soil samples. Air-dried soil was remoistened with 1% methanol and placed in a thin layer in a closed plastic chamber until sclerotia germinated. The number of germinating sclerotia could easily be quantified by counting the developing colony around each sclerotium on the soil surface. The advantage of this assay is the stimulation of viable sclerotia by the methanol, which allows counts of only viable sclerotia in a sample. Direct counting procedures are labor-intensive, but even more so if propagules must be counted microscopically.

1.4.2 Bioassays

Soilborne plant pathogens for which semiselective and selective isolation media are not available may be quantified by bioassays. Adaptation of the most probable number method can make bioassays quantitative. For instance, Pfender et al. (1981) developed a most probable number method based on the proportion of aliquots of each dilution of a soil naturally infested with *Aphanomyces euteiches* in which pea seedlings were infected. A computer program was developed to estimate inoculum density based on actual counts of infected peas at each dilution. The method was time-consuming and required considerable space and facilities for growth of bioassay host plants, but was able to distinguish forms of *A. euteiches* pathogenic to pea from those pathogenic only to bean. Although bioassays may be the only way to quantify a particular pathogen, differences in environmental conditions affecting host growth and inoculum efficiency among laboratories would no doubt lead to different estimates of inoculum densities for the same soil sample at different laboratories.

1.4.3 Soil Assays

Numerous semiselective and selective media have been developed for quantification of most soilborne plant pathogens from soil. Constituents of isolation media inhibit growth of microorganisms that would obscure colonies of a specific pathogen. Procedures to enumerate specific pathogens oftentimes involve pretreatment of soil samples to eliminate fractions that do not contain pathogen propagules. The most common procedure for this purpose is wet-sieving wherein a series of nested sieves separate soil samples by size fraction. Residues collected on sieves that correspond in size to pathogen propagules are then placed on an isolation medium. Wet-sieving is particularly useful for pathogens such as *Phytophthora* spp. that occur at low inoculum densities in soil. For pathogens that occur at higher inoculum densities, such as *Fusarium* spp., soil dilution is more effective.

In addition to wet-sieving, baits and specialized sampling equipment can be effective for pathogens that have low inoculum densities in soil. Baiting procedures normally require the incubation of a suitable bait, such as a stem segment, or leaf tissue, etc. in a soil sample for a given period of time. The tissue may then be recovered by sieving, and cultured for the presence of the pathogen. Pathogens with high competitive saprophytic ability can be enumerated in this manner. In the case of zoosporic pathogens such as *Aphanomyces* sp. and *Phytophthora* spp., soil samples are flooded with water and suitable tissue baits floated over the soil sample for colonization by zoospores produced from inoculum in the soil sample.

A multiple pelleter sampler was developed for enumerating propagules of *Rhizoctonia solani* directly from soil samples (Henis et al. 1978). The pelleter collects a known volume of soil from the sample and deposits it on a semiselective agar medium contained in a petri dish. The sampler allows the operator to process a large number of samples in a short period of time. Pellets from which hyphae of *R. solani* emerge are counted and counts adjusted with the multiple infection correction.

1.4.4 Chemical Assays

Although specific chemical assays have not been used to quantify individual soilborne pathogens, microbial activity associated with disease-suppressive soils has been quantified chemically. Chen et al. (1988) modified a method described by Schnurer and Rosswall (1982) that measures the rate of hydrolysis of fluorescein diacetate by microbes in soil. Enzymatic activity of microorganisms in soil releases water-soluble fluorescein that can be quantified by measuring absorbance at 490 nm. In bark composts suppressive to Pythium damping-off, microbial activity as measured by hydrolysis of fluorescein diacetate was highest in the most suppressive bark composts (Chen et al. 1988).

1.4.5 Serological Assays

In the future, serology and DNA probes may be used widely for detection of pathogens in tissue and soil. MacDonald and Duniway (1979) used fluorescent antibodies to quantify propagules of *Phytophthora megasperma* and *P. cinnamomi* from soil. Although the method required soil dilutions on a selective medium, the fluorescent antibody method allowed distinctions to be made between zoospore cysts and hyphae of the two *Phytophthora* spp. from other fungi that could grow on the selective medium.

Recently, immunoassay kits for several soilborne pathogens based on poly- and monoclonal antibodies have been developed by Agri-Diagnostics Associates (Cinnaminson, New Jersey, USA). Schmitthenner (1988) has developed procedures to quantify the number of oospores of *P. megasperma* in soil with the immunoassay kits. The enzyme-linked immunosorbent assay or ELISA test is quite sensitive for detection of the pathogen, but the spatial pattern of inoculum in the field requires extensive sampling to obtain representative estimates of inoculum density. The same ELISA test was used for detection of *Phytophthora* zoospores in irrigation water (Ali-Shitayeh et al. 1991). A positive correlation was found for the number of zoospores and absorbance reading. The main limitation of the technique was not the sensitivity for the ELISA test but rather the amount of irrigation water that could be sampled through a 0.47-μm membrane before suspended particles restricted filtration. Root rot severity of azaleas infected with *P. cinnamomi* was directly correlated with absorbance reading of the ELISA (Benson 1991). Positive test results were found in azalea root samples with as little as 1% of the roots infected. Further research with ELISA is warranted to increase and expand detection of pathogen inoculum in soil and irrigation water.

Another method of detecting pathogen inoculum in soil resulting from the application of molecular biology techniques is the use of DNA probes that hybridize with the DNA of the target pathogen. Goodwin et al. (1990) were able to detect colonies of *P. parasitica* growing on soil-dilution plates with a DNA probe. The limitation of DNA probe technology, at present, is the need to amplify pathogen biomass so that enough target DNA is available for visualization of the resultant hybridization.

Advances in our understanding of root diseases and their management depend on quantification of inoculum in soil. Traditional methods involving selective and semiselective media and bioassays will continue to be important in the future but the development of serological and molecular techniques for estimating inoculum may improve the precision of estimates and at the same time may reduce the sample cost.

1.5 Inoculum Density-Disease Incidence Relationships

That inoculum density is related to disease is not questioned, but how to describe the direct relationship in quantitative terms has been the subject of considerable controversy since Ralph Baker and colleagues first proposed mathematical models based on three-dimensional relations of inoculum in the infection court (Baker et al. 1967; Baker 1971b). Like inoculum efficiency, inoculum density-disease incidence relations give information about disease development. Unlike disease progress curves, however, inoculum density-disease curves assume that the disease asymptote was reached at the time disease is assessed for analysis. Inoculum density-disease incidence analyses also assume that inoculum of the pathogen is monocylic during disease development.

1.5.1 Inoculum Density-Infection Court Relations

As inoculum density increases in soil, disease incidence increases in proportion, other things being equal. The shape of the inoculum density-disease incidence curve changes with increase in inoculum density as host tissues respond to infection. Baker (1971b) described four regions in a generalized inoculum density-disease incidence curve (Fig. 1.3). At low inoculum densities, increase in disease follows a normal distribution with increasing inoculum density. As inoculum density continues to increase, propagules acting independently cause proportionally more disease. This true logarithmic region of the inoculum density-disease curve is adequately described by the multiple infection correction (Gregory 1948) and results in a straight line with slope of 1.0 when plotted on a log-log basis (Baker et al. 1967). For a few pathogens such as *R. solani*, propagules may not act independently at higher inoculum densities. As propagules "pool" energy, the resultant number of successful infections is greater than the effect of propagules acting independently. Thus, the slope in the synergistic region is greater than would be expected on a log-log basis for inoculum acting independently. A fourth region, the transitional region, occurs as all infection courts are infected or all plants are killed by the pathogen. Further increases in inoculum density result in a plateau of disease incidence as inoculum is much less efficient in this region.

1.5.2 Transformation of Inoculum Density-Disease Curves

Comparisons of untransformed inoculum density-disease curves are difficult because the response is nonlinear. Both the multiple infection transformation (Gregory 1948) and probit transformation (Dimond and Horsfall 1965) have been proposed for linearizing inoculum density-disease curves. A log-log plot of inoculum density versus disease corrected for multiple infections

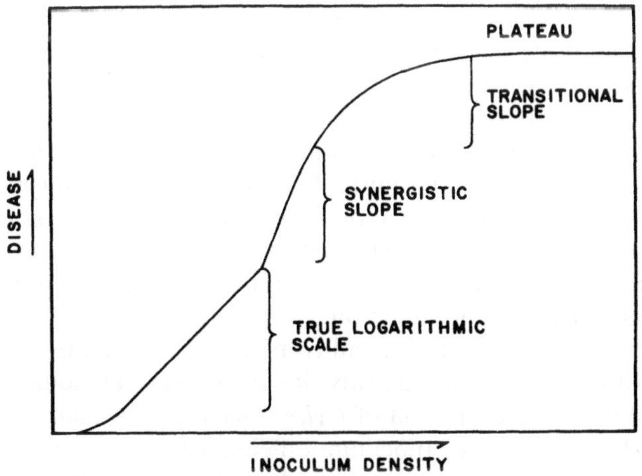

Fig. 1.3. Idealized picture of complexities involved in the inoculum density-disease curve. (Reprinted with permission of R. Baker and Phytopathology)

is predicted to result in a slope of 1.0 for a rhizosphere effect and 0.67 for a rhizoplane effect by the models of Baker et al. (1967). On a log-probit basis slopes should be 2.0 and near 1.3 for the two effects. Numerous examples wherein a slope of 0.67 is observed have been reported for various host-pathosystems since the rhizoplane model was first developed (Baker 1969; Ashworth et al. 1972; Taylor et al. 1981).

1.5.3 Usefulness of Transformed Inoculum Density-Disease Curves

Inoculum density-disease curves can provide information on the effects of environment, host resistance, cultural practices, etc. on disease development and offer comparisons among effects. The effect of temperature on disease development was assessed by analysis of inoculum density-disease incidence curves. Benson and Baker (1974b) observed severe preemergence damping off of radish caused by *R. solani* at 26 °C. Disease was also severe at 22 °C and less at 20 and 30 °C, whereas 15 °C was suboptimal for disease development. Plots of inoculum density versus disease incidence on a log-log basis resulted in estimated inoculum densities required for 50% damping off (ID_{50}) of 0.8 propagules/g soil at 26 °C, rising to 7.0 propagules/g at the least favorable temperature of 15 °C. Inoculum efficiencies were calculated as the slope of the regression line in a plot of inoculum density versus disease incidence corrected for multiple infections [i.e., $\ln 1/(1 - y)$]. At 15 °C inoculum efficiency was estimated at 0.19 (5.3 sclerotia/g soil per infection) and increased to 0.40 (2.5 sclerotia/g soil per infection) at 20 °C, 0.42 (2.4 sclerotia/g soil per infection) at 30 °C, 0.75 (1.3 sclerotia/g soil per infection) at 22 °C, and 1.35 (0.74 sclerotia/g soil per infection) at 26 °C. The high inoculum efficiency for *R. solani* at 22 and 26 °C suggests that inoculum of this soilborne plant pathogen is very effective at these temperatures. The fact that inoculum efficiency was greater than 1.0 at 26 °C may be accounted for by the production of multiple germ tubes from a single sclerotium that incites independent infections.

Likewise the effect of a soil fungicide in suppression of Rhizoctonia damping-off of radish could be evaluated by comparing ID_{50} values that ranged from 5.2 propagules/g soil for the untreated to 46 propagules/g at 400 μg/g PCNB, an eightfold increase in suppression (Benson and Baker 1974a). Presumably, inoculum efficiency decreased as fungicide concentration increased because fewer and fewer propagules at a distance away from the infection court were participating in infection.

Differences in genetic resistance among cultivars of a given crop can also be followed by analysis of inoculum density-disease curves. Hanounik et al. (1977) infested soil with microsclerotia of *Cylindrocladium crotalariae*, then planted peanut in the greenhouse. Inoculum efficiency estimated from the slope of the regression line in plots of inoculum density versus estimated successful infections was 0.0084 (119 microsclerotia/g soil per infection) for the highly susceptible cultivar Florigiant but only 0.00067 (1493 microsclerotia/g soil per infection) for the more resistant, Spancross. Thus, there was a more than tenfold decrease in inoculum efficiency with the resistant cultivar. In experiments by Phipps and Beute (1977) with microsclerotia of *C. crotalariae* differences in inoculum efficiency of 70- to 190-fold were found based on plots of inoculum density-disease incidence curves between Florigiant, and two resistant cultivars: Argentine and NC3033, respectively.

Propagule size may affect inoculum efficiency of soilborne plant pathogens. Microsclerotia of *C. crotalariae* were sieved into size classes then used to infest a Norfolk sandy loam soil at various inoculum densities (Black and Beute 1984). Florigiant peanut was planted and root rot severity assessed after 5 weeks in the greenhouse. Inoculum efficiency increased as size of microsclerotia increased. Inoculum efficiency was calculated from a plot of root rot severity data of Black and Beute (1984) transformed to successful infections by converting the severity scale to a percentage basis versus inoculum densities of 0.8, 2.5, and 10 microsclerotia per gram soil. Microsclerotia between 52 and 38 μm had an inoculum efficiency of 0.0087 (115 microsclerotia/g soil per infection) compared to 0.013 (77 microsclerotia/g soil per infection), 0.031 (32 microsclerotia/g soil per infection), and 0.052 (19 microsclerotia/g soil per infection) for microsclerotia between 105 and - 53 μm, 149 and 106 μm, and larger than 150 μm, respectively. Nutrient status of larger microsclerotia or the ability to form multiple germ tubes may account for the increased inoculum efficiency as propagule size increases.

Sclerotia of *R. solani* smaller than 150 μm were not able to infect bean seedlings (Henis and Ben-Yephet 1970). As sclerotial size increased, a larger percentage of seedlings was infected. Although the sclerotial size class data of Henis and Ben-Yephet (1970) are not suitable for calculation of inoculum efficiency based on inoculum density-disease plots, the number of hyphal extensions was sixfold greater for sclerotia that were 500 to 1000 μm compared to those less than 150 μm (19 vs. 3 per sclerotia, respectively) and colony diameter in a moist chamber was more than twofold greater for the larger sclerotia. At least for sclerotial fungi capable of multiple germ tube formation, propagule size and inoculum efficiency are directly related.

Wilkinson et al. (1985) collected data on the effect of inoculum size and concentration on infection of wheat seedlings by *G. graminis* var. *tritici*. The authors suggest that inoculum efficiency is not related to propagule size once the threshold size is achieved. In a Ritzville silt loam, the threshold size for colonized wheat crowns was larger than 0.25–0.50 mm, whereas the threshold for colonized crowns in a Shano silt loam was 0.15–0.25 mm. Additionally, a threshold size of 0.10–0.15 mm was found for particles of colonized oat grains in the two soils. Although the relation between concentration of inoculum particles of colonized oat grains and lesions per wheat seedling does appear similar for the various particle size classes tested, there seem to be differences in inoculum efficiency for colonized wheat crown inoculum. Inoculum efficiency determined from regression lines in plots of inoculum density versus disease for the Shano silt loam soil for data between 0.5 and 10.0 mg of inoculum per gram of soil (see Fig. 2A in Wilkinson et al. 1985) increased as particle size increased. At a median particle size of 0.375, maximum inoculum efficiency was estimated at 1.513 (0.7 mg crown inoculum/g soil per infection), whereas 31.2 mg crown inoculum/g soil was required per infection at an inoculum size of 0.053 mm.

Inoculum density-disease curves can be used to generate estimates of inoculum efficiency over fields in a growing season. Plotting the data of Ashworth et al. (1972) for infection of cotton (*Gossypium hirsutum*) in 22 different fields by *Verticillium dahliae* as microsclerotia per gram of soil versus the multiple infection correction estimate for the percentage of plants infected results in a mean inoculum efficiency of 0.456 (2.2 microsclerotia/g soil per infection). Estimates from inoculum density-disease curves for inoculum efficiency in cases where data are combined over fields may be subject to large errors because factors such as soil types, pH, moisture content, etc. are not uniform (i.e., constant) among fields. Given these constraints, however, it appears that microsclerotial inoculum of *V. dahliae* is much more efficient on tomato, as Grogan et al. (1979) calculated an inoculum efficiency of 2.219 (0.5 microsclerotia/g soil per infection), one of the highest found in the data examined, when data from five fields in California were combined. Given the unknown differences in virulence of tomato isolates versus cotton isolates of *V. dahliae*, as well as differences in susceptibility of infection courts on tomato and cotton, the differences in estimated inoculum efficiency on tomato may be due to a much more extensive root system in tomato that encounters a greater proportion of the total microsclerotial population of *V. dahliae* over the growing season.

The use of inoculum density-disease curves to evaluate inoculum efficiency can be a useful way to describe the role of inoculum in diseases caused by soilborne plant pathogens. Changes in environmental effects, and differences in resistance and inoculum, can be assessed by a comparison of curves. Mechanistically, the need to transform disease severity and disease incidence data with the multiple infection correction (Gregory 1948) to estimate successful infections results in estimates of inoculum efficiency. Actual root lesion number yields a more accurate assessment of inoculum efficiency, but such data are quite difficult to obtain, particularly from field studies. As illustrated in the above examples, comparisons of inoculum efficiency for pathogens under different test conditions or comparisons of different pathogens are limited at present.

1.5.4 Probability Models for Inoculum Density-Disease Incidence Relations

The controversy over the models proposed by Baker et al. (1967) has developed from an attempt to explain successful infection on the basis of the probability of a propagule to infect the host when it lies at a specific distance from the infection court. The argument by Baker et al. (1967) and Baker (1971b) that differences in rhizosphere- and rhizoplane-effective propagules, based on the theory of tetrahedral packing of solids and surface density, could be measured by comparison of the slope of log-log transformed data was challenged by Gilligan (1985), Ferriss (1981), and Leonard (1980). Criticism of the surface density models was based on the assumption that propagules of inoculum were considered as a mathematical point rather than as three-dimensional objects with volume.

Gilligan (1983, 1985) and Gilligan and Simons (1987) proposed probability models as alternatives to the surface density models to account for host infection. The models are based on an assumption that the center of a propagule must lie within a certain distance from the infection court for infection to occur. Gilligan (1985) defined this distance as the pathozone. The concept has an advantage over rhizosphere, spermosphere, etc., because the distance is not limited by the host exudate pattern and diffusion of nutrients to propagules. Pathozone also defines the competence distance proposed by Grogan et al. (1980), although mathematical models describing infection based on competence distance have not been proposed. The concept of pathozone shifts emphasis from host and root exudate effects to pathogen propagules. The propagule becomes the "center" of the disease triangle in the pathozone concept rather than the entity responding to host and environmental effects.

Gilligan (1985) derived models for predicting success of pathogen infection based on the binomial distribution of pathogens with randomly distributed inoculum and on the negative binomial distribution of pathogens with clumped inoculum. Methods for determining inoculum distribution in soil are discussed in Chapter 7 on spatial pattern analysis. Gilligan (1979) originally developed a probability (displacement) model to study the width of the rhizosphere and inoculum efficiency effects. In calculating rhizosphere width, the original model included the volume of the root. This displacement model was challenged by Leonard (1980) and Ferriss (1981) who proposed probability (shell) models that excluded the root volume. Differences in the two approaches, however, are insignificant when the pathogen propagule is large such as a sclerotium, but are important when the propagule is smaller like a fragmented pycnidium (Gilligan 1985). Use of the probability models is limited by the need to estimate the proportion of the pathogen population that is infective and the assumption that the entire host surface is uniformly susceptible. Refinements in probability models will no doubt continue as new techniques become available.

1.6 Concluding Remarks

Progress in research on epidemiology and management of root diseases has advanced based on the understanding of factors contributing to soil fungistasis, root exudation patterns, survival, improvement in techniques to observe and quantify inoculum in soil, and transformations and models to relate inoculum to disease in susceptible hosts. Advances will continue as new techniques become available to improve studies of population dynamics of soilborne pathogens and as models to relate infection with inoculum density are refined. Management of root diseases will only be successful when effects of environment and host can be understood and integrated in such a way as to predict the consequences of a given management strategy before that strategy is actually applied.

References

Ali-Shtayeh MS, MacDonald JD, Kabashima J (1991) A method for using commercial ELISA tests to detect zoospores of *Phytophthora* and *Pythium* species in irrigation water. Plant Dis 75:305–311

Ashworth LJ Jr, McCutcheon OD, George AG (1972) *Verticillium albo-atrum*: the quantitative relationship between inoculum density and infection in cotton. Phytopathology 62:901–903

Baker R (1965) The dynamics of inoculum. In: Baker KF, Snyder WC (eds) Ecology of soil-borne plant pathogens. Univ California Press, Berkeley, pp 395–419

Baker R (1969) Use of population studies in research on plant pathogens in soil. In: Toussoun TA, Bega RV, Nelson PE (eds) Root diseases and soil-borne pathogens. Univ California Press, Berkeley, pp 11–15

Baker R (1971a) Simulators in epidemiology – can a simulator of root disease be built? Epidemiology of plant diseases, vol 1. Adv Study Inst, Wageningen, pp 1–2

Baker R (1971b) Analyses involving inoculum density of soil-borne plant pathogens in epidemiology. Phytopathology 61:1280–1292

Baker R (1978) Inoculum potential. In: Horsfall JG, Cowling EB (eds) Plant diseases: an advanced treatise, vol II. Academic Press, New York, pp 137–157

Baker R (1991) Biological control: eradication of plant pathogens by adding organic amendments to soil. In: Pimentel D (ed) Handbook of pest management in agriculture, vol II, 2nd edn. CRC Press, Boca Raton, pp 317–327

Baker R, Maurer CL, Maurer RA (1967) Ecology of plant pathogens in soil. VIII. Mathematical models and inoculum density. Phytopathology 57:662–666

Benson DM (1991) Detection of *Phytophthora cinnamomi* in azalea with commercial serological assays kits. Plant Dis 75:478–482

Benson DM, Baker R (1974a) Epidemiology of *Rhizoctonia solani* preemergence damping-off of radish: influence of pentachloronitrobenzene. Phytopathology 64: 38–40

Benson DM, Baker R (1974b) Epidemiology of *Rhizoctonia solani* preemergence damping-off of radish: inoculum potential and disease potential interaction. Phytopathology 64:957–962

Benson DM, Baker R (1974c) Epidemiology of *Rhizoctonia solani* preemergence damping-off of radish: survival. Phytopathology 64:1163–1168

Beute MK, Rodriquez-Kabana R (1979) Effect of volatile compounds from remoistened peanut tissues on growth and germination of sclerotia of *Sclerotium rolfsii*. Phytopathology 69:802–805

Black MC, Beute MK (1984) Relationship among inoculum density, microsclerotium size, and inoculum efficiency of *Cylindrocladium crotalariae* causing root rot on peanuts. Phytopathology 74:1128–1132

Bouhot D (1979) Estimation of inoculum density and inoculum potential: techniques and their value for disease prediction. In: Schippers B, Gams W (eds) Soil-borne plant pathogens. Academic Press, New York, pp 21–34

Bowers JH, Mitchell DJ (1991) Relationship between inoculum level of *Phytophthora capsici* and mortality of pepper. Phytopathology 81:178–184

Bruehl GW (1987) Soilborne plant pathogens. MacMillian, New York, 368 pp

Bristow PR, Lockwood JL (1975) Soil fungistasis: role of the microbial nutrient sink and of fungistatic substances in two soils. J Gen Microbiol 90:147–156

Brown JF, Wyllie TD (1970) Ultrastructure of microsclerotia of *Verticillium albo-atrum*. Phytopathology 60:538–542

Campbell CL, Nelson LA (1986) Evaluation of an assay for quantifying populations of sclerotia of *Macrophomina phaseolina* in soil. Plant Dis 70:645–647

Campbell CL, Powell NT (1980) Progression of diseases induced by soilborne pathogens: tobacco black shank. Prot Ecol 2:177–182

Campbell CL, Jacobi WR, Powell NT, Main CE (1984) Analysis of disease progression and the randomness of occurrence of infected plants during tobacco black shank epidemics. Phytopathology 74:230–235

Chen W, Hoitink HAJ, Schmitthenner AF, Tuovinen OH (1988) The role of microbial activity in suppression of damping-off caused by *Pythium ulitmum*. Phytopathology 78:314–322

Coley-Smith JR, Cooke RC (1971) Survival and germination of fungal sclerotia. Annu Rev Phytopathol 9:65–92

Cook RJ, Schroth MN (1965) Carbon and nitrogen compounds and germination of chlamydospores of *Fusarium solani* f. *phaseoli*. Phytopathology 55:254–256

Cook RJ, Snyder WC (1965) Influence of host exudates on growth and survival of germlings of *Fusarium solani* f. *phaseoli* in soil. Phytopathology 55:1021–1025

Dimond AE, Horsfall JG (1960) Inoculum and the diseased population. In: Horsfall JG, Dimond AE (eds) Plant pathology, an advanced treatise. Academic Press, New York, pp 1–22

Dimond AE, Horsfall JG (1965) The theory of inoculum. In: Baker KF, Snyder WC (eds) Ecology of soil-borne plant pathogens. Univ California Press, Berkeley, 571 pp

Dobbs CG, Hinson WH (1953) A widespread fungistasis in soils. Nature 172:197–199

English JT, Mitchell DJ (1988) Relationships between the development of root systems of tobacco and infection by *Phytophthora parasitica* var. *nicotianae*. Phytopathology 78:1478–1483

Farley JD, Wilhelm S, Snyder WC (1971) Repeated germination and sporulation of microsclerotia of *Verticillium albo-atrum* in soil. Phytopathology 61:260–264

Ferriss RS (1981) Calculating rhizosphere size. Phytopathology 71:1229–1231

Foster RC (1985) The biology of the rhizosphere. In: Parker CA, Rovira AD, Moore KJ, Wong PTW (eds) Ecology and management of soilborne plant pathogens. Am Phytopathol Soc, St Paul, p 358

Fradkin A, Patrick ZA (1985) Interactions between conidia of *Cochliobolus sativus* and soil bacteria as affected by physical contact and exogenous nutrients. Can J Plant Pathol 7:7–18

Garrett SD (1956) Biology of root-infecting fungi. Cambridge Univ Press, London, 294 pp

Garrett SD (1970) Pathogenic root-infecting fungi. Cambridge Univ Press, Cambridge, 294 pp

Gilligan CA (1979) Modeling rhizosphere infection. Phytopathology 69:782–784

Gilligan CA (1983) Modeling of soilborne pathogens. Annu Rev Phytopathol 21: 45–64
Gilligan CA (1985) Probability models for host infection by soilborne fungi. Phytopathology 75:61–67
Gilligan CA, Simons SA (1987) Inoculum efficiency and pathozone width for two host-parasite systems. New Phytol 107:549–566
Goodwin PH, English JT, Neher DA, Duniway JM, Kirkpatrick BC (1990) Detection of *Phytophthora parasitica* from soil and host tissue with a species-specific DNA probe. Phytopathology 89:277–281
Grainger J (1956) Host nutrition and attack by fungal parasites. Phytopathology 46: 445–456
Gregory PH (1948) The multiple-infection transformation. Ann Appl Biol 35:412–417
Griffin GJ (1969) *Fusarium oxysporum* and *Aspergillus flavus* spore germination in the rhizosphere of peanut. Phytopathology 59:1214–1218
Griffin GJ, Baker R (1991) Population dynamics of plant pathogens and associated organisms in soil in relation to infectious inoculum. In: Katan J, DeVay JE (eds) Soil solarization. CRC Press, Boca Raton, pp 3–21
Grogan RG, Ioannou N, Schneider RW, Sall MA, Kimble KA (1979) Verticillium wilt on resistant tomato cultivars in California: virulence of isolates from plants and soil and relationship of inoculum density to disease incidence. Phytopathology 69:1176–1180
Grogan RG, Sall MA, Punja ZK (1980) Concepts for modeling root infection by soilborne fungi. Phytopathology 71:361–363
Hakeem A, Ghaffar A (1977) Reduction of the number of sclerotia of *Macrophomina phaseolina* in soil by organic amendments. Phytopathol Z 88:272–275
Hanounik SB, Pirie WR, Osborne WW (1977) Influence of soil chemical treatment and host genotype on the inoculum density-disease relationship of *Cylindrocladium* black rot of peanut. Plant Dis Rep 61:431–435
Henis Y, Ben-Yephet Y (1970) Effect of propagule size of *Rhizoctonia solani* on saprophytic growth, infectivity, and virulence of bean seedlings. Phytopathology 60:1351–1356
Henis Y, Ghaffer A, Baker R, Gillespie SL (1978) A new pellet-soil sampler and its use for study of population dynamics of *Rhizoctonia solani* in soil. Phytopathology 68:371–376
Hiltner L (1904) Über neuere Erfahrungen und Probleme auf dem Gebiet der Bodenbakteriologie und unter besonderer Berücksichtigung der Gründüngung und Brache. Arb Deutsch Landw Ges 98:59–78
Ho WC, Ko WH (1982) Characteristics of soil microbiostasis. Soil Biol Biochem 14: 589–593
Hora TS, Baker R (1970) Volatile factor in soil fungistasis. Nature 225:1071–1072
Hora TS, Baker R (1972a) Extraction of a volatile factor from soil inducing fungistasis. Phytopathology 62:1475–1476
Hora TS, Baker R (1972b) Soil fungistasis microflora producing a volatile inhibitor. Trans Br Mycol Soc 59:491–500
Hora TS, Baker R (1974) Abiotic generation of a volatile fungistatic factor in soil by liming. Phytopathology 64:624–629
Hora TS, Baker R, Griffin GJ (1977) Experimental evaluation of hypotheses explaining the nature of soil fungistasis. Phytopathology 67:373–379
Horsfall JG (1932) Dusting tomato seed with copper sulphate monohydrate for combating damping-off. NY Agric Exp Stn Tech Bull 198:1–34
Horsfall JG, Dimond AE (1963) A perspective on inoculum potential. Indian Bot Soc 42:46–57

Hsu SC, Lockwood JL (1973) Soil fungistasis: behaviour of nutrient-independent spores and sclerotia in a model system. Phytopathology 63:334–337
Katznelson H (1965) Nature and importance of the rhizosphere. In: Baker KF, Snyder WC (eds) Ecology of soil-borne plant pathogens. Univ California Press, Berkeley, pp 187–209
King JE, Coley-Smith JR (1968) Effects of volatile products of *Allium* species and their extracts on germination of sclerotia of *Sclerotium cepivorum* Berk. Ann Appl Biol 61:407–414
Ko WH, Lockwood JE (1967) Soil fungistasis: relation to fungal spore nutrition. Phytopathology 57:894–901
Ko WH, Hora FK (1972) The nature of a volatile inhibitor from certain alkaline soils. Phytopathology 62:573–575
Ko WH, Hora FK, Herlicska E (1974) Isolation and identification of a volatile fungistatic substance from alkaline soil. Phytopathology 64:1398–1400
Koch DO, Jeger MJ, Gerik TJ, Kenerley CM (1987) Effects of plant density on progress of Phymatotrichum root rot in cotton. Phytopathology 77:1657–1662
Leach LD, Davey AE (1938) Determining the sclerotial population of *Sclerotium rolfsii* by soil analysis and predicting losses of sugar beets on the basis of these analyses. J Agric Res 56:619–631
Leonard KJ (1980) A reinterpretation of the mathematical analysis of rhizoplane and rhizosphere effects. Phytopathology 70:695–696
Linderman RG, Gilbert RG (1973) Behavior of sclerotia of *Sclerotium rolfsii* produced in soil or in culture regarding germination stimulation by volatiles, fungistasis, and sodium hypochloride treatment. Phytopathology 63:500–504
Lingappa BT, Lockwood JL (1964) Activation of soil microflora by fungus spores in relation to soil fungistasis. J Gen Microbiol 35:215–227
Lockwood JL (1975) Quantitative evaluation of a leaching model system for soil fungistasis. Phytopathology 65:460–464
Loria R, Lacy ML (1979) Mechanism of increased susceptibility of bleached pea seeds to seed and seedling rot. Phytopathology 69:573–575
MacDonald JD, Duniway JM (1979) Use of fluorescent antibodies to study the survival of *Phytophthora megasperma* and *P. cinnamomi* zoospores in soil. Phytopathology 69:436–441
Mitchell JE (1979) The dynamics of inoculum potential of populations of soil-borne plant pathogens in the soil ecosystem. In: Schippers B, Gams W (eds) Soil-borne plant pathogens. Academic Press, New York, pp 3–20
Papavizas GC, Davey CB (1961) Extent and nature of the rhizosphere of *Lupinus*. Plant Soil 14:215–236
Park D (1965) Survival of microorganisms in soil. In: Baker KF, Snyder WC (eds) Ecology of soil-borne plant pathogens. Univ California Press, Berkeley, pp 82–98
Pavlica DA, Hora TS, Bradshaw JJ, Skogerboe RK, Baker R (1978) Volatiles from soil influencing activities of soil fungi. Phytopathology 68:758–765
Pfender WF, Rouse DI, Hagedorn DJ (1981) A "most probable number" method for estimating inoculum density of *Aphanomyces euteiches* in naturally infested soil. Phytopathology 71:1169–1172
Pfender WF, Hagedorn DJ (1983) Disease progress and yield loss in Aphanomyces root rot of peas. Phytopathology 73:1109–1113
Phipps PM, Beute MK (1977) Sensitivity of susceptible and resistant peanut cultivars to inoculum densities of *Cylindrocladium crotalariae* microsclerotia in soil. Plant Dis Rep 61:300–303
Punja ZK (1986) Relationships among soil depth, soil texture, and inoculum placement in infection of carrot roots by eruptively germinating sclerotia of *Sclerotium rolfsii*. Phytopathology 76:976–980

Punja ZK, Grogan RG (1981) Eruptive germination of sclerotia of *Sclerotium rolfsii*. Phytopathology 71:1092–1099
Punja ZK, Jenkins SF, Grogan RG (1984) Effect of volatile compounds, nutrient, and source of sclerotia on eruptive sclerotial germination of *Sclerotium rolfsii*. Phytopathology 74:1290–1295
Reynolds KM, Benson DM, Bruck RI (1985) Epidemiology of Phytophthora root rot of Fraser fir: rhizosphere width and inoculum efficiency. Phytopathology 75: 1010–1014
Ribeiro OK (1983) Physiology of asexual sporulation and spore germination in *Phytophthora*. In: Erwin DC, Bartnicki-Garcia S, Tsao PH (eds) *Phytophthora*, its biology, taxonomy, ecology, and pathology. Am Phytopathol Soc, St Paul, p 392
Rodriquez-Kabana R, Beute MK, Backman PA (1980) A method for estimating numbers of viable sclerotia of *Sclerotium rolfsii* in soil. Phytopathology 70: 917–919
Romine M, Baker R (1972) Properties of a volatile fungistatic factor in soil. Phytopathology 62:602–605
Rovira AD, Davey CB (1974) Biology of the rhizosphere. In: Carson EW (ed) The plant root and environment. University Press of Virginia, Charlottesville, p 691
Schippers B, Palm LC (1973) Ammonia a fungistatic volatile in chitin-amended soil. Neth J Plant Pathol 79:279–281
Schmitthenner AF (1988) ELISA detection of *Phytophthora* from soil. Phytopathology 78:1576 (Abstr)
Schnurer J, Rosswall T (1982) Fluorescein diacetate hydrolysis as a measure of total microbial activity in soil and litter. Appl Environ Microbiol 43:1256–1261
Schroth MN, Snyder WC (1961) Effect of host exudates on chlamydospore germination of the bean root rot fungus, *Fusarium solani* f. *phaseoli*. Phytopathology 51:389–393
Schroth MN, Cook RJ (1964) Seed exudation and its influence on pre-emergence damping-off in bean. Phytopathology 54:670–673
Schuster ML, Coyne DP (1974) Survival mechanism of phytopathogenic bacteria. Annu Rev Phytopathol 12:199–221
Short GE, Lacy ML (1974) Germination of *Fusarium solani* f. *pisi* chlamydospores in the spermosphere of pea. Phytopathology 64:558–562
Short GE, Lacy ML (1976a) Carbohydrate exudation from pea seeds: effect of cultivar, seed age, seed color, and temperature. Phytopathology 66:182–187
Short GE, Lacy ML (1976b) Factors affecting pea seed and seedling rot in soil. Phytopathology 66:188–192
Singleton LL, Mihail JD, Rush CM (1992) Methods for research on soilborne phytopathogenic fungi. Am Phytopathol Soc, St Paul, 266 pp
Taylor JD, Griffin GJ, Garren KH (1981) Inoculum pattern, inoculum density-disease incidence relationships, and populations fluctuations of *Cylindrocladium crotalariae* microsclerotia in peanut field soil. Phytopathology 71:1297–1302
Tomimatsu GS, Griffin GJ (1982) Inoculum potential of *Cylindrocladium crotalariae*: infection rates and microsclerotial density-root infection relationships on peanut. Phytopathology 72:511–517
Toussoun TA (1970) Nutrition and pathogenesis of *Fusarium solani* f. *phaseoli*. In: Toussoun TA, Bega RV, Nelson PE (eds) Root diseases and soil-born pathogens. Univ California Press, Berkeley, p 252
Vanderplank JE (1975) Principles of plant infection. Academic Press, New York, pp 84–87
Watson AG, Ford EJ (1972) Soil fungistasis – a reappraisal. Annu Rev Phytopathol 10:327–347

Wilkinson HT, Cook RJ, Alldredge JR (1985) Relation of inoculum size and concentration to infection of wheat roots by *Gaeumannomyces graminis* var. *tritici*. Phytopathology 75:98–103

Yarwood CE, Sylvester ES (1959) The half-life concept of longevity of plant pathogens. Plant Dis Rep 43:125–128

2 Host Roots

J.T. English and D.J. Mitchell

2.1 Introduction

Soilborne disease results from interactions among a host, pathogen, and environment, and roots obviously are the most important component of the host in diseases involving soilborne pathogens. A great deal of insight has been gained over the years, particularly in a qualitative sense, into the contributions that roots make to epidemic development. However, much remains to be done to quantify these contributions. This lag in quantitative insight derives to a great extent from the plasticity of root system development, which reflects temporal and spatial changes in production of specific root tissues in response to host genetics and diverse environmental signals.

Root system plasticity imposes severe constraints on the abilities of investigators to sample and quantify the contributions of these host tissues to disease. Ultimately, however, plasticity will have to be addressed because it reflects the summation of processes which determine the manner in which roots interact with pathogen populations. This is important both in the many pathosystems in which the opportunities for host-pathogen interaction are directly related to movement of susceptible root tissues to nonmotile pathogen propagules, and in pathosystems in which the pathogens themselves are motile (Baker et al. 1967). Within this framework of plasticity, this chapter focuses on the following aspects of root systems in relation to soilborne epidemics: (1) root systems as driving or constraining forces on opportunities for host-pathogen interactions; (2) methods used to assess or quantify the contributions of roots to epidemic processes; and (3) factors which influence the susceptibility of host roots to disease.

2.2 Root Growth and Energy Availability

Root systems have evolved along a variety of pathways to accommodate specific plant requirements for water and nutrients under a broad range of

C.L. Campbell/D.M. Benson (Eds.)
Epidemiology and Management of Root Diseases

environmental settings. The morphology and physiology of a root system of a specific species reflect the strategy of that species to accommodate the demands which shoot and root tissues place on each other (Troughton 1977), the requirements of shoot and root tissues themselves resulting from interactions with their enviroments. In the course of these tissue interactions, roots play a fundamental role in epidemic development through their contribution of the carbon and mineral nutrients that are required for the function of soilborne pathogens. A great deal of research on roots and soil has focused on the quantity and characteristics of carbon compounds which are derived from carbon dioxide fixed by photosynthetic processes, translocated through the shoot to the root system, and ultimately transferred into the soil. In the soil these carbon compounds provide a major source of energy for growth of microbial populations, including those which are pathogenic.

An indication of the importance of roots in energy availability to soilborne pathogens is derived from research on productivity of plant roots in annual and perennial agroecosystems. Fogel (1985) has summarized a number of investigations which estimate the annual production of fine roots as an indicator of the relative partitioning of carbon resources between shoots and roots. Many of these estimates were made for forest ecosystems and other systems dominated by perennial species. For instance, estimates of annual fine root production in forest ecosystems have ranged from 3.5 to 11.0 Mg of fine roots produced per hectare for 5-year-old Scots pine (Persson 1978) and 55-year-old Douglas fir stands (Fogel and Hunt 1983), respectively. Annual production of fine roots of herbaceous plants in both annual and perennial systems has been estimated to be intermediate or less prolific than in forest ecosystems. For example, Gregory et al. (1978) estimated that about 5.1 Mg of fine roots were produced per hectare annually in a perennial tall grass prairie, while only 1.2 Mg of fine roots were produced per hectare in an annual planting of winter wheat. In annual systems, fine root production as low as 0.6 Mg/ha has been reported for soybean plantings (Sivakumar et al. 1977). In contrast, up to 4.2 Mg of fine roots of corn have been produced per hectare (Barber 1971).

The great quantity of carbon that is introduced into soil via roots is underestimated in these investigations because additional organic material is introduced through rhizodeposition processes associated with normal root function. For example, Barber and Martin (1976) and Martin (1977) estimated that between 12 and 23% of photosynthetically fixed carbon is released by roots of wheat and barley into soil. Although the energy resources contributed by roots are obviously considerable, the spatial and temporal availability to resident pathogens is quite variable and complex. Cook and Baker (1983; Chap. 4, this Vol.) have described qualitatively the dynamics of nutrient availability in relation to root system development. The complexities of these developmental processes ultimately define the structure and dynamics of rhizosphere processes.

2.2.1 Specific Tissue Contributions

The nature of nutrients introduced into soil varies with the stage of root tissue development. The influences of these nutrients on surrounding pathogen populations are at the level of the microsite which includes as few as one or two plant cells and the adjacent soil. At any microsite, nutrients are derived from either functioning or decomposing tissues (Foster 1982; Foster and Bowen 1982; Foster et al. 1983; Newman 1985), and can include components of the tissues themselves or materials released from them into soil. Because a root system at any moment is composed of tissues which differ in age and function, there are great variations in the types and quantities of nutrients present, and in the abilities of pathogens to utilize them. The responses of pathogens to these energy sources vary among species. Consequently, an important step in evaluating the contributions of roots to epidemic development in a specific pathosystem is the characterization of tissues which are relevant to defined host-pathogen interactions of infection, tissue colonization, and pathogen reproduction and survival.

Differentiating Tissues. The importance of compounds released from differentiating tissues proximal to apical meristems in providing environmental cues to surrounding pathogenic organisms is well known. These are tissues which are predominantly involved in the uptake of water and nutrients from soil. A diverse array of organic materials, including amino acids, organic acids, nucleotides and proteins, are released from these and other root tissues. Cook and Baker (1983), Rovira (1969), and Schroth and Hildebrand (1964) summarized the numerous important roles ascribed to these materials in stimulating the development and growth of many types of pathogenic organisms. For instance, one of the most well-known and best-characterized effects of released materials is the attraction of zoospores of *Phytophthora* and *Pythium* species to susceptible tissues in the region of elongation behind the root tip. Carlile (1983) and Royle and Hickman (1964a,b) discussed possible mechanisms for the tactic responses of pathogens within these genera. Organic materials released from this region also influence nonmotile pathogens by stimulating the germination of dormant propagules. Griffin (1969), for example, characterized the germination response of chlamydospores of *Fusarium oxysporum* relative to their position along roots of 3-day-old peanut plants. In this pathosystem, chlamydospore germination began within 9.5 mm of the root tip and increased through distances of 13 mm from the tip.

It is important to recognize that the success of infection depends on the balance of root extension and tissue maturation rates relative to germination and tactic responses of pathogens. Cook and Baker (1983) described qualitatively the possible scenarios involved in pathogen responses to tissue differentiation.

In addition to stimulatory and tactic influences on pathogens, the dynamics of differentiation may affect the success of tissue colonization after infection. In the case of wheat infected by *Gaeumannomyces graminis* var. *tritici*, more severe disease can develop after infection of newly formed and relatively undifferentiated tissues than after infection of differentiated tissues (Deacon and Henry 1980). Although initial infection occurs readily in both cases, subsequent vascular colonization can be limited in differentiated tissues by the presence of an endodermis.

Mature Tissues. Organic materials are released not only from differentiating tissues located near root tips, but from mature tissues at greater distances from growing points (McDougall and Rovira 1970). These regions of roots are involved to a large extent in the translocation of resources between roots and shoots. Foster (1982), Foster and Bowen (1982), and Foster et al. (1983) enumerated processes of cell maturation and senescence, and cell sloughing and lysis by which these materials are released. Complex rhizosphere microbial communities typically develop in response to the availability of the nutrients. Cell lysis, in particular, may occur at wound sites in mature tissues. Organic materials released after lysis stimulate the germination and tactic responses of dormant propagules. For example, Marx and Davey (1969) described this scenario as a means by which pine roots can be infected by *Phytophthora cinnamomi*, and Cook and Baker (1983) reported wound sites as infection points for species of *Fusarium* and *Verticillium*.

Senescing cortical cells, which appear either in response to microbial degradation or under the control of the plant, provide additional sites of pathogen stimulation. Cortical cell senescence may be the first step in general root senescence which occurs in many plant species as they shift from vegetative to reproductive phases of development (Huisman 1982). Senescence may begin much earlier, however, as has been reported in cereals, and the rate and pattern of occurrence are influenced by both plant genetics and environment (Holden 1975, 1976; Henry and Deacon 1981; Lewis and Deacon 1982). For example, Deacon and Lewis (1982) demonstrated this possibility for wheat in relation to infection by *Cochliobolus sativus*, a major component of the common root rot complex. These investigators worked with the wheat cultivars Rescue, which exhibits a high rate of early cortical cell senescence, and Cadet, which exhibits a low rate of early senescence. Coincidentally, Rescue is more susceptible than Cadet to disease caused by *C. sativus* (Larson and Atkinson 1970). Deacon and Lewis (1982) demonstrated that the rate of senescence was controlled by genes on specific chromosome pairs of the two cultivars, and when the chromosome pairs were switched between cultivars, both the senescence and susceptibility responses were reversed. From these observations, the authors suggested that small differences in the rate of senescence processes may strongly influence the success of infection and disease development. The extent to which cortical cell senescence influences disease progress in other plant species is as yet uncertain; however, in addition to wheat and barley, the phenomenon has been observed in roots of rape (MacLeod et al. 1986).

Decomposing Tissues. Root tissues begin to decompose after plant death in annual species, or after active growth has ceased in perennial species. As decomposition begins, root tissues may already have been colonized to some degree by one or more pathogenic organisms. This is particularly likely for wilt-inducing species of *Fusarium* and *Verticillium* which can colonize vascular tissues of host plants extensively.

Because of prior colonization, decomposing tissues may provide limited resources for colonization by pathogens in surrounding soil. However, the tissues are important in providing protection for those pathogens already within the tissues. Numerous investigators have described the occurrence of hyphae or survival structures, such as chlamydospores, oospores, or microsclerotia, in undefined root debris (Boosalis and Scharen 1959; Evans et al. 1966; Ashworth et al. 1974; Mitchell and Kannwischer-Mitchell 1983). Although fewer investigators have quantified the impact of these occurrences on pathogen survival, its potential importance has been demonstrated for survival of *Verticillium albo-atrum* in colonized cotton tissues. In field-based studies, Evans et al. (1966) observed that microsclerotia survived for longer periods of time in basal stem tissue than freely in soil. Although root tissues were not examined as a reservoir of survival propagules, it is likely that a significant proportion of the residual population of *V. albo-atrum* at the time of planting also represents the microsclerotia originally in root tissues.

In addition to pathogens, decomposing tissues support a diverse array of saprophytic organisms which can utilize components of dead tissues. Antagonistic members of this developing microbial community may strongly influence the survival of pathogens within the decomposing tissues. Wilkinson et al. (1982) suggested that this was the case for antagonistic strains of *Pseudomonas* spp., which colonized root lesions on live roots caused by *G. graminis* var. *tritici*. Colonization in this manner reduced infections of plants established in soil containing these infected roots.

Although quantification of the microbial community dynamics in decomposing tissues is a daunting task, much could be gained from the effort. Pfender (1988a) and Pfender and Wootke (1988b) examined these same processes in relation to succession of decomposing organisms on wheat debris and derived some insight into the potential management of community components for reducing surviving pathogen populations.

2.2.2 Characterization of Relevant Root Tissues

Although plant pathologists historically have acknowledged the contribution of tissue status to host-pathogen interactions, relatively few have quantified this component of the interaction (Gilligan 1983; Jeger 1987). As investigators move towards the development of a more biologically based understanding of epidemic processes, greater consideration should be given to the quantification of the roles of specific tissues in the host-pathogen interaction. It should be possible to more clearly define the importance of microsite events as biotechnological methods become more readily available. Deacon and Lewis (1982), for instance, showed the potential for more clearly defining the role of specific tissues by direct genetic manipulation. With the rapid development of biotechnological procedures for even more precise genetic manipulation, research on host-pathogen interactions at the microsite level should become more approachable.

In addition to biotechnological methods, a variety of other approaches, including some of which are relatively simple, are available to characterize host-pathogen interactions associated with specific root tissues, and to define organic materials of importance in these interactions. For instance, to determine the germination efficiency of chlamydospores of *F. oxysporum* along the length of peanut roots, as described earlier, Griffin (1969) used the root slide method of Jackson (1960) in which peanut root growth was directed along slides coated with water agar that had been embedded with macroconidia and chlamydospores of the pathogen. After a period of growth, germination of pathogen propagules could be evaluated in relation to root position and tissue status. English and Mitchell (1989) used a simple point inoculation method to quantify susceptibility of tobacco root tissues to infection by *Phytophthora parasitica*. When root tips of these plants were inoculated with zoospores, over 70% of inoculations led to root infection. In contrast, only 10% of root tissues 2 cm behind root tips, and less than 5% of root tissues behind the first emerged lateral root were infected after inoculation. The rates of lateral root extension in these experiments varied between 0.3 and 0.7 cm/day. It was apparent, therefore, that changes in tissue structure, physiology, and susceptibility occurred very rapidly over a period of just days. The type of information developed by these approaches should prove useful in understanding the timing of host-pathogen interactions that define the efficiency of initial infection processes.

By another simple method, that of direct observation, Reynolds et al. (1985b) evaluated the dynamics of secondary inoculum production by *Phytophthora cinnamomi* on roots of Fraser fir. By this approach, these investigators determined that the majority of sporangia of this pathogen were produced within 5 mm of actively growing root tips, and that the number produced was affected by soil moisture and time. By combining this information with evaluations of numbers of susceptible infection sites, it should be possible to estimate the potential for secondary inoculum formation in relation to root system development.

Direct and indirect assays are available to associate various organic materials with specific tissues (Curl and Truelove 1986). For example, McDougall and Rovira (1970) characterized the distribution of carbon exudates of all forms by exposing wheat seedlings to radiolabeled carbon dioxide. Exudates containing labeled compounds were collected on filter paper in contact with growing wheat roots. As visualized by autoradiography, the most intense exudation occurred at the apices of lateral roots where tissues were differentiating rapidly. Curl and Truelove (1986) also summarized studies by Frenzel (1957, 1960) who used growth of nutritional mutants of *Neurospora crassa* to determine which root tissues of sunflower released specific amino acids. In other experiments, Pearson and Parkinson (1961) evaluated the distribution of protein exudates from broad bean seeds germinated on filter paper. After ninhydrin staining, these investigators observed that most proteinaceous materials were associated with the region of tissue just behind root tips. Schroth and Snyder (1961) used this same approach to describe similar regions of high exudation of ninhydrin-staining materials and reducing sugars from roots of developing seedlings of pinto bean. Ultimately, enhanced characterization of root tissues and their influence on epidemics depend on the creativity of the individual investigator to effectively combine these types of approaches.

2.3 Root System Plasticity

The distribution of nutrient sources depends on the patterns of growth of specific root tissues in soil. These distributions, in turn, influence the course of disease development on root systems of individual plants. This becomes apparent if disease progress is viewed as repeated cycles of root tissue infection and progressive tissue colonization (Huisman 1982; Gilligan 1983). Infection of a root system results initially from the contact of susceptible tissues with residual pathogen propagules produced during a preceding cycle of plant growth. Subsequently, reinfection of the same root system occurs when susceptible tissue develop adjacent to colonized roots which support the production of pathogen inoculum, or when tissues develop adjacent to infected roots of nearby plants. Root tissues may be colonized progressively from these infection points, and the specific tissues colonized are characteristic of the association which has evolved between a specific host and pathogen.

The arrangement of differentiating, mature, and decomposing tissues in relation to one another will influence the manner in which discrete processes are coupled to affect the dynamics of disease development. Although the basic patterns of root system branching, and thus, tissue distribution for any plant species is controlled genetically, root systems also respond to the heterogeneity of a variety of soil factors (Zobel 1975, 1986). The interactions of genetic control and environment permit a great deal of plasticity in root system morphology and architecture. For instance, Wiersum (1958) observed that the length and density of specific lateral roots of pea were affected differentially by the type and concentration of ions examined in sand. Investigations by Drew (1975) and Drew and Saker (1978) provided evidence of root system plasticity in barley, which resulted from differential branching in regions of high or low concentrations of nitrogen and phosphorus in soil.

The degree of plasticity expressed by plants is species-dependent. Crick and Grimes (1987), for example, differentiated morphological plasticity of root systems of *Agrostis stolonifera* and *Scripus sylvaticus* in response to fluctuating soil fertility levels. *Agrostis stolonifera* responded to spatial and temporal variability in soil fertility by producing numerous lateral roots, which increased the surface area of absorptive tissues. This change in root system structure enabled efficient utilization of temporarily available nutrients. In contrast, *S. sylvaticus* maintained a less plastic root system which effectively utilized flushes of nutrients through modification of uptake rates.

2.3.1 Conceptual Frameworks for Root System Structure

The plasticity and size of root systems are factors that have discouraged evaluations of root system branching and tissue distribution in epidemiological studies. In fact, because of labor and time constraints, investigators often use soil coring methods to quantify the abundance of roots as average root density or total root length per unit volume of soil. Roots collected in cores can be characterized by categories such as "fine roots" or "rootlets," and such characterizations have proven useful for comparing relative infection patterns of different pathogens (Burke and Barker 1966; Dryden and Van Alfen 1984; Hancock 1985; Huisman 1988a; Hancock and Grimes 1990). Soil cores have also proven useful in characterizing root density in relation to pathogen infection and important soil factors. For instance, by using soil cores, Dryden and Van Alfen (1984) were able to elucidate the response of dry beans to infection by *Fusarium oxysporum*. Host response depended on the vertical distribution of roots, pathogen propagules, and soil water. In contrast, Kotcon et al. (1984) were not able to detect differences between root densities of healthy potatoes and those which exhibited symptoms of early dying.

Although soil coring methods provide information useful for characterizing root density patterns in relation to plant growth and soil characteristics, they destroy the integrity of tissue connections and preclude evaluation of the spatial arrangement of specific tissues. Incorporation of root system plasticity into evaluations of epidemics will not be achieved until an appropriate conceptual framework has been established. Several models which have been devised to describe root system growth and development may assist in establishing this framework.

Developmental Model. Historically, a chronologically based, developmental model has been used to describe root system structure (Fig. 2.1). In the developmental model, root tissues are characterized by the time of lateral

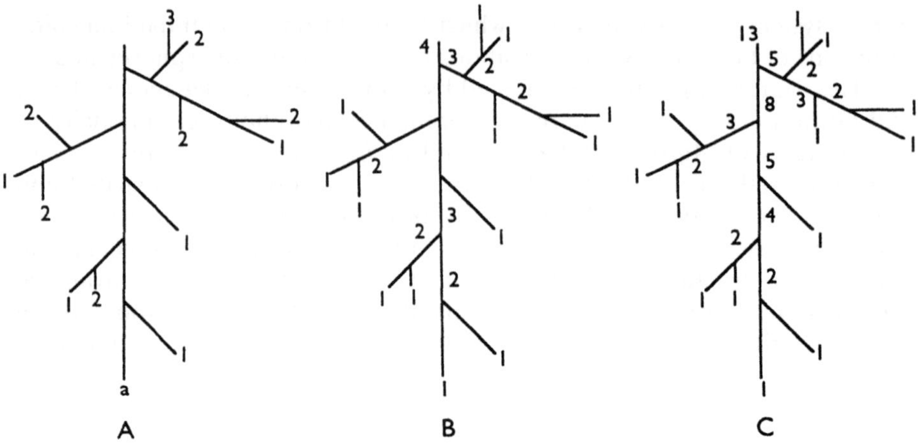

Fig. 2.1. Diagrammatic root systems classified by the developmental scheme **A** in which *a* represents the axis and *numbers* refer to lateral root classes (Barley 1970); by the morphometric scheme **B** in which *numbers* refer to root-order classes (Fitter 1982); and by the topological scheme **C** in which each *number* refers to the magnitude of a specific link whose value is equal to the sum of magnitudes of two daughter links. (Fitter 1986)

root initiation (Barley 1970; Rose 1983). In this scheme, roots which emerge initially from the plant crown are defined as axes. Any root which emerges from an axis is defined as a primary lateral root, and any root which emerges from a primary lateral is defined as a secondary lateral root, and so forth. By this system, all tissues which make up the length of a root are considered as part of the same lateral group regardless of physiological status and function of these tissues.

The developmental model has been used in studies of plant responses to soil factors and infection. Tennant (1976) used this model to describe the differential sensitivity of wheat lateral roots to alterations in levels of nitrogen, potassium, and phosphorus. May et al. (1967) employed the model to evaluate changes in root length extension rates as higher order lateral roots begin to emerge in developing seedlings of barley. More recently, Belford et al. (1987), Klepper et al. (1984), and Porter et al. (1986) described forms of the developmental model which were useful in elucidating the balance among the development of leaves, tillers, and roots of wheat.

The developmental model has seen much less use in quantitative pathological studies. However, its potential utility was shown by Bloomberg (1979a,b) who used the model to effectively quantify the likelihood of infection of Douglas-fir seedlings by *Fusarium oxysporum* in relation to the rate of production and extension of lateral roots. The model proved useful in evaluating this pathosystem because the process of concern was root tip infection. The developmental model has more limited utility for delimiting tissues in other parts of a root system in relation to susceptibility to infection or production of secondary inoculum.

Morphometric Model. An alternative model for describing root system structure, developed by Fitter (1982), is based on morphometric analysis, and provides an alternative view of the heterogeneity of root tissues (Fig. 2.1). This model was derived from a system of analytical methods used to describe watershed geomorphology (Strahler 1957). In the morphometric framework, root systems are classified in a direction opposite to that of the developmental approach. Accordingly, roots which terminate in apical meristems are defined as first-order lateral roots. Where two first-order roots merge, there begins a second-order lateral root, and so forth. An advantageous, but coincidental, characteristic of this scheme is that each lateral root is broken into regions which describe more realistically tissue developmental state and function than in the developmental model. Specifically, the length of a root is classified by regions which are more closely related to activities such as water uptake proximal to the apical meristem, and translocation functions and cell sloughing at points distal to the meristem. This type of tissue division should be useful for characterization of host tissues in relation to specific epidemic events.

The utility of the morphometric framework was demonstrated by Fitter (1982) in evaluations of the complexities of root branching patterns of *Poa annua* and *Rumex crispis* in relation to soil fertility. Fitter observed greater changes in the number of second-order roots than first-order roots relative to soil fertility level for both species. Thus, the resulting changes in proportions of root tissues involved in uptake and transport were reflected in alterations in the relative proportions of root categories, or branching ratios.

The morphometric model has been used rarely in epidemiological investigations. In the root infection studies of English and Mitchell (1989) described previously, root tissues were classified according to this framework. In other experiments, English and Mitchell (1984) demonstrated changes in the branching of *Catharanthus roseus* in response to infection by *Phytophthora parasitica*.

Analogous parameters have been described within the developmental and morphometric models to characterize morphological aspects of root system development. For example, descriptive parameters for root diameter, length, rate of branching, and rate of extension are available in both models. Direct comparisons of parameter estimates in the two models should be made with caution, however, because the orientation of root classification differs greatly (Fitter 1982; Rose 1983). The morphometric model provides additional capability for describing the relative proportions of roots within specific categories through parameters of branching and length ratios. These descriptors provide some estimate of root system architecture.

Topological Model. Fitter (1987) argued recently that a major weakness of the developmental and, to a lesser extent, the morphometric model is the segment-based approach used for root classification in which tissues of different function are included in the same root element category. The inability to classify segment function into mutually exclusive categories makes it difficult to evaluate and describe the manners in which root systems explore and exploit soil resources. To address this concern, Fitter (1985, 1986, 1987, 1991) established an architectural root model which is composed of topological and geometric components (Fig. 2.1). As indicated by Hetrick et al. (1988, 1991), this framework is derived from concepts developed in graph theory. Root elements within the architectural model are defined quite differently from those of the earlier models. Link characterization provides the topological orientation of the model. An interior link is a length of root bound by junctions with other links, and an exterior link is a length of root bound by a meristematic region and a root junction. The geometric components of this model include descriptive parameters of root tissue structure analogous to those described in earlier models. These include root length, root width, and branch angle of the root elements of interest.

In the architectural model, exterior links represent tissues involved in resource uptake, and interior links represent tissues involved in translocation functions. The manner in which links are joined determines the efficiency of resource uptake and efficiency of transport to the plant shoot. The manner in which links are joined to achieve these functions is described by parameters such as magnitude, or total number of exterior links feeding into a link of interest; total exterior path length, or the sum of all pathways from exterior links to the shoot base; and altitude, or the longest individual path length. Fitter (1985, 1986, 1987) described the theoretical structures that root systems should exhibit on the basis of idealized parameter values and their resulting utilities for soil resource utilization. Fitter (1986) also used these parameters to elucidate the response of root system development of *Trifolium pratense* to soil moisture. Specifically, he associated alterations in root system architecture of *T. pratense* with changes in link assembly as reflected in altered values of altitude. Soil moisture had no effect on geometric parameters such as mean length of internal links.

The architectural model has been used only once in evaluations of host-pathogen interactions. Larkin et al. (1992) evaluated the impact of infection of alfalfa seedlings by *Pythium* spp. using parameters from this model. The authors observed that infected seedlings had shorter and less branched root systems than did noninfected or metalaxyl-treated seedlings; the reductions in root system branching were reflected in reductions in root system path length. Hetrick et al. (1988, 1991) also used the topological model to describe the plasticity of root systems of a warm-season grass in relation to mycorrhizal associations. Again, root system plasticity was reflected in infection-dependent changes in path length. In contrast, phosphorus concentration had no influence on root system architecture.

Further application of this model to studies of plant infection under defined soil conditions should enhance efforts to explore the types of interactions which can occur among roots and pathogens, and to interpret whole plant responses in epidemics. As greater consideration is given to understanding genetic control of rooting strategies, this model should provide fresh insights into management strategies based on whole root system manipulations.

2.4 Evaluation of Root Systems

Epidemiologists are hampered by technical difficulties involved in sampling root systems in a manner in which relevant tissue arrangements are maintained. Historically, destructive root-sampling methods have been used to assess root growth. Many of these involve soil cores, trenches, monoliths, etc., and have been described in detail by Böhm (1979). Although labor and time intensive, these methods have proven useful for characterizing the spatial arrangement of roots of undefined type relative to soil characteristics and plant community structure.

Radiotracer methods have been combined with soil core procedures to evaluate the distribution of roots of individual plants in soil. For example, Bookman and Mack (1982) used a double-labeling radioisotope technique to monitor root competition of two grass species, *Bromus tectorum* and *Poa pratensis*. A different radioisotope was applied to the leaves of each species, and after a period of growth in various combinations of plant densities, soil cores were taken at defined depths and distances from labeled plants. The pattern of occurrence of one or both isotopes in roots removed from cores reflected the extent of interspecific root competition.

Radiotracer methods have also been used as a nondestructive technique of evaluating root distribution in soil. Abbot and Fraley (1991) described several methods by which radioisotopes can be injected into soil and evaluated for uptake by actively growing roots. In the central injection method, for instance, a radioisotope is injected into soil at a defined depth in the center of a monoculture or mixed-species planting. After sufficient time for uptake of the isotope by roots, foliage of surrounding plants are evaluated in situ or in the laboratory for the level of isotope activity. The pattern of radioactivity in surrounding plants defines the soil volume occupied by individual plants. This approach can be applied periodically to evaluate changes in root distribution during plant growth. In epidemiological studies, specifically, the approach should help to define the distribution of susceptible, differentiating root tissues involved in the uptake of radioisotope.

Other nondestructive techniques may also prove useful for detailed examination of epidemic processes. For example, rhizotrons have been used for many years to quantify root system productivity in terms of seasonal rates of lateral root production, extension, and senescence (Atkinson 1985). More recently, mini- and microrhizotrons have been used to gather this type of information under field production conditions (Upchurch and Ritchie 1983; Carpenter et al. 1985; Itoh 1985). These devices may prove useful for estimating spatial and temporal changes in total root density. However, their utility for observing root development in relation to disease development is tempered by questions of growth artifacts associated with root development along various glass or plastic surfaces (Taylor and Böhm 1976; Atkinson 1985), and by limitations for usage in nonshrinking soil conditions (Rush et al. 1984).

Another newly developed approach of potential utility for nondestructive observation of root growth and interactions with pathogens is magnetic resonance imaging (MRI). This type of imaging uses the interaction between magnetic fields and imposed radio frequencies to detect and visualize the concentration of protons associated predominantly with water (Brown et al. 1986; Cooper et al. 1989). It is possible with this approach to visualize root tissues in situ potentially with high resolution. The utility of this method is its nondestructive method of visualizing the physiological status and spatial distribution of roots in soil while maintaining information on tissue arrangement. Additionally, such visualization can be made over time with precise relocation of specific tissues in soil. A major drawback of this approach is currently the great expense and limited availability of imaging equipment. Furthermore, imaging equipment is not portable, and plants must be brought into the laboratory for analysis. This constrains seriously the size of root systems which can be evaluated by this method.

A final method of analysis of root system growth and structure, which has become more useful, is that of mathematical models. This approach employs both nondestructive and destructive methods. Destructive methods, such as described previously, are used to develop initial estimates of biologically important parameters included in a model. Nondestructive simulation methods then are used to examine rapidly scenarios of root growth and subsequent root system architecture in relation to factors of interest. Destructive or nondestructive methods are used, once again, to validate the accuracy of model predictions in controlled experiments.

Models often prove to be useful in developing and testing hypotheses of root system development as influenced by interacting environmental variables. This approach has been used often by investigators to describe inherent patterns of root growth (Hackett and Rose 1972; Gerwitz and Page 1974; Hayhoe 1981; McCoy and Boersma 1986; Bowers et al. 1991); such models have utilized both empirical and process-oriented approaches. Modeling techniques have been used to predict root system responses to soil moisture, temperature, and bulk density (Lungley 1973; Dexter 1978, 1987; Hewitt and Dexter 1979; Stone et al. 1983; Thornley 1985; Protopapas and Bras 1987; Simmons and Pope 1988).

Although modeling as an investigative tool in epidemiological studies has been used only rarely, its utility in this regard is evident. For example, Bloomberg (1979a,b) simulated the growth of root systems of Douglas-fir and subsequent infection by *Fusarium oxysporum*. By considering the host growth as a component of the epidemic, he was able to predict the occurrence of root rot in relation to soil environmental conditions. Reynolds et al. (1986) also incorporated root growth as a factor of importance in simulating the secondary spread of *Phytophthora cinnamomi* in a population of 2-year-old Fraser-fir seedlings. Root growth, which was estimated as a function of shoot development, was incorporated into a model describing infection and reproduction of *P. cinnamomi*. The model allowed for meaningful investigation of disease spread from root system to root system as a function of plant density and soil environmental factors. In a different approach, Jeger (1987) developed a theoretical model, based on differential equations, to evaluate the contribution of root growth to disease. Specifically, this approach provided a means to evaluate possible scenarios for the dynamic properties of disease increase on individual root systems. To date, few root growth models have been validated under field conditions (Bloomberg 1979a; Hayhoe 1981; Stone et al. 1983). However, this is something that will have to be done before their utility is accepted.

The combination of relevant conceptualizations of root system structure and function with additional innovative experimental approaches to visualization of root systems should stimulate interest in considering realistically the important roles that roots play in epidemics. It also should help focus attention on the development of innovative sampling strategies that allow for consideration of the complexities of root systems, and the relationships of these complexities to the potential for infection and the response of the host to infection and colonization.

2.5 Host Susceptibility

Alteration in susceptibility to pathogens is the most important influence that the host has on altering the rate of disease development (Burdon 1987; Seem 1988; Hau and Kranz 1990). Thus, as increasing emphasis is placed on host growth in the analysis of root disease epidemics, it becomes more important to evaluate factors that alter susceptibility of the host, and, in turn, to determine the influence of changes in host susceptibility on disease development and pathogen dynamics. Overviews of impacts of alterations in host susceptibility on host-pathogen interactions, particularly of foliar diseases, have been provided by Yarwood (1959), Populer (1978), and Seem (1988). The influence of the host and environment on alterations in susceptibility has been treated under headings such as predisposition, environmentally conditioned susceptibility, acquired susceptibility, physiological susceptibility, or induced susceptibility (Yarwood 1959; Schoeneweiss 1975). Plant pathologists historically have emphasized factors that predispose plants toward greater susceptibility. However, it is important to integrate knowledge of induced changes toward both lesser and greater susceptibility into the current concepts of epidemiology that incorporate decreases, as well as increases, of plant diseases in a host population in time or space (Gilligan 1987; Kranz 1990a). Although a myriad of host-pathogen interactions, as moderated by a complex physical and biological environment, influence the genetic control of the plant's defenses against disease, the emphasis in this section will be limited to the roles of the host and important environmental factors as moderators of root susceptibility to soilborne pathogens. Space limitations preclude discussion of static host influences, such as genetic susceptibility and stand density of a crop.

2.5.1 Host Influences on Susceptibility Alterations

Susceptibility of plants to soilborne diseases may be altered as plants age. Four general classes of susceptibility alterations with plant age, as presented by Yarwood (1959) and discussed by Populer (1978) and Seem (1988), include: susceptibility decreasing with the age of the plant, particularly following juvenile susceptibility; susceptibility increasing with age; and susceptibility of plants of intermediate age either greater or less than that of young or old plants. Yarwood (1959), Populer (1978), and Burdon (1987) provide examples of diseases fitting these categories, but few critical studies have been conducted with root diseases. It is difficult to critically assess susceptibility alterations in roots to soilborne pathogens, especially in plants past the seedling stage, because of the need for multiple disease evaluations over the physiological ages of plants which are exposed to uniform inoculum and maintained under standardized environmental conditions. The problem is compounded by the current limitations of nondestructive methods for observing growth and development of intact root systems in soil.

In general, decreases in susceptibility to soilborne pathogens with host age may be rapid, as typified by damping-off of various hosts by species of *Pythium* (Mellano et al. 1970), or gradual, as exemplified by the progressive expression of resistance of pepper plants to *Phytophthora capsici* after the susceptible one to eight-leaf stages (Kim et al. 1989). Altered susceptibility of hosts of intermediate ages may be associated with physiological changes related to shifts to reproductive stages. Symptoms of Verticillium wilt, for example, appeared in susceptible cultivars of potato, but not resistant cultivars, after tuber initiation. Symptom appearance corresponded with a sharp drop in phenol contents of vascular elements to a level below that required to inhibit extracellular enzymes of *Verticillium albo-atrum* (Huber 1989).

Changes in susceptibility with ontogenetic differentiation over time also can be attributed to the formation of different plant organs with varying levels of susceptibility (Last 1971; Populer 1978; Seem 1988). Although alterations in susceptibility associated with diseases of specialized organs, such as tubers, are typically presented as examples (Yarwood 1959; Populer 1978; Seem 1988), the detection of alterations of susceptibility in roots of different physiological or architectural stages should be of paramount importance to epidemiologists working on root diseases.

Structure and function of different types of roots vary with position within the architecture of the root system (Fitter 1985, 1991; Zobel 1991a,b), and, because individual roots can respond differently than the entire root system to stress (Waisel and Eshel 1991), susceptibility may vary with the position and type of root. Even in the few cases in which infection in individual root systems at defined inoculum densities has been quantified, susceptibility of different types of roots could not be assessed because estimates have been based on infections per unit length of root and the types of root infected rarely have been defined (Tomimatsu and Griffin 1982; Dryden and Van Alfen 1984; Hancock 1985). By assessing patterns of infection by *Phytophthora parasitica* var. *nicotianae* over time in entire root systems differentiated into root orders, English and Mitchell (1989) demonstrated that more than 70% of all infections per tobacco plant occurred on first-order (root tips) roots. Thus, if the pathogen infects primarily first-order roots, the proportion of tissue susceptible to soilborne pathogens at different times or stages of host development will depend on the ratio of susceptible first-order roots to nonsusceptible second- or higher-order roots. Pathogens, such as species of *Cylindrocladium*, *Fusarium*, and *Verticillium*, may colonize different types or orders of roots uniformly (Gerik and Huisman 1988; Huisman 1988a,b; Tomimatsu and Griffin 1988). However, point inoculations of different types of roots over time are needed to define clearly the different susceptibilities of root tissues to primary infection and, subsequently, to differentiate factors that induce colonization to progress from the cortex to the vascular tissue. Additional studies are needed for the evaluation of the likelihood of infection and subsequent disease development in different types of roots from different locations in the root system over time. Only by carefully controlling environmental conditions will host-mediated changes in susceptibility be differentiated from environmental predisposition.

Many other components of host development may contribute indirectly to alterations of susceptibility to root diseases. The differential responses of the host to various pathogens or beneficial microorganisms may result in alterations of susceptibility to other pathogens. Plants may alter the soil environment by changing soil temperature through shading, mulch formation, or soil water manipulations; by influencing soil pH; by differential utilization of nutrients; by increasing bulk density; or by quantitatively or qualitatively changing rhizodeposition. These host-induced environmental changes can influence the composition, physiological activity, and spatial patterns of microbes associated with hosts to favor either beneficial or deleterious microorganisms, which in turn may result in altered expressions of host susceptibility. With the integration of quantitative information on root colonization by microbes into root growth and response models, new insight should be provided on interrelationships of the host and various microbes in altering susceptibility. Advances in techniques for marking microorganisms, such as the tracing of transposon mutants by the most-probable-number-DNA hybridization procedure and antibiotic resistance (Bazin et al. 1990; Lam et al. 1991) should make such quantitative information available.

Future studies should provide insight into daily, seasonal, and yearly fluctuations in host susceptibility. For example, daily quantitative or qualitative changes in root exudates, which have been shown to occur in resistant and susceptible plants (Funck-Jensen and Hockenhull 1984), could alter root susceptibility. Susceptibility may change with physiological and anatomical responses of individual roots and root systems to seasonal changes or to modifications of plant growth over many years. For example, changes in susceptibility are expressed in the patterns of seasonal or perennial containment or breakout of pathogens compartmentalized in bark or wood of older roots (Shigo 1984).

Changes in the composition or structure of host populations in plant communities (Burdon 1987; Alexander 1990; May 1990; Paul 1990; Harper 1990) may have important impacts on proportions or distributions of susceptible plants. Populer (1978) proposed evaluating changes in susceptibility in populations of plants by comparing over time the proportions or accumulations of susceptible plant parts found at different stages. Evaluations of the influences of alterations in susceptibility in host populations within communities have been of interest in plant ecology studies. Although disputed by Kranz (1990b), it has been generalized that epidemics rarely occur in natural plant communities (Burdon 1987; Augspurger 1990). The possibility that epidemics are restricted in natural communities because diseases are only important in specific phases of population development led to the prediction that newly established, expanding plant populations, with high densities and homogeneous stands, are more susceptible to epidemic "outbreaks" by invading pathogens than are old, established populations (Harper 1977). This prediction has been tested by following the incidence of disease in individual populations of hosts over time in long-term field studies (Harper 1977), and recently in a unique study on islands on which approximate ages of host populations could be estimated by correlation to island ages (Carlsson et al. 1990). In the latter study, susceptibilities of three hosts to smuts and a rust were largely restricted to early-intermediate phases of host population development within communities. Similar studies with root diseases in natural communities could provide significant insight into susceptibility alterations in host populations within plant communities.

2.5.2 Environmental and Cultural Factors in Susceptibility Alterations

Biotic and abiotic environmental factors may affect susceptibility through alterations in the host prior to infection (predisposition) or in the host-pathogen association after infection has occurred (Levitt 1972; Colhoun 1973; Schoeneweiss 1975; Burdon 1987). A multitude of complex interactions of the soil microflora, microfauna, and macrofauna occur within and around plant roots (Beute and Benson 1979; Coleman 1985; Curl and Truelove 1986; Bazin et al. 1990; Curl and Harper 1990; Lynch 1990), and alterations in plant-microbe interactions may result in increased or decreased susceptibility to soilborne pathogens. Recognition of components of susceptibility to root diseases will depend on new approaches or techniques for studies of interactions in the more carefully defined endorhizosphere, rhizoplane, ectorhizosphere, and mycorrhizosphere (Lynch 1990; Reid 1990). Much of the information on the impact of plant-microbe influences on reductions of susceptibility comes from studies on symbionts, particularly mycorrhizae, or on biological control of root diseases (Cook and Baker 1983; Perrin 1990; Reid 1990). Conversely, increased susceptibility has been reported typically from work on interactions of soilborne organisms, such as nematodes or mites, with fungi (Powell 1971; Beute and Benson 1979; Curl and Truelove 1986).

Plants generally exist as symbionts with various microorganisms, and, since the physiology of the entire plant is altered when symbiosis occurs, studies on susceptibility should be conducted with plants in representative symbiotic associations whenever possible (Linderman 1988; Campbell and Greaves 1990). Vesicular-arbuscular endomycorrhizae (VAM) and ectomycorrhizae are among the most common symbiotic associations (Perrin 1990; Wilcox 1991), and variations in their root and fungal components or stages of development may critically influence susceptibility. Many studies with ectomycorrhizae and several studies with VAM indicate decreased susceptibility of mycorrhizal plants to diseases caused by soilborne fungi or nematodes; however, disease severity may also be increased or unaffected in mycorrhizal plants as compared to nonmycorrhizal plants (Perrin 1990). While ectomycorrhizae may provide physical or chemical protection against pathogens, vesicular-arbuscular mycorrhizae generally influence disease by plant growth stimulation (Perrin 1990; Reid 1990). It is interesting to speculate that intra- and interspecific connections of host roots by hyphae of both ecto- and vesicular-arbuscular mycorrhizal fungi (Read et al. 1985; Reid 1990) may allow translocation within plant populations of factors important in altering susceptibility.

Studies on biological control have shown that disease caused by one microorganism may be lessened by prior inoculation with a different microorganism (Cook and Baker 1983). Thus, reduced susceptibility may result from cross-protection or induced resistance. Virulent strains of several forma speciales of *Fusarium oxysporum*, for example, may cause little or no wilt if their hosts have been inoculated previously with avirulent strains of *F. oxysporum* (Louter and Edgington 1990). Although sound evidence for microbially induced alterations in susceptibility to soilborne pathogens is limited, van Peer et al. (1991) recently demonstrated induced resistance of carnation plants to Fusarium wilt based on phytoalexin accumulation in response to bacterization of roots with a *Pseudomonas* sp. Additional clarification of the influence of plant-microbe interactions on susceptibility should come from future studies on the manipulation of microbes, such as bacterial antagonists in the xylem tissue or fungal antagonists in the cortex, to formulate internal, as well as external, layers of protection (Harman and Lumsden 1990).

The most important abiotic factors that influence changes in susceptibility of hosts to root diseases include light and soil water, temperature, aeration, mineral nutrition, hydrogen ion concentration, and bulk density. The influences of many of these environmental components on both the pathogen and the host in plant disease development are covered in depth by MacDonald (Chap. 4, this Vol.) and Sumner (Chap. 11, this Vol.) and in various reviews (Duniway 1979; Burdon 1987; Engelhard 1989; Huber 1989, 1990; Drew 1990; Rovira et al. 1990). Several environmental factors may be of unique importance to changes in susceptibility to root diseases (Burdon 1987). For example, flooded soil may increase disease severity by predisposing plants to infection through adverse effects on host physiology (Drew and Lynch 1980; Kuan and Erwin 1980; Blaker and MacDonald 1981; Matheron and Mircetich 1985; Wilcox and Mircetich 1985; Fraedrich and Tainter 1989; Schaffer et al. 1992); by depletion of soil oxygen which restricts root regeneration from necrotic roots (Stolzy and Sojka 1984; Schaffer et al. 1992); and by direct influences on pathogens, such as the promotion of the release of zoospores and mediation of the infection processes by species of *Phytophthora* and *Pythium* (Duniway 1979, 1983). Periods of drought may also predispose plants to various soilborne diseases (Schoeneweiss 1975; Burdon 1987).

Since the metabolic expression of genetically controlled defenses against disease is regulated by mineral ions, root diseases are critically influenced by applications of fertilizers or the manipulation of nutrients through cultural practices, such as liming for pH adjustment, adding organic amendments, altering crop sequences, and mainpulating tillage or irrigation (Huber 1989; Rovira et al. 1990). Changes in disease resistance may be consequences of these cultural practices that directly supply nutrients or change their availability through altered microbial activity. Huber (1989, 1990) summarized the susceptibility-altering roles of macro- and minor nutrients, exemplified by, respectively, the increased resistance of cell wall pectates to degradation by soilborne pathogens with applications of calcium and the increased severity of take-all or Verticillium wilt with the immobilization of manganese.

It is difficult to assess the impact of the infinitely varied forms of nutrients on root disease because of the influences of abiotic and biotic factors in the environment, other diseases, time of application, stage of crop growth, previous crop history, or ratios with other nutrients. In general, however, susceptibility to root disease increases with deficiencies in phosphorus and potassium; varies with ammonium and nitrate nitrogen sources, which generally have opposite effects on disease levels; and increases with deficiencies in minor nutrients (Burdon 1987). The mechanisms of control are also largely unknown. However, susceptibility may be preconditioned; for example, tomato seedlings grown on nitrate as a nitrogen source are preconditioned to be more resistant to Fusarium wilt than are seedlings grown on ammonium nitrogen sources (Jones et al. 1989). Because nutrition influences host predisposition and tolerance or resistance to disease, as well as pathogen growth, virulence, reproduction, and survival, more information on cultural control of diseases needs to be incorporated into crop management programs.

Various other cultural management practices may have important direct or indirect roles, other than in nutrient manipulation, in altering host susceptibility (Rovira et al. 1990; Chap. 11, this Vol.). Predisposition of roots to disease in various monocultures or crop rotations under different tillage practices may be a consequence of autotoxicity or allelopathy (Rovira et al. 1990). If root development is slowed or roots are damaged by compaction (Drew 1990), susceptibility to disease may be altered. The restrictive influences of soil compaction or poor drainage on root growth and tolerance to disease have been reviewed by Allmaras et al. (1988). Timing of crop management practices can greatly influence host susceptibility. Planting dates may be selected to avoid the predisposition of roots to disease under seasonal environmental influences, such as extremes in temperature or available soil water. Some *Pythium* spp., for example, infect roots of pasture grasses at temperatures favorable for root growth but do not cause disease except at higher temperatures; consequently, pasture crops can be planted later in the fall to retain low susceptibility at the low temperatures (Freeman et al. 1966; Mitchell 1975). Various components of integrated pest management systems may have important influences on nontarget organisms or on host susceptibility. Many instances of changes in host susceptibility with herbicide applications have been reported (Altman and Rovira 1989; Campbell and Greaves 1990; Rovira et al. 1990). For example, Trifluralin, a widely used herbicide, reduces Aphanomyces root rot of pea and Rhizoctonia damping-off of cowpea, but it can predispose soybeans to infection by *Phytophthora megasperma* f. sp. *glycinea* and cotton by *Rhizoctonia* sp. Trifluralin appeared to predispose soybeans to Fusarium root rot by causing hypocotyl swelling and cracking, which provided favorable sites for penetration by *Fusarium oxysporum* (Carson et al. 1991). Indirect influences on susceptibility to pathogens may result from alterations caused by herbicide damage in the balances in invasion and colonization of tissues by protective and deleterious microbes (Altman and Rovira 1989; Campbell and Greaves 1990).

2.6 Measuring Susceptibility Alterations

The influences of altered host susceptibility on disease and pathogen development include, as summarized by Seem (1988): reductions or increases in total numbers of infections or lesions; alterations in sizes or expansion rates of lesions or colonized areas; lengthening or shortening the incubation or latent periods; and changing the infectious period or the reproductive capacity of the pathogen. Additionally, changes in host susceptibility may alter maximum disease development. Measurements of these major components of epidemic development would allow the generation of quantitative information on the precise impact of changes of host susceptibility on disease development, as well as information on the impact of various host and environmental components on alteration of susceptibility. Uniform inoculations at specific times over the physiological development of the host maintained under standardized environmental conditions will be required to differentiate age-dependent susceptibility and environmentally induced predisposition or susceptibility alterations (Populer 1978; Seem 1988). Assessment during the various series of events in root diseases, however, may require much different approaches than those employed for foliar diseases. In contrast to the polycyclic chain of latent periods and infectious periods with nearly instantaneous dispersal that is typical of many foliar diseases, the infection chain of many soilborne pathogens may be heterogeneous. Thus, the infection chain in soilborne systems may consist of several nonuniform or dissimilar cycles or phases of nonparasitic pathogen activity, primary infections from surviving inoculum, inoculum formation, and cycles of secondary infections (Reynolds et al. 1986; Gilligan 1987; Brassett and Gilligan 1988). Also, dispersal events for soilborne pathogens are usually more limited in duration than for foliar pathogens.

Very little quantitative information is available on the influence of changes in root susceptibility on the lengths of the incubation, latent, or infectious periods and on the reproductive capacity of pathogens (Gilligan 1987). However, models of the relationships of inoculum density to infection of roots by pathogens and of efficiency of inoculum in infection or disease (Chap. 1, this Vol.; Tomimatsu and Griffin 1982; Gilligan 1987) could be modified to investigate the impact of alterations in susceptibility on disease development. Such modifications require the maintenance of uniform environmental conditions, inoculations over progressive host ages, and sampling of host or pathogen responses over the physiological development of host roots. More information on variation in intrinsic susceptibility of different morphotypes of roots will be required to standardize models; for example, the general assumption in pathozone analyses that susceptibility of roots along the root axis is uniform (Gilligan 1987) may be valid for young radicles (Reynolds et al. 1985a), but many factors, such as the emergence of lateral roots from the main root axis, may alter susceptibility in more mature tissue. Analytical approaches that allow quantification of infections, subsequent disease, and signs of pathogens in different ages and types of roots over time should be particularly useful (Rush et al. 1984; Reynolds et al. 1985b; English and Mitchell 1988; Gerik and Huisman 1988). This type of information can then be integrated into root development models (Bloomberg 1979a,b, 1988; Gutierrez et al. 1983; Reynolds et al. 1986; Bowers et al. 1991). The mechanistic model of Bowers et al. (1991), for example, allows the determination of relationships among root growth, infection by soilborne pathogens, and colonization by various microorganisms. The model, which is based on processes of initiation and development of root segments of morphometric root orders, would be informative in studies on susceptibility changes over time because time delays between successive stages of root components are major determinatives of the dynamics of the model. Other approaches, such as path analysis or path coefficient analysis (Bowers et al. 1990; Hau and Kranz 1990) should be beneficial in elucidating the indirect as well as the direct effects of environmental factors or host components, such as age and morphology, on predisposition or susceptibility alterations.

Ultimately, any of a myriad of factors may be involved in altering host susceptibility to soilborne diseases. It is the responsibility of the epidemiologist to be aware of important factors when conceptualizing experiments, formulating procedures, especially for timing exposures of host roots to inoculum or preparing sampling schedules, and interpreting results. Although there are more cases of predisposition of resistant plants to susceptibility than of susceptible plants to resistance (Yarwood 1959), it is the latter case that should receive increasing emphasis in the management of soilborne diseases by biological or cultural control.

2.7 The Future

The challenge for the future of host root evaluations is to develop useable sampling and root manipulation technologies to allow assessment of the roles that root development plays in epidemic processes. Although progress towards this end has been made in discrete areas in recent years, the advancements have yet to be brought together. In addition to newly developed conceptual frameworks of root system structure, advances have been made in biotechnological, computer, and visualization methods.

It is time to integrate these advancements with consideration of host susceptibility to synthesize a clarified view of the roles that roots play relative to specific steps in epidemic development. With greater emphasis on roots as a controlling component of root diseases, there should come fresh approaches towards disease management based on root system manipulation.

References

Abbot ML, Fraley LF Jr (1991) A review: radiotracer methods to determine root distribution. Environ Exp Bot 31:1–10

Alexander HM (1990) Dynamics of plant-pathogen interactions in natural plant communities. In: Burdon JJ, Leather SR (eds) Pests, pathogens and plant communities. Blackwell Scientific, Boston, pp 31–45

Allmaras RR, Kraft JM, Miller DE (1988) Effects of soil compaction and incorporated crop residue on root health. Annu Rev Phytopathol 26:219–243

Altman J, Rovira AD (1989) Herbicide-pathogen interactions in soil-borne root diseases. Can J Plant Pathol 11:166–172

Ashworth LL, Huisman OC, Harper DM, Stromberg LK (1974) Free and bound microsclerotia of *Verticillium albo-atrum* in soils. Phytopathology 64:563–564

Atkinson D (1985) Spatial and temporal aspects of root distribution as indicated by the use of a root observation laboratory. In: Fitter AH, Atkinson D, Read DJ, Usher MB (eds) Ecological interactions in soil. Blackwell Scientific, Boston, pp 43–65

Augspurger CK (1990) Spatial patterns of damping-off disease during seedling recruitment in tropical forests. In: Burdon JJ, Leather SR (eds) Pests, pathogens and plant communities. Blackwell Scientific, Boston, pp 131–144

Baker R, Maurer CL, Maurer RA (1967) Ecology of plant pathogens in soil. VII. Mathematical models and inoculum density. Phytopathology 57:662–666

Barber SA (1971) Effect of tillage practice on corn (*Zea mays* L.) root distribution and morphology. Agron J 63:724–726

Barber DA, Martin JK (1976) The release of organic substances by cereal roots into soil. New Phytol 76:69–80

Barley KP (1970) The configuration of the root system in relation to nutrient uptake. Adv Agron 22:159–201

Bazin MJ, Markham P, Scott EM, Lynch JM (1990) Population dynamics and rhizosphere interactions. In: Lynch JM (ed) The rhizosphere. John Wiley & Sons, New York, pp 99–127

Belford RK, Klepper B, Rickman RW (1987) Studies of intact shoot-root systems of field-grown winter wheat. II. Root and shoot developmental patterns as related to nitrogen fertilizer. Agron J 79:310–319

Beute MK, Benson DM (1979) Relation of small soil fauna to plant disease. Annu Rev Phytopathol 17:485–502

Blaker NS, MacDonald JD (1981) Predisposing effects of soil moisture extremes on the susceptibility of rhododendron to Phytophthora root and crown rot. Phytopathology 71:831–834

Bloomberg WJ (1979a) A model of damping-off and root rot of Douglas-fir seedlings caused by *Fusarium oxysporum*. Phytopathology 69:74–81

Bloomberg WJ (1979b) Model simulations of infection of Douglas-fir seedlings by *Fusarium oxysporum*. Phytopathology 69:1072–1077

Bloomberg WJ (1988) Modeling control strategies for laminated root rot in managed Douglas-fir stands: model development. Phytopathology 78:403–409

Böhm W (1979) Methods of studying root systems. Springer, Berlin Heidelberg New York

Bookman PA, Mack RN (1982) Root interaction between *Bromus tectorum* and *Poa pratensis*: a three-dimensional analysis. Ecology 63:640–646

Boosalis MG, Scharen AL (1959) Methods for microscopic detection of *Aphanomyces euteiches* and *Rhizoctonia solani* and for isolation of *Rhizoctonia solani* associated with plant debris. Phytopathology 49:192–198

Bowers JH, Smerage GN, Mitchell DJ (1991) The development of a conceptual model of root growth. In: Keister DL, Cregan PB (eds) The rhizosphere and plant growth. Kluwer Academic, Dordrecht, pp 73–80
Bowers JH, Sonoda RM, Mitchell DJ (1990) Path coefficient analysis of the effect of rainfall variables on the epidemiology of Phytophthora blight of pepper caused by *Phytophthora capsici*. Phytopathology 80:1439–1446
Brassett PR, Gilligan CA (1988) A discrete probability model for polycyclic infection by soil-borne plant parasites. New Phytol 109:183–191
Brown JM, Johnson GA, Kramer PJ (1986) In vivo magnetic resonance microscopy of changing water content in *Pelargonium hortorum* roots. Plant Physiol 82:1158–1160
Brugge R, Thornely JHM (1985) A growth model of root mass and vertical distribution, dependent on carbon substrate from photosynthesis and with non-limiting soil conditions. Ann Bot 55:563–577
Burdon JJ (1987) Diseases and plant population biology. Cambridge Univ Press, New York, 208 pp
Burke DW, Barker AW (1966) Importance of lateral roots in Fusarium root rot of beans. Phytopathology 56:292–294
Campbell R, Greaves MP (1990) Anatomy and community structure of the rhizosphere. In: Lynch JM (ed) The rhizosphere. John Wiley & Sons, New York, pp 11–34
Carlile MJ (1983) Motility, taxis, and tropism in *Phytophthora*. In: Erwin DC, Bartnicki-Garcia S, Tsao PH (eds) *Phytophthora*: its biology, taxonomy, ecology, and pathology. APS Press, St Paul, pp 95–108
Carlsson U, Elmqvist T, Wennstrom A, Ericson L (1990) Infection by pathogens and population age of host plants. J Ecol 78:1094–1105
Carpenter A, Cherrett JM, Ford JB, Thomas M, Evans E (1985) An inexpensive rhizotron for research on soil and litter-living organisms. In: Fitter AH, Atkinson D, Read DJ, Usher MB (eds) Ecological interactions in soil: plants, microbes and animals. Blackwell Scientific, Oxford, pp 67–71
Carson ML, Arnold WE, Todt PE (1991) Predisposition of soybean seedlings to Fusarium root rot with trifluralin. Plant Dis 75:342–346
Coleman DC (1985) Through a ped darkly: an ecological assessment of root-soil-microbial-faunal interactions. In: Fitter AH, Atkinson D, Read DJ, Usher MB (eds) Ecological interactions in soil: plants, microbes and animals. Blackwell Scientific, Oxford, pp 1–21
Colhoun J (1973) Effects of environmental factors on plant diseases. Annu Rev Phytopathol 11:343–364
Cook RJ, Baker KF (1983) The nature and practice of biological control of plant pathogens. Am Phytopathol Soc, St Paul, pp 1–539
Cooper GP, Brown JM, Johnson GA (1989) In vivo magnetic resonance microscopy at $5\,\mu$m. J Mag Resonan 83:608–616
Crick JC, Grimes JP (1987) Morphological plasticity and mineral nutrient capture in two herbaceous species of contrasted ecology. New Phytol 107:403–414
Curl EA, Harper JD (1990) Fauna-microflora interactions. In: Lynch JM (ed) The rhizosphere. John Wiley & Sons, New York, pp 369–388
Curl EA, Truelove B (1986) The rhizosphere. Springer, Berlin Heidelberg New York
Deacon JW, Henry, CM (1980) Age of wheat and barley roots and infection by *Gaeumannomyces graminis* var. *tritici*. Soil Biol Biochem 12:113–118
Deacon JW, Lewis SJ (1982) Natural senescence of the root cortex of spring wheat in relation to susceptibility to common root rot (*Cochliobolus sativus*) and growth of a free-living nitrogen-fixing bacterium. Plant Soil 66:13–20

Dexter AR (1978) A stochastic model for the growth of roots in tilled soil. J Soil Sci 29:102–116
Dexter AR (1987) Mechanics of root growth. Plant Soil 98:303–312
Drew MC (1975) Comparison of the effects of a localized supply of phosphate, nitrate, ammonium and potassium on the growth of the seminal root system, and the shoot, of barley. New Phytol 75:479–490
Drew MC (1990) Root function, development, growth and mineral nutrition. In: Lynch JM (ed) The rhizosphere. John Wiley & Sons, New York, pp 35–57
Drew MC, Lynch JM (1980) Soil anaerobiosis, microorganisms and root function. Annu Rev Phytopathol 18:37–66
Drew MC, Saker LR (1978) Nutrient supply and the growth of the seminal root system in barley. III. Compensatory increases in growth of lateral roots, and in rates of phosphate uptake, in response to a localized supply of phosphate. J Exp Bot 109:435–451
Dryden P, Van Alfen NK (1984) Soil moisture, root system density, and infection of roots of pinto beans by *Fusarium solani* f. sp. *phaseoli* under dryland conditions. Phytopathology 74:132–135
Duniway JM (1979) Water relations of water molds. Annu Rev Phytopathol 17:431–460
Duniway JM (1983) Role of physical factors in the development of Phytophthora diseases. In: Erwin DC, Bartnicki-Garcia S, Tsao PH (eds) *Phytophthora*: its biology, taxonomy, ecology, and pathology. Am Phytopathol Soc, St Paul, pp 175–187
Engelhard AW (ed) (1989) Soilborne plant pathogens: management of diseases with macro- and microelements. APS Press, St Paul, pp 1–217
English JT, Mitchell DJ (1984) Patterns of root growth and inoculum production in the *Catharanthus roseus – Phytophthora parasitica* pathosystem. Phytopathology 74:813
English JT, Mitchell DJ (1988) Relationships between the development of root systems of tobacco and infection by *Phytophthora parasitica* var. *nicotianae*. Phytopathology 78:1478–1483
English JT, Mitchell DJ (1989) Use of morphometric analysis for characterization of tobacco root growth in relation to infection by *Phytophthora parasitica* var. *nicotianae*. Plant Soil 113:243–249
Evans G, Snyder WC, Wihelm S (1966) Inoculum increase of the Verticillium wilt fungus in cotton. Phytopathology 56:590–593
Fitter AH (1982) Morphometric analysis of root systems: application of the technique and influence of soil fertility on root system development in two herbaceous species. Plant Cell Environ 5:313–322
Fitter AH (1985) Functional significance of root morphology and root system architecture. In: Fitter AH, Atkinson D, Read DJ, Usher MB (eds) Ecological interactions in soil. Blackwell Scientific, Boston, pp 87–106
Fitter AH (1986) The topology and geometry of plant root systems: influence of watering rate on root system topology in *Trifolium pratense*. Ann Bot 58:91–101
Fitter AH (1987) An architectural approach to the comparative ecology of plant root systems. New Phytol 106:61–77
Fitter AH (1991) Characteristics and functions of root systems. In: Waisel Y, Eshel A, Kafkafi U (eds) Plant roots: the hidden half. Marcel Dekker, New York, pp 3–25
Fogel R (1985) Roots as primary producers in below-ground ecosystems. In: Fitter AH, Atkinson D, Read DJ, Usher MB (eds) Ecological interactions in soil. Blackwell Scientific, Boston, pp 23–36
Fogel R, Hunt G (1983) Contribution of mycorrhizae and soil fungi to nutrient cycling in a Douglas-fir ecosystem. Can J For Res 13:219–232

Foster RC (1982) The fine structure of epidermal cell mucilages of roots. New Phytol 91:727–740
Foster RC, Bowen GD (1982) Plant surfaces and bacterial growth: the rhizosphere and rhizoplane. In: Mount MS, Lacy GH (eds) Phytopathogenic prokaryotes. Academic Press, New York, pp 159–185
Foster RC, Rovira AD, Cook TW (1983) Ultrastructure of the root-soil interface. APS Press, St Paul, pp 1–56
Fraedrich SW, Tainter FH (1989) Effect of dissolved oxygen concentration on the relative susceptibility of shortleaf and loblolly pine root tips to *Phytophthora cinnamomi*. Phytopathology 79:1114–1118
Freeman TE, Luke HH, Sechler DT (1966) Pathogenicity of *Pythium aphanidermatum* on grain crops in Florida. Plant Dis 50:292–294
Frenzel B (1957) Zur Abgabe von Aminosäuren und Amiden an das Nährmedium durch die Wurzeln von *Helianthus annuus* L. Planta 49:210–234
Frenzel B (1960) Zur Ätiologie der Anreicherung von Aminosäuren und Amiden im Wurzelraum von *Helianthus annuus* L. Ein Beitrag zur Klärung der Probleme der Rhizosphäre. Planta 55:169–207
Funck-Jensen D, Hockenhull J (1984) Root exudation, rhizosphere microorganisms and disease control. Vaxtskyddsnotiser 48:49–54
Gerik JS, Huisman OC (1988) Study of field-grown cotton roots infected with *Verticillium dahliae* using an immunoenzymatic staining technique. Phytopathology 78:1174 - 1178
Gerwitz A, Page ER (1974) An empirical mathematical model to describe plant root systems. J Appl Ecol 11:773–781
Gilligan CA (1983) Modeling of soilborne pathogens. Annu Rev Phytopathol 21: 45–64
Gilligan CA (1987) Epidemiology of soil-borne plant pathogens. In: Wolfe MS, Caten CE (eds) Populations of plant pathogens: their dynamics and genetics. Blackwell Scientific, Boston, pp 119–133
Gregory PJ, McGown M, Biscoe PV, Hunt B (1978) Water relations of winter wheat. 1. Growth of the root system. J Agric Sci 91:91–102
Griffin GJ (1969) *Fusarium oxysporum* and *Aspergillus flavus* spore germination in the rhizosphere of peanut. Phytopathology 59:1214–1218
Gutierrez AP, DeVay JE, Pullman GS, Friebertshauser FE (1983) A model of a Verticillium wilt in relation to cotton growth and development. Phytopathology 73:89–95
Hackett C, Rose DA (1972) A model of the extension and branching of a seminal root of barley, and its use in studying relations between root dimensions. I. The model. Aust J Biol Sci 25:669–679
Hancock JG (1985) Fungal infection of feeder rootlets of alfalfa. Phytopathology 75:1112–1120
Hancock JG, Grimes DW (1990) Colonization of rootlets of alfalfa by species of *Pythium* in relation to soil moisture. Phytopathology 80:1317–1322
Harman GE, Lumsden RD (1990) Biological disease control. In: Lynch JM (ed) The rhizosphere. John Wiley & Sons, New York, pp 259–280
Harper JL (1977) Plant population biology. Academic Press, London
Harper JL (1990) Pests, pathogens and plant communities: an introduction. In: Burdon JJ, Leather SR (eds) Pests, pathogens and plant communities. Blackwell Scientific, Boston, pp 3–14
Hau B, Kranz J (1990) Mathematics and statistics for analyses in epidemiology. In: Kranz J (ed) Epidemics of plant diseases: mathematical analysis and modeling, 2nd edn. Springer, Berlin Heidelberg New York, pp 12–52
Hayhoe H (1981) Analysis of a diffusion model for plant root growth and application to plant soil-water uptake. Soil Sci 131:334–343

Henry CM, Deacon JW (1981) Natural (non-pathogenic) death of the cortex of wheat and barley seminal roots, as evidenced by nuclear staining with acridine orange. Plant Soil 60:255–274

Hetrick BAD, Leslie JF, Wilson GT, Kitt DG (1988) Physical and topological assessment of effects of a vesicular-arbuscular mycorrhizal fungus on root architecture of big bluestem. New Phytol 110:85–96

Hetrick BAD, Wilson GWT, Leslie JF (1991) Root architecture of warm- and cool-season grasses: relationship to mycorrhizal dependence. Can J Bot 69:112–118

Hewitt JS, Dexter AR (1979) An improved model of root growth in structured soil. Plant Soil 52:325–343

Holden J (1975) Use of nuclear staining to assess rates of cell death in cortices of cereal roots. Soil Biol Biochem 7:333–334

Holden J (1976) Infection of wheat seminal roots by varieties of *Phialophora radicicola* and *Gaeumannomyces graminis*. Soil Biol Biochem 8:109–119

Huber DM (1989) Introduction. In: Engelhard AW (ed) Soilborne plant pathogens: management of diseases with macro- and microelements. APS Press, St Paul, pp 1–8

Huber DM (1990) Fertilizers and soil-borne diseases. Soil Use Manag 6:168–172

Huisman OC (1982) Interrelations of root growth dynamics to epidemiology of root-invading fungi. Annu Rev Phytopathol 20:303–327

Huisman OC (1988a) Colonization of field-grown cotton roots by pathogenic and saprophytic soilborne fungi. Phytopathology 78:716–722

Huisman OC (1988b) Seasonal colonization of roots of field-grown cotton by *Verticillum dahliae* and *V. tricorpus*. Phytopathology 78:708–716

Itoh S (1985) In situ measurement of root density by micro-rhizotron. Soil Sci Plant Nutr 31:653–656

Jackson RM (1960) Soil fungistasis and the rhizosophere. In: Parkinson D, Waid JS (eds) The ecology of soil fungi. Liverpool Univ Press, Liverpool, pp 168–181

Jeger MJ (1987) The influence of root growth and inoculum density on the dynamics of root disease epidemics: theoretical analysis. New Phytol 107:459–478

Jones JP, Engelhard AW, Wolz SS (1989) Management of Fusarium wilt of vegetables and ornamentals by macro- and microelement nutrition. In: Engelhard AW (ed) Soilborne plant pathogens: management of diseases with macro- and microelements. APS Press, St Paul, pp 18–32

Kim YJ, Hwang BK, Park KW (1989) Expression of age-related resistance in pepper plants infected with *Phytophthora capsici*. Plant Dis 73:745–747

Klepper B, Belford RK, Rickman RW (1984) Root and shoot development in winter wheat. Agron J 76:117–122

Kotcon JB, Rouse DI, Mitchell JE (1984) Dynamics of root growth in potato fields affected by early dying syndrome. Phytopathology 74:462–467

Kranz J (ed) (1990a) Epidemics of plant diseases: mathematical analysis and modeling, 2nd edn. Springer, Berlin Heidelberg New York, 268 pp

Kranz J (1990b) Fungal diseases in multispecies plant communities. New Phytol 116:383–406

Kuan TL, Erwin DC (1980) Predisposition effect of water saturation of soil on Phytophthora root rot of alfalfa. Phytopathology 70:981–986

Lam ST, Ellis DM, Ligon JM (1991) Genetic approaches for studying rhizosphere colonization. In: Keister DL, Cregan PB (eds) The rhizosphere and plant growth. Kluwer Academic, Norwell, pp 43–50

Larkin RP, English JT, Mihail JD (1992) The influence of infection by *Pythium* spp. on root system morphology of alfalfa seedlings. Phytopathology 82:1114

Larson RL, Atkinson TG (1970) A cytogenetic analysis of reaction to common root rot in hard red spring wheats. Can J Bot 48:2059–2067

Last FT (1971) The role of the host in the epidemiology of some nonfoliar pathogens. Annu Rev Phytopathol 9:341–362
Levitt J (1972) Responses of plants to environmental stresses. Academic Press, New York, 697 pp
Lewis SJ, Deacon JW (1982) Effects of shading and powdery mildew infection on senescence of the root cortex and coleoptile of wheat and barley seedling, and implications for root- and foot-rot fungi. Plant Soil 69:401–411
Linderman RG (1988) Mycorrhizal interactions with the rhizosphere microflora: the mycorrhizosphere effect. Phytopathology 78:366–371
Louter JH, Edgington LV (1990) Indications of cross-protection against fusarium crown and root rot of tomato. Can J Plant Pathol 12:283–288
Lungley DR (1973) The growth of root systems – a numerical computer simulation model. Plant Soil 38:145–159
Lynch JM (1990) Introduction: some consequences of microbial rhizosphere competence for plant and soil. In: Lynch JM (ed) The rhizosphere. John Wiley & Sons, New York, pp 1–10
MacLeod WJ, Robson AD, Abbott LK (1986) Effects of phosphate supply and inoculation with vesicular-arbuscular mycorrhizal fungus on the death of the root cortex of wheat, rape, and subterranean clover. New Phytol 103:349–357
Martin JK (1977) Factors influencing the loss of organic carbon from wheat roots. Soil Biol Biochem 9:1–7
Matheron ME, Mircetich SM (1985) Relative resistance of different rootstocks of English walnut to six *Phytophthora* spp. that cause root and crown rot in orchard trees. Plant Dis 69:1039–1041
Marx DM, Davey CB (1969) The influence of ectotrophic mycorrhizal fungi on the resistance of pine roots to pathogenic infections. IV. Resistance of naturally occurring mycorrhizae to infections by *Phytophthora cinnamomi*. Phytopathology 59:559–565
May LH, Randles FH, Aspinall D, Paleg LG (1967) Quantitative studies of root development. II. Growth in the early stages of development. Aust J Biol Sci 20:273–283
May RM (1990) Population biology and population genetics of plant-pathogen associations. In: Burdon JJ, Leather SR (eds) Pests, pathogens and plant communities. Blackwell Scientific, Boston, pp 309–325
McCoy EL, Boersma L (1986) Dynamic equations of axial root growth from the empirical solution of a mechanistic model. Bot Gaz 147:383–391
McDougall BM, Rovira AD (1970) Sites of exudation of ^{14}C-labelled compounds from wheat roots. New Phytol 69:999–1003
Mellano HM, Munnecke DE, Endo RM (1970) Relationship of seedling age to development of *Pythium ultimum* on roots of *Antirrhinum majus*. Phytopathology 60:935–942
Mitchell DJ (1975) Density of *Pythium myriotylum* oospores in soil in relation to infection of rye. Phytopathology 65:570–575
Mitchell DJ, Kannwischer-Mitchell ME (1983) Relationship of inoculum density of *Phytophthora* species to disease incidence in various hosts. In: Erwin DC, Bartnicki-Garcia S, Tsao PH (eds) *Phytophthora*: its biology, taxonomy, ecology, and pathology. APS Press, St Paul, pp 95–108
Newman EI (1985) The rhizosphere: carbon sources and microbial populations. In: Fitter AH, Atkinson D, Read DJ, Usher MB (eds) Ecological interactions in soil: plants, microbes, and animals. Blackwell Scientific, Boston, pp 107–121
Paul ND (1990) Modification of the effects of plant pathogens by other components of natural ecosystems. In: Burdon JJ, Leather SR (eds) Pests, pathogens and plant communities. Blackwell Scientific, Boston, pp 81–96

Pearson R, Parkinson D (1961) The sites of excretion of ninhydrin-positive substances by broad bean seedlings. Plant Soil 4:391–396
Perrin R (1990) Interactions between mycorrhizae and diseases caused by soil-borne fungi. Soil Use Manag 6:189–194
Persson H (1978) Root dynamics in a young Scots pine stand in central Sweden. Oikos 30:508–519
Pfender WF (1988a) Suppression of ascocarp formation in *Pyrenophora tritici-repentis* by *Limonomyces roseipellis*, a basidiomycete from reduced-tillage wheat straw. Phytopathology 78:1254–1258
Pfender WF, Wootke SL (1988b) Microbial communities of *Pyrenophora*-infested wheat straw as examined by multivariate analysis. Microb Ecol 15:95–113
Populer C (1978) Changes in host susceptibility with time. In: Horsfall JG, Cowling EB (eds) Plant disease: an advanced treatise, vol 2. Academic Press, London, pp 239–262
Porter JR, Klepper B, Belford RK (1986) A model (WHTROOT) which synchronizes root growth and development with shoot development for winter wheat. Plant Soil 92:133–145
Powell NT (1971) Interactions between nematodes and fungi in disease complexes. Annu Rev Phytopathol 9:253–274
Protopapas AL, Bras RL (1987) A model for water uptake and development of root systems. Soil Sci 144:352–366
Read DJ, Francis R, Finlay RD (1985) Mycorrhizal mycelia and nutrient cycling in plant communities. In: Fitter AH (ed) Ecological interactions in soil: plants, microbes and animals. Blackwell Scientific, Boston, pp 193–217
Reid CPP (1990) Mycorrhizas. In: Lynch JM (ed) The rhizosphere. John Wiley & Sons, New York, pp 281–315
Reynolds KM, Benson DM, Bruck RI (1985a) Epidemiology of Phytophthora root rot of Fraser fir: rhizosphere width and inoculum efficiency. Phytopathology 75:1010–1014
Reynolds KM, Benson DM, Bruck RI (1985b) Epidemiology of Phytophthora root rot of Fraser fir: root colonization and inoculum production. Phytopathology 75:1004–1009
Reynolds KM, Gold HJ, Bruck RI, Benson DM, Campbell CL (1986) Simulation of the spread of *Phytophthora cinnamomi* causing a root rot of Fraser fir in nursery beds. Phytopathology 76:1190–1201
Rose DA (1983) The description of the growth of root systems. Plant Soil 75: 405–415
Rovira AD (1969) Plant root exudates. Bot Rev 35:35–57
Rovira AD, Elliot LF, Cook RJ (1990) The impact of cropping systems on rhizosphere organisms affecting plant health. In: Lynch JM (ed) The rhizosphere. John Wiley & Sons, New York, pp 389–436
Royle DJ, Hickman CJ (1964a) Analysis of factors governing in vitro accumulation of zoospores of *Pythium aphanidermatum* on roots. I. Behavior of zoospores. Can J Microbiol 10:151–162
Royle DJ, Hickman CJ (1964b) Analysis of factors governing in vitro accumulation of zoospores of *Pythium aphanidermatum* on roots. II. Substances causing response. Can J Microbiol 10:201–219
Rush CM, Upchurch DR, Gerik TJ (1984) In situ observations of *Phymatotrichum omnivorum* with a borescope mini-rhizotron system. Phytopathology 74:104–105
Schaffer B, Andersen PC, Ploetz RC (1992) Responses of fruit crops to flooding. Hort Rev 13:257–313
Schoeneweiss DF (1975) Predisposition, stress, and plant disease. Annu Rev Phytopathol 13:193–211

Schroth MN, Hildebrand DC (1964) Influence of plant exudates on root-infecting fungi. Annu Rev Phytopathol 2:101–132
Schroth MN, Snyder WC (1961) Effect of host exudates on chlamydospore germination of the bean root rot fungus, *Fusarium solani* f. *phaseoli*. Phytopathology 51:389–393
Seem RC (1988) The measurement and analysis of the effects of crop development on epidemics. In: Kranz J, Rotem J (eds) Experimental techniques in plant disease epidemiology. Springer, Berlin Heidelberg New York, pp 51–68
Shigo AL (1984) Compartmentalization: a conceptual framework for understanding how trees grow and defend themselves. Annu Rev Phytopathol 22:189–214
Simmons GL, Pope PE (1988) Development of a root growth model for yellow-poplar and sweetgum seedlings grown in compacted soil. Can J For Res 18:728–732
Sivakumar MVK, Taylor HM, Shaw RH (1977) Top and root relations of field-grown soybeans. Agron J 69:470–473
Stolzy LH, Sojka RE (1984) Effects of flooding on plant disease. In: Kozlowski TT (ed) Flooding and plant growth. Academic Press, New York, pp 221–264
Stone JA, Kaspar TC, Taylor HM (1983) Predicting soybean rooting depth as a function of soil temperature. Agron J 75:1050–1054
Strahler AN (1957) Quantitative analysis of watershed geomorphology. Trans Am Geophy Union 38:913–920
Taylor HM, Böhm W (1976) Use of acrylic plastic as rhizotron windows. Agron J 68:693–694
Tennant D (1976) Root growth of wheat. I. Early patterns of multiplication and extension of wheat roots including effects of levels of nitrogen, phosphorus and potassium. Aust J Agric Res 27:183–196
Tomimatsu GS, Griffin GJ (1982) Inoculum potential of *Cylindrocladium crotalariae*: infection rates and microsclerotial density-root infection relationships in peanut. Phytopathology 72:511–517
Tomimatsu GS, Griffin GJ (1988) Infection levels and extent of colonization by *Cylindrocladium crotalariae* of roots of cylindrocladium black rot-susceptible and -resistant peanut cultivars in naturally infested soil. Can J Plant Pathol 10: 11–14
Troughton A (1977) Relationship between the root and shoot systems of grasses. In: Marshall JK (ed) The belowground ecosystem: a synthesis of plant-associated processes. Range Sci Dep Sci Ser 26:39–51
Upchurch DR, Ritchie JT (1983) Root observations using a video recording system in mini-rhizotrons. Agron J 75:1009–1015
van Peer R, Niemann GJ, Schippers B (1991) Induced resistance and phytoalexin accumulation in biological control of Fusarium wilt of carnation by *Pseudomonas* sp. strain WCS417r. Phytopathology 81:728–734
Waisel Y, Eshel A (1991) Multiform behavior of various constituents of one root system. In: Waisel Y, Eshel A, Kafkafi U (eds) Plant roots: the hidden half. Marcel Dekker, New York, pp 39–52
Wiersum LK (1958) Density of root branching as affected by substrate and separate ions. Acta Bot Neerl 7:174–190
Wilcox HE (1991) Mycorrhizae. In: Waisel Y, Eshel A, Kafkafi U (eds) Plant roots: the hidden half. Marcel Dekker, New York, pp 731–765
Wilcox WF, Mircetich SM (1985) Effects of flooding duration on the development of Phytophthora root and crown rots of cherry. Phytopathology 75:1451–1455
Wilkinson HT, Weller DM, Alldredge JR (1982) Enhanced biological control of wheat take-all when inhibitory *Pseudomonas* strains are introduced on inoculum or seed as opposed to directly into soil. Phytopathology 72:948–949

Yarwood CE (1959) Predisposition. In: Horsfall JG, Diamond AE (eds) Plant pathology: an advanced treatise, vol 1. Academic Press, London, pp 521–562
Zobel RW (1975) The genetics of root development. In: Torrey JG, Clarkson DF (eds) The development and function of roots. Academic Press, London, pp 261–275
Zobel RW (1986) Rhizogenetics (root genetics) of vegetable crops. Hortscience 21:956–959
Zobel RW (1991a) Genetic control of root systems. In: Waisel Y, Eshel A, Kafkafi U (eds) Plant roots: the hidden half. Marcel Dekker, New York, pp 27–38
Zobel RW (1991b) Root growth and development. In: Keister DL, Cregan PB (eds) The rhizosphere and plant growth. Kluwer Academic, Boston, pp 61–71

3 Mycorrhizae

F. Schönbeck, G. Grunewaldt-Stöcker, and H. von Alten

3.1 Distribution and Ecological Impact of the Mycorrhizal Symbiosis

The formation of a mycorrhizal symbiosis is considered a widespread strategy of plants to obtain advantages under selection pressure in natural ecosystems. Most cultivated plant species can be infected by mycorrhizal fungi. Depending on the host plant and the infecting fungus, different mycorrhizae are formed: ectomycorrhizae, ericaceous, orchidaceous, and vesicular-arbuscular mycorrhizae (VAM). The main importance for plant production within this group has to be attributed to ectomycorrhizae (forestry) and VAM (most agricultural and horticultural plants).

The symbiosis is necessary for the successful development of many plants, at least in certain stages of development (Koide 1991). However, plants do not always respond to mycorrhizal infection. The effects depend on:

1. Plant species or cultivar.
2. Properties of the given soil (e.g., nutrient availability).
3. Presence of special stress situations which can be compensated by symbiosis.
4. Frequency and efficacy of the infecting mycorrhizal fungi.

There are significant differences between host species in the dependency on symbiosis, due to root morphology (Baylis 1970, 1972) or other plant traits. For example, *Allium* species are known to be highly mycotrophic even without severe P-deficiency, whereas members of the Gramineae seem rather independent of symbiosis. Nevertheless, even within the latter group considerable differences have been reported (Hetrick et al. 1990).

In nature, mycorrhizal symbiosis not only allows the host plant to grow particularly at the borders of its ecological niche, but the wide range of this symbiosis implies that it goes far beyond that. In practical plant production growth conditions sometimes occur which allow the mycorrhizal fungi to exceed the natural mycotrophy of the plants, e.g., fumigated or artificial soils in nurseries or micropropagation of plants.

C.L. Campbell/D.M. Benson (Eds.)
Epidemiology and Management of Root Diseases

With regard to the fact that agricultural soils from low to high fertility contain mycorrhizal inoculum (Land et al. 1990), all considerations on plant nutrition, rhizosphere ecology, or root health in the field have to include mycorrhizal aspects – or they would be incomplete.

3.2 Productivity of Cultivated Mycorrhizal Plants

Agricultural and horticultural plant production is found to a large extent on fertilized soils where no severe nutrient deficiency should occur. However, mycorrhiza researchers have focused most attention on the striking nutritional effects of this symbiosis, especially regarding phosphate. Now mycorrhizal growth responses are understood more and more as multifactorial effects. Besides nutrient deficiency or pathogens, many possible stress factors can limit plant development and yield. Abiotic factors that mycorrhizal plants can better withstand than nonmycorrhizal ones are drought (Allen and Boosalis 1983), salinity (Hirrel and Gerdemann 1980), heavy metals (Schuepp et al. 1987), and transplantation (Williams et al. 1992). However, it will be necessary to evaluate the significance of mycorrhiza-induced stress tolerance in practical field situations.

Without doubt, poor nutrient availability is important in many regions of the world, particularly in developing countries. On the other hand, in many regions of intense agriculture, the amount of P-fertilization has often been several times higher than necessary for decades. High P-availability reduces VA-mycorrhizal infection (Koide and Li 1990). However, host plants have often shown growth reductions in cases where intense mycorrhizal colonization could develop in spite of high P-levels (Kiernan et al. 1983). Thus, VAM was thought to have vanished in normal (= fertilized) agricultural soils. However, evaluations of the mycorrhizal status of field crops showed that VAM infection is extensive (Baltruschat and Dehne 1982). Land (1990) described the mycorrhizal populations in arable, normally cultivated soils (e.g., tillage, fertilization up to 300 kg N/ha per year, and chemical disease control) of different fertility in northern Germany. She found about 500 mycorrhizal spores and up to 20 000 infective units (most probable number, MPN) in 100 g soil, even in fields where 10 t wheat/year was harvested.

Land (1990) transferred VAM isolates from these soils to greenhouse cultures to examine whether a more fertile soil contains less effective VAM fungi. The efficiency was tested as the ability to increase plant growth under conditions of P-deficiency. In these tests all soils, regardless of fertility, proved to have very efficient to totally inefficient VAM fungi.

The question remains as to the role mycorrhizal fungi may play in arable, well-fertilized soils of temperate climates – particularly for cereal crops. Here, they should not be essential for nutrient uptake – at least in regard to major nutrients. The biological function of VA-mycorrhizae in such soils might be to enhance the tolerance of host plants to biotic and abiotic stress (Dehne 1987b). However, the intensity and impact of stress are often unpredictable. Thus, the benefit from mycorrhizae is very difficult to calculate compared to the carbon costs of the endophytes. Considering the differences in mycotrophy of plant species and cultivars, it has to be expected that certain host plants show positive responses to mycorrhizal infection even under apparent optimum conditions.

The plant response to mycorrhizae strongly depends on the properties of the soil in which it is grown. Therefore, it is impossible to generalize about the importance of the symbiosis in the field. Plants are also grown in nutrient-deficient soils without sufficient fertilization. Here, P-acquisition is a major problem. Most phosphorus is bound and has to be released by the mineralization of organic matter or desorption from soil surfaces in order to contribute to plant nutrition.

The amount of phosphorus that a plant can acquire is not only determined by the soil, but also by the plant properties. When the growth rate of shoots is high, the plant root system must provide large quantities of phosphorus. Thus, fast-growing plants produce long root systems which can bypass the buffering action of the soil by exploiting new soil regions. Even shoot parts can contribute to P-acquisition: plants can remove about 50% of the leaf phosphorus prior to senescence (Chapin 1980). If this ability to supplement phosphorus from the roots by transferring P from more or less senescent plant parts to reproductive tissues (fruit and seed production) is increased in new cultivars, then the ability to transfer P can make a remarkable difference in yield. Specific root activities, such as rhizosphere acidification, and exudation of organic compounds and phosphatases, can also influence P-nutrition, but will only be significant with high P-availability (Chapin 1980).

When roots are colonized by VAM fungi, the endophytes can exploit new P-sources with extraradical hyphae (Sanders and Tinker 1973). Additionally, symbiosis alters plant traits involved in P-nutrition (Koide 1991). According to this role of the endophytes as nutrient servers, plants in P-deficient soils often have entirely colonized root systems.

The positive effect of the mycorrhizal fungi on their hosts are not only expressed as increases in the quantity of total dry matter, but also as increases in qualitative aspects, e.g., the balance between plant parts or the health of the total plant. Due to better growth or the physiological effect of symbiosis, mycorrhizal plants in any soil prove to be more vigorous than nonmycorrhizal ones, thus damage or loss of plant parts caused by environmental stress factors might be compensated for much faster. Ecto- (Clowes 1981) or VA-mycorrhizal (Berta et al. 1990) development can also effect meristematic cell division, thus altering root branching and reducing root/shoot ratios. As shoot parts are mostly harvested, this effect can contribute to yield increases and to the better use of resources if the relative deficit in absorbing tissue is compensated by a higher efficacy of mycorrhizal root systems.

3.3 Mycorrhizae – Root Disease Relationships

There have been several reviews on the protective influence of mycorrhizal fungi on root diseases as well as on the potential use of these symbionts in plant protection (e.g., Marx 1972; Schönbeck 1979; Bagyaraj 1984; Jeffries 1987; Duchesne et al. 1989b).

It is well documented that mycorrhizal fungi can reduce disease severity to a tolerable minimum, but they do not achieve complete protection. This is similar to other biological agents used in disease control. Mycorrhizal fungi may affect the development of root pathogens and limit the plant damage caused by them (Dehne 1982; Caron 1989; Kope and Fortin 1989). Moreover, the ability to increase plant tolerance (i.e., enhanced efficiency despite damage) offers the interesting aspect of compensation.

Previously, research has focused on VA mycorrhizae as possible biocontrol agents, but recently the focus has shifted to ectomycorrhizae (Duchesne et al. 1989b). However, the ubiquitous distribution of VAM fungi in nearly all the main annual and perennial crop plants of agriculture and horticulture (except the cruciferous and chenopodial genera) supports the strategy of making use of the beneficial traits of these fungi. In numerous experiments selected isolates of *Glomus caledonium*, *G. epigaeum*, *G. etunicatum*, *G. fasciculatum*, *G. intraradices*, *G. manihotis*, *G. mosseae*, *Gigaspora gigantea*, *G. margarita*, and *Entrophospora columbiana* were demonstrated to lower the severity of diseases caused by soilborne pathogens (Table 3.1).

Since the reports of Dehne (1982) and Hussey and Roncadori (1982), the range of root pathogens affected adversely by mycorrhizae has not been extended greatly. More recent results involve pathogenic bacteria (Garcia-Garrido and Ocampo 1989), some fungi (e.g., *Sclerotinia* spp., *Gaeumannomyces graminis*, *Fusarium* spp.; Krishna and Bagyaraj 1983; Dehne 1987a,b), and some nematode species (Diederichs 1987; Umesh et al. 1989; Baghel et al. 1990). It should be emphasized, however, that individual VAM-fungal species are not pathogen-specific. The specificity of the interactions, including the host plant, must still be determined.

The actual success of a biocontrol organism is not only determined by the general efficiency of the control mechanism involved, but the control effects strongly depend on the temporal and spatial distribution of the organisms. The spatial distribution of mycorrhizal fungi in soil involves both the mycorrhizal roots and the hyphal nets. The rate of mycorrhizal root colonization in some instances has been proven to be dependent on the host variety and is not necessarily linked to a positive stimulation of plant growth

Table 3.1. Plant pathogens impaired by VA-mycorrhizal fungi in biocontrol of soilborne pathogens

Fungi	Nematodes
Olpidium brassicae	*Heterodera cajani*
Pythium ultimum	*H. solanacearum*
Phythophthora capsici	*Meloidogyne hapla*
P. cinnamomi	*M. incognita*
P. megasperma	*M. javanica*
P. palmivora	*Radopholus similis*
P. parasitica	*Rotylenchus reniformis*
Fusarium avenaceum	*Pratylenchus brachyurus*
F. culmorum	*P. penetrans*
F. graminearum	*Tylenchulus semipenetrans*
F. oxysporum	Bacteria
F. solani	*Erwinia carotovora* pv. *carotovora*
Verticillium dahliae	*Pseudomonas solanacearum*
Rhizoctonia solani	*P. syringae*
Macrophomina phaseolina	
Cochliobolus sativus	
Cylindrocladium scoparium	
Gaeumannomyces graminis	
Phoma terrestris	
Pyrenochaeta terrestris	
Sclerotium rolfsii	
Sclerotinia sclerotiorum	

or with the ability to increase plant resistance or inhibit pathogen development (Schönbeck and Dehne 1989). In addition, to investigate the mechanisms of interaction, not only are the culture substrates used and the duration of the experiment important considerations, but also the sequence of inoculation.

The often described necessity of a well-established mycorrhiza association in the root system for disease control (e.g., Dehne 1987a) is based on many results in various test combinations. However, there are also reports on the protective effects in simultaneous inoculations, e.g., with plant pathogenic bacteria and nematodes, or even after pathogen infection (Garcia-Garrido and Ocampo 1989; Thomas et al. 1989). Caron et al. (1986a) showed that the effect of *G. intraradices* for limiting tomato root rot as well as the population of *F. oxysporum* f. sp. *radicis-lycopersici* was independent of the sequence of inoculation. However, the sequence of inoculation can have either a negative or positive influence on VAM-root colonization. Often a suppression of the root-fungi association or fungal sporulation is caused by pathogenic fungi or nematodes (Krishna and Bagyaraj 1983; Diederichs 1987; Carling et al. 1989; Caron 1989). Infection by *F. oxysporum* f. sp. *radicis-lycopersici*, however, unexpectedly increased the root-fungi association when *G. intraradices* was applied after infection by the pathogen (Caron et al. 1986a). The reasons for such unusual results may lie in the unique performance of each test combination and its specific mode of interaction.

In this connection, the inoculum density of the symbionts and the pathogens remains a weak point that is often neglected when the efficiency of mycorrhizae is considered. This could be due to the difficulties regarding the estimation of infectious propagules. At least in experiments, varied initial inoculum densities should be tested to establish threshold values for a minimal colonization rate of mycorrhizal fungi and for a maximum infection rate of the pathogen, where mycorrhizae could still be effective.

In plant production systems, therefore, the disease-protecting potential of mycorrhizal fungi might be influenced in several ways. To make use of the beneficial traits of mycorrhizal fungi, knowledge of the mechanisms controlling disease suppression is essential.

3.3.1 Mechanisms Decreasing the Impact of Soilborne Pathogens in Mycorrhizal Root Systems

The protective role of mycorrhizae against soilborne plant pathogens and the adverse effects on plants caused by the latter depends upon several mechanisms that have been postulated previously, but have not yet been completely clarified. This chapter will refer to recent contributions that give more substantiated evidence for interaction processes.

3.3.1.1 Direct Interactions of Mycorrhizae and Soilborne Pathogens

Direct mechanisms that act against soilborne pathogens are mainly attributed to ectomycorrhizal fungi (Table 3.2).

A mechanical barrier provided by the sheath of the symbiont protects the mycorrhizal root from infection by an unknown range of pathogens. This has often been reviewed, exemplified in the experiments of Marx and Davey (1969) with *Leucopaxillus cerealis* var. *piceina* and *Phytophthora cinnamomi* on *Pinus echinata*, and is a rather simple way of protection. Moreover, the mantle supports a protective microbial rhizosphere population, leading to direct (antibiotic) or indirect inhibition of pathogens (Garbaye and Bowen 1989).

Antibiotic or other inhibitory substances that affect root pathogens are produced by more than 90 ectomycorrhizal fungi (Marx 1972). However, as this antibiosis has mostly been based on in vitro tests (e.g., Kope et al. 1991), the efficacy of antibiosis in situ under more or less natural mycorrhizosphere conditions is not well documented. However, there are two examples for the production of inhibitors in situ: *L. cerealis* var. *piceina* (polyacetylene, Marx 1969) and *Paxillus involutus* (oxalic acid, Duchesne et al. 1989a). The stimulation of oxalic acid synthesis by root exudates of *Pinus* sp. gives evidence that the often supposed antibiosis effect appears to

Table 3.2. Examples of recent investigations on the mechanisms of protection by ectomycorrhizal fungi against soilborne plant pathogenic fungi

Ectomycorrhizal fungus (active principle or substance)	Pathogen	Reference
Tested in situ:		
Paxillus involutus (oxalic acid; antibiosis)	*Fusarium oxysporum* f.sp. *pini*	Duchesne et al. (1989a,b) Farquhar and Peterson (1990)
Pisolithus arhizus (p-hydroxybenzoyl-formic acid, (R)-(−)-p-hydroxy-mandelic acid; antibiosis)	*Truncatella hartigii* *Rhizoctonia solani* *Sclerotinia sclerotiorum* *Fusarium solani* *Cylindrocarpon destructans*	Kope et al. (1991)
Laccaria laccata (phenols; induced resistance)	*Fusarium oxysporum* *Pythium irregulare*	Sylvia and Sinclair (1983)
Tested in vitro:		
Pisolithus tinctorius (extracellular metabolites; antibiosis)	*Truncellata* sp. *Rhizoctonia* sp. *Sphaerosporella* sp. *Fusarium solani* *Brunchorstia* sp. *Cochliobolus* sp.	Kope and Fortin (1989, 1990)
Laccaria bicolor	*Rhizoctonia solani*	Zhao and Guo (1989)
Lactarius deliciosus		
Gomphidius roseus (hyperparasitism)		
Suillus brevipes (antibiosis)	*Armillaria mellea*	Natarayan and
	Cylindrocladium spp. *Fusarium oxysporum* *Fusarium solani* *Rhizoctonia solani*	Govindasamy (1990)

be more sensitive to environmental changes and more difficult to assess in situ. This may explain why in vitro selected mycorrhizal strains were frequently ineffective under practical test conditions. The production of inhibitory substances by plants in response to ectomycorrhizal and VA-mycorrhizal fungi will be discussed below in Section 3.3.1.2.

The hyperparasitism of several ectomycorrhizal fungi is reported in in vitro studies, but its relevance to disease control is still open.

Up to now, the competition for infection sites has been a more theoretical than practically proven mode of action of the symbionts. There could be a limitation to those pathogens that infect identical roots parts or act in physiological development stages in the root system for which mycorrhizal fungi are specialized. The methods of investigation must ensure the identity of each infection structure in order to distinguish those from the pathogenic fungus and those from the mycorrhizal fungi. Thus, new methodological approaches by special labeling techniques are needed.

3.3.1.2 Indirect Interaction of Mycorrhizae and Soilborne Pathogens

Most of the literature concerning mechanisms of interaction with soilborne plant pathogens deals with disease protection that is dependent on the changed metabolism of the mycorrhizal host plant. Therefore, the summarizing principle of these interactions is induced resistance.

Indirectly, plant health and defense vigor can be augmented by a general increase in tolerance, i.e., better adaptation to and compensation of adverse abiotic stress factors. Mycorrhizae generally enhance the uptake and utilization of mineral nutrients, especially P, in addition to K, Ca, Mg, Mn, and Zn. Also, a more profitable uptake and partitioning of water in the symplast (Auge and Stodola 1990) render mycorrhizal plants more vigorous.

Even without affecting pathogen development, these effects can contribute to a general increase in tolerance to pathogen attack, as measured in increased plant yield or biomass production. Examples of interactions with various pathogenic fungi in intensive crop production as well as with parasitic nematodes are given below.

Morphogenetic alterations of the root systems (Berta et al. 1990) and morphological changes in the vascular tissue (increased number of vessels) due to the symbiont (Dehne and Schönbeck 1979a) may support the compensation effects in mycorrhizal plants.

Disease resistance mediated by the host plant can be induced by ectomycorrhizal and VA-mycorrhizal fungi. Due to various localized and/or translocated effects, the consequences of the root-fungi association for shoot and root systems are mostly considered as a complex of growth and health stimulation. However, experimental designs frequently do not allow a clear distinction of modifying parameters and interdependencies. Induced resistance seems to be due to: (1) inhibitory chemicals synthesized or accumulated by symbiontically infected host tissue; (2) biochemical reactions of mycorrhizal plants (e.g., enhanced lignification, chitinase activity, changed hormonal balance, or other reactions that have adverse effects on pathogen infection and spread); and (3) improved phosphorus content in the tissue.

1. Synthesis and Accumulation of Inhibitory Substances by the Host Plant. Chemical defense reactions have already been considered a beneficial consequence of mycorrhizae in orchids (Gäumann and Kern 1959). The accumulation of phenolic substances in cortical cells of Douglas-fir roots, induced by *Laccaria laccata* (Sylvia and Sinclair 1983) and leading to disease resistance to *Fusarium* and *Rhizoctonia* infection, confirms these early observations of induced resistance through different types of mycorrhizae.

Other reports of increased amounts of phenolic compounds in VA-mycorrhizal root parts of infected peanut (*Sclerotium rolfsii*, Krishna and Bagyaraj 1983) or tomato (*Fusarium oxysporum*, Dehne and Schönbeck 1979b) underline the fact that mycorrhizae can stimulate the phenol metabolic pathway known in plant defense reactions.

However, investigations on isoflavonoid phytoalexin synthesis and accumulation in mycorrhizal plants could not establish a correlation with disease suppression or prove that the induced glyceollin accumulation caused disease resistance. Also, the role and relevance of scopoletin and cyanoglycosides (HCN precursors) in VA-mycorrhizal root systems for disease reduction remained indistinct.

With regard to quantitative changes in amino acids, an increased arginine content has often been cited as an example that a single substance may be responsible for the direct inhibition of chlamydospore production of *Thielaviopsis basicola* in mycorrhizal tobacco roots (Baltruschat and Schönbeck 1975). However, this hypothesis is not sufficiently documented.

2. Metabolic Reactions with a Disease-Reducing Effect. Increased activity of chitinolytic enzymes has been described as an induced metabolic change in mycorrhizal roots with disease-protecting effects, e.g., on wilt diseases (Dehne et al. 1978). A study on chitinase activity in leek roots during the stages of colonization and establishment of *G. versiforme* revealed interesting aspects of the time course of synthesis and the specificity of that plant enzyme (Spanu et al. 1989).

The penetration of mycorrhizal fungi in the root caused an early defense reaction (increased chitinase activity), as also shown earlier for peroxidase. Later, in established symbiosis, the defense reaction is suppressed, due to an altered hormonal balance. Localization studies with immunolabeling techniques showed that in vivo the chitin in the fungal walls did not come into contact with the plant chitinase and so was inaccessible.

These results should encourage further investigations on the effectiveness of this induced hydrolase in preventing subsequent fungal infections. Following research trends in molecular biology, the mode of induction of chitinase increase as a possible resistance factor is highly interesting.

Enhanced lignification was shown to inhibit the spread of the Fusarium wilt pathogen in tomato and cucumber (Dehne and Schönbeck 1979a,b). The change in cell wall structure led to the fortification of vascular tissue and endodermis cells, consequently limiting the pathogen. There have also been observations of extensive lignituber formation as a cellular defense reaction to penetration attempts of *T. basicola* in mycorrhizal tomato plants (Dehne 1987b). Surprisingly, detailed histological and histochemical studies on these mechanisms are scarce.

Competition for nutrients is another mode of interaction often mentioned, but with minor emphasis, perhaps due to the rare evidence for this cause of pathogen impairment. For instance, Priestel (1980) showed competition between *Meloidogyne incognita* and *G. mosseae* for carbohydrates (glucose and fructose) in the infected tissue, which may have inhibited nematode development and spread.

The utilization of particular nutrients, which may be disadvantageous to the pathogen, is speculated as an additional mechanism in combination with others (e.g., Caron et al. 1986b; Garcia-Garrido and Ocampo 1989). As the preoccupation of root cells by VAM fungi is coupled with biochemical changes, it is difficult to determine what really renders the colonized tissue unfavorable as a food source for parasitic nematodes and pathogenic microorganisms.

3. Improved Phosphorus Nutrition. Smith (1988) and Graham (1988) extensively discussed the relevance of increased P-nutrition in VA-mycorrhizal plants for the interaction with soilborne pathogens. They defined test conditions indispensable for such an appraisal. Due to the improved capability of the mycorrhizal root system to absorb and translocate P and minor elements through the network of external hyphae, enhanced P-nutrition has indirect effects on different pathogens.

Although the variability of interactions is great, regarding sedentary nematodes, an increased tolerance of mycorrhizal root parts is often reported. Especially in field experiments with a high soil P-level, VAM fungi can alleviate Zn-deficiency induced by nematodes, thus increasing host tolerance. Their influence on nematode infection and reproduction must be measured in an appropriate way in order to avoid misinterpreting changes in population densities due to dilution or disease escapes in enlarged root systems dependent on the improved P-supply. Mycorrhizal fungi seem to affect nematode penetration and infection processes in the same way as P-fertilization. There are several reports on the adverse effects of higher P-content in mycorrhizal roots on the reproduction of various *Meloidogyne* species (Hussey and Roncadori 1982; Diederichs 1987; Carling et al. 1989), however, this is not consistently observed as a general defense mechanism. Apparently, P-increase and the consequent enhancement of plant growth and vigor represent only one effect; the reduction in nematode populations and disease severity may be linked to this growth promotion, but must not necessarily be dependent on the improved P-status.

Numerous combinations of soilborne fungal pathogens with VAM fungi have also been correlated with the improved P-nutrition. Graham (1988) pointed out, in detail, how the VA root-fungi association interferes with increasing root P-content, the reduction in membrane permeability, and the increasing allocation of carbon to the mycorrhizal fungus in the root, resulting in less exudation out of the root. This P-dependent effect of limited root exudation indirectly alters the activities of microorganisms that respond to quantitative changes in root metabolites in and around roots. The so established mycorrhizosphere exerts its own selective influence on microbial activities in the surrounding soil and also may lower the attractiveness of root infection sites for pathogens.

Regarding these interactions with soilborne fungal pathogens, Fusarium crown and root rot on tomato is one example where host response differed between mycorrhizal and P-fertilized nonmycorrhizal plants (Caron et al. 1986b). Root necrosis and the pathogen population were unaffected by P-fertilization but significantly reduced by *G. intraradicis*. This supports the supposition that induced mechanisms, independent of the increased P-status, control the impairment of fungal infections in roots, and confirms several other results where P-fertilization could not replace the symbiont activity. Also, opposite effects of VA-mycorrhizae on shoot and leaf diseases (increased susceptibility) conflict with the hypothesis of a predominant influence of the altered P-status in the host on disease resistance (Schönbeck 1979).

Overall, the actual knowledge about mechanisms controlling disease reduction is still unsatisfactory. Also, studies on the specificity within the host-symbiont-pathogen system are rather rare. The enumeration of possible modes of interaction should be replaced by more detailed evidence of direct or induced disease reduction. This demand must be fulfilled if mycorrhizal fungi are indeed to attain the importance which many scientists are convinced that they can have in future concepts of crop protection.

3.3.2 Role of Mycorrhizal Fungi as Biocontrol Organisms – Significance of the Known Biological Effects in Practical Plant Production

In contrast to the amount of literature on laboratory experiments, field or greenhouse trials with mycorrhizae involving practical plant production lag far behind. Field studies mostly consider indigenous populations of mycorrhizal fungi and their interaction with plant nutrition and yield, while field trials with inoculations of selected fungal strains for biocontrol purposes are more seldom. Presently, the use of ectomycorrhizal fungi is already a feasible forestry practice, although the response is one of better plant growth on marginal habitats rather than biocontrol of root diseases. Investigations of biological effects on soilborne diseases under practical conditions are beginning, however. Unfortunately, many inconsistent results confirm the distinct dependence on environmental conditions (climate, soil/substrate).

Interference with indigenous mycorrhizal fungi complicates the analysis and would demand fumigated or chemically treated soils. Nevertheless, in order to select adequate fungal strains for successful biocontrol one must also consider their fitness in the rhizosphere populations and their adaptation to abiotic environmental factors. It is still difficult to attain a sufficient degree of efficiency and certainty of a positive response. More and better knowledge about the complex influence on the hosts by the symbionts could help to improve and support the efficiency of the symbionts regarding disease reduction or increase in tolerance. As yet, a specific mechanism predominantly involved in disease control has not been determined.

In summary, the possible beneficial effects of mycorrhizae include: increased crop yields, protection against certain root pathogens, increased tolerance to environmental stress, and a reduction in the use of fertilizers. However, the ability to reduce soilborne diseases represents only one trait that cannot, in general, be attributed to every fungal strain.

Therefore, it seems that a separate consideration of the disease-reducing or pathogen-inhibiting ability of mycorrhizal fungi will not describe fully the complex benefits that they may give to the crop plants. Here, we will focus on the practical application of mycorrhizal fungi (indigenous or even inoculated). Their usefulness as biocontrol agents has been less convincingly proved and demonstrated than their general and often striking promotion of plant vigor and yield under conditions of biotic or abiotic stresses.

Thus, up to now, the value of mycorrhizal fungi seems to lie more in their health-promoting activities rather than resulting from single, specific control mechanisms. Consequently, the high expectancies placed on the disease-controlling efficiency of mycorrhizal fungi should be brought to a realistic level, without diminishing the importance of their beneficial potential for plant health.

3.3.3 Mycorrhizae in Integrated Plant Production

As opposed to classic biocontrol agents, mycorrhizal fungi can be used in two different ways to support plant health and success in plant production: (1) by maintaining and promoting advantageous conditions for indigenous populations of mycorrhizal fungi at certain production sites and (2) by inoculating selected fungal strains.

The application of mycorrhizal fungi requires the development and production of a distributable and storable product. To guarantee sufficient efficiency, screening at various stages is necessary. However, this will require further research. Nevertheless, the introduction of mycorrhizal fungi into soil or onto seeds has markedly improved, since inorganic carrier material or fermenter cultures (here, only ectomycorrhizal fungi) simplified inoculum production and large-scale formulation (see Jeffries 1987).

The selection of fungal strains having sufficient effectivity and combining adequately with host and/or target pathogens is a quite more complex and uncertain task. An economically acceptable application procedure can only be developed successfully under actual conditions for the cultivation of host plants.

Taking these aspects into consideration, the suitable range of application is apparently restricted to:

1. Forest seedling nurseries (ectomycorrhizal fungi);
2. Nurseries and horticultural crops in artificial or fumigated substrates; especially pot cultures and transplanted ex vitro material from micropropagation;
3. Seed treatment of agricultural crops.

To profit from the possible benefits of introduced or indigenous symbionts, the influence of plant production measurements, such as fertilization, tillage, and soil amendments with green manure, on the establishment of mycorrhizal fungi and on their performance in the host plant must be taken into consideration. In the extensive literature on the effects of plant protection chemicals applied to roots/soils or shoots, reports show various influences of these chemicals: from detrimental to no effect or even, stimulatory to mycorrhizae.

In order to support a balanced mycorrhizosphere and to avoid interference with crop production practices, much more must be learned about disturbing factors.

The role and importance of the mycorrhizae present in conventional, intensive crop production have not as yet been clarified. However, in discussions on future farming concepts, mycorrhizal fungi are attracting progressively more attention. Thus, the estimation of their beneficial properties might be increased in sustainable agriculture (use of less fertilization, plant protection, and reduced tillage). These aspects favor conditions under which the symbionts can function either as indigenous or introduced populations in promoting plant health.

3.4 Future Research

To acquire more detailed information about the mechanisms involved in disease control phenomena reported with regard to VAM and to attain the goal of using VA-mycorrhizal fungi as health-promoting agents in plants, future activities should focus on the following:

1. Experiments conducted under conditions that allow the investigation of disease control mechanisms in mycorrhizal plants – independently of alterations in nutrient uptake.
2. Cultivar specificity of the plant responses to VAM.
3. Long-term field trials to test the stability and reliability of the effects of mycorrhizae on plant health. These experiments should include the complex interactions of factors such as soil structure and fertility, crop rotation, inoculum density of VA-mycorrhizal fungi and pathogens, and mixed populations of symbiotic and deleterious microbes.
4. Determination of the role of mycorrhizal symbiosis in either intensive plant production or LISA (low input sustainable agriculture).

5. Competitiveness of isolates to be inoculated in unsterile soil.
6. In vitro culture of VA-mycorrhizal fungi.
7. Improvement of the usefulness of mycorrhizal strains by genetic methods, including breeding of host plants that show extraordinary qualities with regard to symbiosis.

References

Allen MF, Boosalis MG (1983) Effects of two species of vesicular-arbuscular mycorrhizal fungi on drought tolerance of winter wheat. New Phytol 93:67–76

Auge RM, Stodola AJW (1990) An apparent increase in symplastic water contributes to greater turgor in mycorrhizal roots of droughted *Rosa* plants. New Phytol 115:285–295

Baghel PPS, Bhatti DS, Jalali BL (1990) Interaction of VA mycorrhizal fungus and *Tylenchulus semipenetrans* on citrus. In: Jalali BL, Chand H (eds) Current trends in mycorrhizal research. Proc Natl Conf on Mycorrhiza, Hisar, India, pp 118–119

Bagyaraj DJ (1984) Biological interactions with VA mycorrhizal fungi. In: Powell CL, Bagyaraj DJ (eds) VA mycorrhiza. CRC Press, Boca Raton, pp 131–153

Baltruschat H, Dehne HW (1982) Zum Auftreten der endotrophen Mykorrhizen in verschiedenen Getreidefruchtfolgen und Düngesystemen. Med Fac Landbouww Rijksuniv Gent 47:831–839

Baltruschat H, Schönbeck F (1975) Untersuchungen über den Einfluß der endotrophen Mycorrhiza auf den Befall von Tabak mit *Thielaviopsis basicola*. Phytopath Z 84:172–188

Baylis GTS (1970) Root hairs and phycomycetous mycorrhizas in phosphorus-deficient soil. Plant Soil 33:713–716

Baylis GTS (1972) Fungi, phosphorus, and the evolution of root systems. Search 3:257–259

Berta G, Fusconi A, Trotta A, Scannerini S (1990) Morphogenetic modifications induced by the mycorrhizal fungus *Glomus* strain E_3 in the root system of *Allium porrum* L. New Phytol 114:207–215

Carling DE, Roncadori RW, Hussey RS (1989) Interactions of vesicular-arbuscular mycorrhizal fungi, root-knot nematode and phosphorus fertilization on soybean. Plant Dis 73:730–732

Caron M (1989) Potential use of mycorrhizae in control of soilborne diseases. Can J Plant Pathol 11:177–179

Caron M, Fortin JA, Richard C (1986a) Effect of inoculation sequence on the interaction between *Glomus intraradices* and *Fusarium oxysporum* f. sp. *radicis-lycopersici* on tomatoes. Can J Plant Pathol 8:12–16

Caron M, Fortin JA, Richard C (1986b) Effect of phosphorus concentration and *Glomus intraradices* on *Fusarium* crown and root rot of tomatoes. Phytopathology 76:942–946

Chapin FS III (1980) The mineral nutrition of wild plants. Annu Rev Ecol Syst 11:233–260

Clowes FAL (1981) Cell proliferation in ectotrophic mycorrhizas of *Fagus sylvatica* L. New Phytol 87:547–555

Cooper KM (1982) Mycorrhizal fungi – their role in plant establishment. Comb Proc Int Plant Prop Soc 32, November 5–6, 1992, Auckland, New Zealand, Int Plant Prop Soc Inc, Boulder, Colorado, pp 396–404

Dehne HW (1982) Interactions between vesicular-arbuscular mycorrhizal fungi and plant pathogens. Phytopathology 72:1115–1119
Dehne HW (1987a) Zur Nutzung der VA-Mykorrhiza als Antistressfaktor. Angew Bot 61:135–143
Dehne HW (1987b) Zur Bedeutung der vesikulär-arbuskulären (VA) Mykorrhiza für die Pflanzengesundheit. Habilitationsschrift, University Hannover, Germany
Dehne HW, Schönbeck F (1979a) Untersuchungen zum Einfluß der endotrophen Mycorrhiza auf Pflanzenkrankheiten. I. Ausbreitung von *Fusarium oxysporum* f. sp. *lycopersici* in Tomaten. Phytopath Z 95:105–110
Dehne HW, Schönbeck F (1979b) Untersuchungen zum Einfluß der endotrophen Mykorrhiza auf Pflanzenkrankheiten. II. Phenolstoffwechsel und Lignifizierung. Phytopath Z 95:210–216
Dehne HW, Schönbeck F, Baltruschat H (1978) Untersuchungen zum Einfluß der endotrophen Mykorrhiza auf Pflanzenkrankheiten. 3. Chitinase-Aktivität und Ornithin-Zyklus. Z Pflanzenkr Pflanzenschutz 85:666–678
Diederichs C (1987) Interaction between five endomycorrhizal fungi and the root-knot nematode *Meloidogyne javanica* on chickpea under tropical conditions. Trop Agric (Trinidad) 64:353–355
Duchesne LC, Ellis BE, Peterson RL (1989a) Disease suppression by the ectomycorrhizal fungus *Paxillus involutus*: contribution of oxalic acid. Can J Bot 67:2726–2730
Duchesne LC, Peterson L, Ellis BE (1989b) The future of ectomycorrhizal fungi as biological control agents. Phytoprotection 70:51–57
Farquhar MC, Peterson RL (1990) Early effects of the ectomycorrhizal fungus *Paxillus involutus* on the root rot organism *Fusarium* associated with *Pinus resinosa*. Can J Bot 68:1589–1596
Garbaye J, Bowen GD (1989) Stimulation of ectomycorrhizal infection of *Pinus radiata* by some microorganisms associated with the mantle of ectomycorrhizas. New Phytol 112:383–388
Garcia-Garrido JM, Ocampo JA (1989) Effect of VA mycorrhizal infection of tomato on damage caused by *Pseudomonas syringae*. Soil Biol Biochem 21:165–167
Gäumann E, Kern H (1959) Über chemische Abwehrreaktionen bei Orchideen. Phytopath Z 36:1–26
Graham JH (1988) Interactions of mycorrhizal fungi with soilborne plant pathogens and other organisms: an introduction. Phytopathology 78:365–366
Hetrick BAD, Wilson GWT, Todd TC (1990) Differential responses of C_3 and C_4 grasses to mycorrhizal symbiosis, phosphorus fertilization, and soil micro-organisms. Can J Bot 68:461–467
Hirrel MC, Gerdemann JW (1980) Improved growth of onion and bell pepper in saline soils by two vesicular-arbuscular mycorrhizal fungi. Proc Soil Sci Am 44:654–655
Hussey RS, Roncadori RW (1982) Vesicular-arbuscular mycorrhizae may limit nematode activity and improve plant growth. Plant Dis 66:9–14
Jain RK, Sethi CL (1988) Interaction between vesicular-arbuscular mycorrhiza, *Meloidogyne incognita* and *Heterodera cajani* on cowpea as influenced by time of inoculation. Ind J Nem 18:263–268
Jeffries P (1987) Use of mycorrhizae in agriculture. CRC Crit Rev Biol 5:319–357
Kiernan JM, Hendrix JW, Maronek DM (1983) Fertilizer-induced pathogenicity of mycorrhizal fungi to sweetgum seedlings. Soil Biol Biochem 15:257–262
Koide RT (1991) Nutrient supply, nutrient demand and plant response to mycorrhizal infection. New Phytol 117:365–386
Koide RT, Li M (1990) On host regulation of the vesicular-arbuscular mycorrhizal symbiosis. New Phytol 114:59–65

Kope HH, Fortin JA (1989) Inhibition of phytopathogenic fungi in vitro by cell free culture media of ectomycorrhizal fungi. New Phytol 113:57–63
Kope HH, Fortin JA (1990) Antifungal activity in culture filtrates of the ectomycorrhizal fungus *Pisolithus tinctorius*. Can J Bot 68:1254–1259
Kope HH, Tsantrizos YS, Fortin JA, Ogilvie KK (1991) p-Hydroxybenzoylformic acid and (R)-(−)-p-hydroxymandelic acid, two antifungal compounds isolated from the liquid culture of the ectomycorrhizal fungus *Pisolithus arhizus*. Can J Microbiol 37:258–264
Krishna KR, Bagyaraj DJ (1983) Interaction between *Glomus fasciculatum* and *Sclerotium rolfsii* in peanut. Can J Bot 61:2349–2351
Land S (1990) Auftreten und Charakterisierung der vesikulär-arbusculären (VA) Mykorrhiza in intensiv genutzten Ackerböden. PhD Thesis, Universität Hannover, Germany
Land S, Dauck H, von Alten H (1990) Evaluation of VA-mycorrhizae in different agricultural soils. Agr Ecosyst Environ 29 (1–4):217–224
Marx DH (1969) The influence of ectotrophic mycorrhizal fungi on the resistance to pathogenic infections. II. Purification, identification and biological activity of antibiotics produced by *Leucopaxillus cerealis* var. *piceina*. Phytopathology 59: 411–417
Marx DH (1972) Ectomycorrhizae as biological deterrents to pathogenic root infections. Annu Rev Phytopathol 10:429–454
Marx DH, Davey CB (1969) The influence of ectotrophic mycorrhizal fungi on the resistance to pathogenic infections. III. Resistance of aseptically formed mycorrhizae to infection by *Phytophthora cinnamomi*. Phytopathology 59:549–558
Mosse B (1973) Plant growth responses to vesicular-arbuscular mycorrhiza. IV. In soil given additional phosphate. New Phytol 72:127–136
Natarajan K, Govindasamy K (1990) Antagonism of ectomycorrhizal fungi to some common root pathogens. In: Jalali BL, Chand H (eds) Current trends in mycorrhizal research. Proc Natl Conf Mycorrhiza, Hisar, India, pp 98–99
Priestel G (1980) Wechselbeziehungen zwischen der endotrophen Mykorrhiza und dem Wurzelnematoden *Meloidogyne incognita* (Kolfoid u. White 1919, Chitwood 1949) an Gurke. PhD Thesis University Hannover, Germany
Sanders FE, Tinker PB (1973) Phosphate flow into mycorrhizal roots. Pestic Sci 4:385–395
Schönbeck F (1979) Endomycorrhiza in relation to plant diseases. In: Schippers B, Gams W (eds) Soil-borne plant pathogens. Academic Press London, pp 271–280
Schönbeck F, Dehne HW (1989) VA-mycorrhiza and plant health. In: Vancura V, Kunc F (eds) Interrelationship between microorganisms and plants in soil. Academia Publ Cz Acad Sci Praha, pp 83–91
Schuepp H, Dehn B, Sticher H (1987) Interaktionen zwischen VA-Mykorrhizen und Schwermetallbelastungen. Angew Bot 61:85–96
Smith GS (1988) The role of phosphorus nutrition in interactions of vesicular-arbuscular mycorrhizal fungi with soilborne nematodes and fungi. Phytopathology 78:371–374
Spanu P, Boller T, Ludwig A, Wiemken A, Faccio A, Bonfante-Fasolo P (1989) Chitinase in roots of mycorrhizal *Allium porrum*: regulation and localization. Planta 177:447–455
Sylvia DM, Sinclair WA (1983) Phenolic compounds and resistance to fungal pathogens induced in primary roots of Douglas-fir seedlings by the ectomycorrhizal fungus *Laccaria laccata*. Phytopathology 73:390–397
Thomas GV, Sundararaju P, Ali SS, Ghai SK (1989) Individual and interactive effects of VA mycorrhizal fungi and root-knot nematode *Meloidogyne incognita* on cardamom. Trop Agric UK 66:21–24

Umesh KC, Krishnappa K, Bagyaraj DJ (1989) Interaction of burrowing nematode *Radopholus similis* (Cobb 1893, Thorne 1949) and VA mycorrhiza *Glomus fasciculatum* (Thaxt Gerd et Trappe) in banana (*Musa acuminata* Colla.). Ind J Nematol 18:6–11

Williams SCK, Vestberg M, Uosukainen M, Dodd JC, Jeffries P (1992) Effects of fertilizers and arbuscular mycorrhizal fungi on the post-vitro growth of micropropagated strawberry. Agronomie 12 (10):851–859

Zhao ZP, Guo XZ (1989) Study on hyphal hyperparasitic relationships between *Rhizoctonia solani* and ectomycorrhizal fungi. Acta Microbiol Pao 29:170–173

4 The Soil Environment

J.D. MacDonald

4.1 Introduction

Many components of the natural environment influence the epidemic spread of plant diseases. Processes critical to pathogens, such as survival, dispersal, germination, host infection, growth within and outside the host, and reproduction, are affected directly and indirectly by the environment. The environment also influences the tissue growth, three-dimensional arrangement, physiological hardiness, and disease proneness of the host plant. Knowledge of the environment and its potential effects in disease epidemiology is critical to disease management strategies.

In the case of aboveground diseases, environmental factors such as temperature, moisture, wind, and radiation are well known for their effects on disease development and pathogen spread (e.g., Zadoks and Schein 1979; Fry 1982; Campbell and Madden 1990). While many of these same environmental factors can influence the epidemiology of root diseases, their relative importance often is quite different. For example, air currents may transport propagules of foliar pathogens great distances before they land on susceptible host tissues. Within the soil matrix, however, wind is not a factor. Host-pathogen contact in root diseases relies largely on the invasive growth of roots through soil or motility on the part of the pathogen.

Another unique feature of the soil environment is the nearly continuous presence of liquid water. However, unlike the water that periodically condenses or splashes on foliar tissues, the chemical potential of soil water can be very low, owing to the capillary forces of soil pores or the presence of dissolved solutes. The soil environment also undergoes unique changes in the gaseous composition of its atmosphere, and is buffered against changes in temperature and water status to an extent rarely duplicated in the aboveground environment. This buffering significantly slows the rate of change of certain environmental factors and points out another distinction between the above- and below-ground environments: the time scale over which conditions may change from restrictive to permissive.

Identifying epidemiologically significant aspects of the soil environment, quantifying them, and manipulating them at either the crop management or research experiment level poses many challenges. The soil environment, and how plants and pathogens respond over time to the direct, indirect, and

C.L. Campbell/D.M. Benson (Eds.)
Epidemiology and Management of Root Diseases

integrated effects of its physical, chemical, and biological components are topics that could fill entire books. In this single chapter, it will not be possible to deal with all relevant aspects of the soil environment or to discuss all potential examples. Thus, I have chosen to confine my discussion to selected aspects of the physical soil environment, how they influence root disease epidemiology, and where some important data gaps exist. Information concerning biotically influenced soil microenvironments (i.e., rhizosphere, spermosphere, etc.) can be obtained in other sections of this book or from other writings (e.g., Lockwood 1964; Garrett 1970; Baker and Cook 1974; Bowen and Rovira 1976; Cook and Baker 1983; Foster et al. 1983; Foster 1986; Weller 1988).

4.2 Components of the Physical Soil Environment

Soil is a complex substance, consisting of much more than just the weathered, decomposed remnants of rock. Actually, soil is a "system" more than a "material". It is simultaneously heterogeneous, polyphasic, particulate, disperse, and porous, with a very large interfacial area per unit volume (Hillel 1980). The term "polyphasic" refers to the common characterization of soil as a three-phase system; one possessing solid, liquid, and gaseous components. The liquid and gaseous phases exist within the void space of the solid phase, and all three phases undergo continuous, often simultaneous, interrelated changes. These changes can have important effects in root disease epidemiology. Some prominent features of the soil system are summarized in the following sections.

4.2.1 The Soil Matrix

The solid component of soil consists largely of variously sized particles derived from the gradual fragmentation and chemical decomposition of primary minerals forming the earth's crust. Depending upon their size, mineral fragments are classified as stones, gravel, sand, or silt. Clay particles, which are colloidal in size, are secondary minerals formed from chemical decomposition products of the parent minerals. Interspersed among these various mineral particles are particles of organic origin. In agricultural soils, the organic component derives largely from plant roots and incorporated crop residues, but also consists of the remains of various soil-inhabiting animals and microorganisms. Humus is the name applied to a class of soil particles that are the semistable, colloidally sized decomposition products of larger organic materials (Hillel 1980).

The proportionate amount of sand, silt, and clay in a soil is a relatively stable feature over time (in the context of human life span), and imparts a certain "texture" to the soil. Texture is one feature by which soils are classified and described. For example, a "loam" soil is one that has roughly equal amounts of sand, silt, and clay. Soils dominated by one or two particle classes are given textural names reflecting their composition (i.e., "sandy loam", "silty clay loam", or "sandy clay"). Soils that contain large amounts of organic matter (15–20%) may have the prefix "mucky" added to the textural classification, whereas soils which contain very large amounts of organic matter (>20%) are simply called muck (if the organic matter is well decomposed) or peat (if the organic matter is still recognizable) (Hausenbuiller 1972).

The mineral and organic particles that comprise the solid phase of soil lie together in a three-dimensional network referred to as the soil "matrix". The matrix usually contains considerable void space since soil particles are irregular in size and shape and do not adjoin in an organized manner. Furthermore, the matrix possesses a higher level of organization, or "structure", resulting from the aggregation of individual particles into larger units.

Soil aggregates form as a result of various chemical, biological, and physical forces that act upon soil particles and cement them together. The smallest unit in the hierarchy of aggregates is the clay domain, or floccule, which may measure 1–2 μm across (Dexter 1988). These subunits combine with other floccules and mineral grains to form clusters (2–20 μm), which combine to form microaggregates (20–250 μm), which combine to form aggregates (>250 μm), which combine to form clods (>25 mm) (Dexter 1988). The cementing of smaller subunits into larger structures is a central process in aggregate formation. Forces that bring the subunits together and facilitate cementing include the process of soil consolidation, pressures resulting from gravity or soil swelling, and the surface tension of capillary water.

Clay particles play an important role in aggregate formation because they tend to coat larger particles and, owing to their negative surface charge, cross-link those particles through di- and trivalent cations (principally Ca^{+2}, Mg^{+2}, and Al^{+3}). Calcium carbonate, silicas, and sesquioxides also play important roles in cementing soil particles into larger aggregates (Hillel 1980). Additionally, the mucilaginous materials secreted by plant roots and soil microorganisms, and organic chemicals released during the decomposition of biological substrates, function to cement aggregates (Dexter 1988). Once formed, aggregates >250 μm are reinforced and stabilized against the dispersive forces of water by fungal mycelium (particularly mycorrhizal fungi) or fine (0.2–1.0 mm diameter) roots (Tisdale and Oades 1982; Oades 1984; Miller and Jastrow 1990).

Thus, the pore system within soil consists not only of interparticle spaces, but also of networks of interaggregate spaces (Fig. 4.1A). Indeed, it is the interaggregate spaces that provide many of the large diameter pores critical for water and air infiltration, as well as unimpeded root growth (Dexter 1988). Large diameter pores also are created by the burrowing

Fig. 4.1. Soil fractures illustrating contrasting structural properties. Both samples are of the 5–15 cm profile from two locations. Sample **A** is a Yolo loam from a nonagricultural area having tree and annual grass cover. Note the abundance of macropores, large soil aggregates, interaggregate fracture lines, and old root channel pores. Sample **B** is a Yolo fine sandy loam from an agricultural area showing the effects of surface compaction. Note that large aggregates are ill-defined and interaggregate pores are essentially nonexistent. Soil properties such as water infiltration, water-holding capacity, air-filled porosity, soil strength, soil heat capacity, gas diffusion rates, etc. are very different between these soils. As described in the text, these differences influence many features of root growth, movement of motile pathogens, and many other processes involved in root disease development. *Vertical bars* indicate 1 cm

activity of animals (e.g., earthworms) and the growth (followed by death and decay) of plant roots. Thus, the soil matrix is permeated by pores which may be as wide as several millimeters or as minute as the spaces between clay lamellae. A typical medium-textured soil may contain approximately 50% solids and 50% pore space (Singer and Munns 1987).

In contrast to soil texture, soil structure is a dynamic, highly mutable feature of the matrix – particularly in agroecosystems. This is because the bonds between aggregate subunits remain as planes of weakness within higher-order aggregates, and become fracture planes if shear or tensile forces exceed aggregate strength (Hillel 1980; Dexter 1988).

Many natural forces act to fragment large aggregates into smaller subunits, including soil shrinkage upon drying, invasive root growth, and the burrowing of soil-inhabiting animals. Many agricultural practices also affect soil structure, sometimes adversely. For example, some cultivated plants may not produce the high root densities or lengths of fine roots (0.2–1.0 mm diameter) critical for aggregate stabilization, or soil treatments with fumigants or certain fungicides may reduce the growth of fungi active in aggregate stabilization. And at the same time aggregates may have reduced biotic stabilization, they may be subjected to severely destructive forces. For example, air slaking (aggregate fracture resulting from internal air compression; Hillel 1980) occurs when air-dry aggregates, at the surface of fallow fields or between crop rows, are wetted rapidly during irrigation. Aggregates also are fragmented by the impact force of rain or sprinkler drops or the abrasive action of sand particles carried in flowing water. At the same time water is providing energy for aggregate fracture, it also is weakening the bonds between aggregate subunits (Dexter 1988). Furthermore, if irrigation water contains sodium ions, Na^+ may eventually become concentrated enough to cause the dispersion of clay floccules and subsequent disintegration of all higher-order aggregates (Hillel 1980).

Fragmentation of aggregates into smaller subunits increases soil density and strength (strength refers to a soil's ability to resist an applied force; Taylor 1974). Structural breakdown is common and readily visible at soil surfaces following irrigation or rainfall, when "crusts" form and inhibit seedling emergence. Under extreme conditions, surface aggregates may become so completely dispersed that a slick, dense layer of mud forms, effectively sealing the surface and severely restricting water and air entry into the profile (Dexter 1988). To some extent, the adverse effects of surface crusts can be moderated by tillage to fracture them. However, tillage also has disruptive effects on soil structure because vehicle traffic causes compression and compaction (Fig. 4.1B). Also, if a soil is too dry when tilled, aggregates are subjected to grinding abrasion, and if it is too wet, aggregates are smeared or the soil puddled into dense, hard clods (Hausenbuiller 1972).

Because soil structure affects virtually all other properties of the soil system, it has many important direct and indirect effects on root disease epidemiology.

4.2.2 The Soil Solution

The liquid and gaseous components of soil coexist within the pores permeating the solid matrix. The water contained within soil is at once a solvent, an ion transport medium, a chemically reactive substance, a physical force reshaping the solid matrix, and a lubricant. Its properties vary greatly from that of pure water and it is most properly referred to as the "soil solution". Chemical properties of the soil solution differ greatly from soil to soil, owing to differing amounts of interfacial area within the matrix ($1000\,cm^2\,g^{-1}$ of sandy soil, to $1\,000\,000\,cm^2\,g^{-1}$ of clay soil; Day et al. 1967) and complex interactions between the solid and liquid phases (e.g., cation and anion exchange, ion dissolution and precipitation, hydrogen and covalent bonding, and surface chelation; Singer and Munns 1987).

Rain or irrigation water enters the soil matrix through cracks and pores that open to the surface. Water infiltration is driven by the capillary and adsorptive forces of soil pores, as well as hydrostatic pressures and gravitational forces acting on the water mass (Holmes et al. 1967; Hausenbuiller 1972). As water enters the matrix, soil pores become saturated and excess water is conducted to deeper parts of the profile. When entry ceases, water continues to drain from the matrix until the gravitational forces pulling it downward are countered by capillary and adsorptive forces holding it within the pores. The amount of water retained by a soil at the point when drainage becomes negligible is defined as "field capacity" and is a function of the soil's texture, pore volume, and pore size distribution.

The soil solution, like all other matter, possesses both kinetic and potential energies. However, its kinetic energy usually is negligible since flow velocities through the matrix are so low. Hence, the energy status of the soil solution is described best by its potential energy (Miller and Klute 1967; Hillel 1980; Singer and Munns 1987). The potential energy of water (water potential) can be expressed as energy per unit volume, which has the dimensions of pressure (Hillel 1980; Papendick and Campbell 1981). The accepted SI unit for pressure is the Pascal (Pa), although many other units have been used to describe water potential in earlier literature – including atmospheres, cm H_2O, and bars ($1\ bar = 1 \times 10^5\ Pa = 100\ kPa$) (Papendick and Campbell 1981).

The water potential of the soil solution (Ψ) is the sum of all forces affecting its energy state and is described by the equation $\Psi = \psi_s + \psi_m + \psi_g + \psi_p$; where ψ_s, ψ_m, ψ_g, and ψ_p are the solute, matric, gravitational, and pressure potentials, respectively. The solute potential (also referred to as osmotic potential, ψ_π) results from dissolved substances in the water and has a negative influence on Ψ. The matric potential reflects capillary and adsorptive forces of the soil matrix, which also have a negative influence on Ψ. Gravitational potential results from differences in elevation relative to a reference point, and can have a positive or negative effect on Ψ. Pressure potential derives from the hydrostatic pressure of water (e.g., ponded surface water) and may have a positive or negative effect on potential, depending upon the reference pressure (Papendick and Campbell 1981). Soil physicists frequently combine the matric and pressure components under ψ_p, since ψ_m is essentially a negative pressure potential (Day et al. 1967; Hillel 1980). However, because ψ_p also describes hydrostatic pressures apart from the soil matrix (e.g., the positive pressure associated with cell turgor), it can be useful to retain a distinction between the terms.

By definition, pure water at atmospheric pressure has a Ψ value of zero (Papendick and Campbell 1981). Relative to this reference value, the Ψ of the soil solution is almost always negative – due to the fact that in unsaturated soil, Ψ is dominated by the ψ_m component (Papendick and Campbell 1981). In some cases, however (particularly in salt-affected soils), the ψ_s component may become significant as evapotranspiration removes water from the soil and concentrates dissolved salts in ever smaller water films (Raney 1970; Papendick and Campbell 1981).

Water potential in the soil-plant-atmosphere system describes the tendency of water to move from areas of high relative potential to areas of lower relative potential (i.e., from an area of negative potential to an area of more negative potential). At field capacity, the Ψ for most soils ranges between -10 to -35 kPa (Singer and Munns 1987). As plants absorb water, the Ψ in the soil nearest the roots drops below that of the bulk soil, forming a gradient that causes water to move toward the roots. However, as the soil dries, its conductivity decreases sharply as the number of water-filled pores decreases, pathways become more tortuous, and friction becomes significant (Hillel 1980). Eventually, the soil dries to a point where plants cannot extract water fast enough from any part of the profile to offset transpirational loss, and leaves wilt. The soil water content at which plants are unable to regain turgor is defined as the "permanent wilting point", and is approximately -1500 kPa (Papendick and Campbell 1981).

The energy status of the soil solution is a highly significant factor in root disease epidemiology because it describes the availability of water to plant roots and soil-inhabiting microorganisms, and strongly influences the relative activity of various organisms. However, the soil solution also influences root disease epidemiology in ways not so directly linked to the water relations of plants or pathogens in soil. For example, water weakens interaggregate bonds, aiding root penetration of high-strength soils and influencing root distribution within the profile. Water also is the medium in which motile pathogens move within soil, and through which chemical substances (nutrient ions, root exudates, etc.) diffuse along concentration gradients. Some of these effects are discussed in following sections.

Finally, water participates in many chemical reactions that occur at the surfaces of soil mineral and organic particles, and which largely determine mineral nutrient availability. The abundance or deficiency of certain macro- and micronutrients influences the occurrence or severity of numerous root diseases (Engelhard 1989). The chemistry of the soil solution, however, is a complex subject that cannot be treated adequately in this chapter, and readers are referred to more detailed accounts (e.g., Chapman 1970; Barber 1984; Bohn et al. 1985).

4.2.3 The Soil Atmosphere

Since the liquid and gaseous phases of soil coexist within the pore volume of the solid matrix, the amount of pore volume occupied by one is inversely related to the amount occupied by the other. Plant roots and most soil-inhabiting organisms require oxygen for respiration so, under optimum conditions, soil contains a mixture of air- and water-filled pores.

In a fully saturated soil, all pores are water-filled and air (except for entrapped bubbles) is excluded from the matrix. However, as described in the preceding section, hydrostatic and gravitational forces cause excess water (that which exceeds the field capacity water content) to drain from the profile. Soil pores with neck diameters $>10\,\mu m$ drain at ψ_m values of $-30\,kPa$ (an approximation of field capacity) (Papendick and Campbell 1981). Thus, at field capacity, the proportion of air-filled pores is a function of pore size distribution, and gives rise to the terms "air capacity" or "air-filled porosity" (the fractional volume of air in soil at the field capacity water content; Hillel 1980).

As a water front moves into a profile, it alters soil air pressures (Stolzy 1974; Hillel 1980). Below the front, air is compressed and displaced from the profile, as evidenced by bubbles escaping at surface cracks or large pores. As the front moves downward, soil air pressure in the upper profile decreases (relative to atmospheric) and air is drawn into the large diameter pores. As more water drains from the profile (or is removed by evapotranspiration) air is drawn into increasingly smaller diameter pores. This mass flow of air, in response to changes in soil water content, reflects the inverse relationship between air and water content, and also contributes to the mixing and specific composition of the soil air.

The principal gases in soil air are the same ones in atmospheric air: N_2, O_2, CO_2, and H_2O. However, their relative concentrations within the soil matrix can be quite different. For example, the relative humidity of atmospheric air fluctuates widely, but in soil (at water contents conducive to plant growth) the relative humidity seldom drops below 99% (Singer and Munns 1987). Also, in the upper 15 cm of a well-structured soil, the soil air contains approximately 79.2% N_2, 20.6% O_2, and 0.25% CO_2 (Stolzy 1974). This is an O_2 concentration slightly lower, and a CO_2 concentration eight times greater than atmospheric air (20.97 and 0.03%, respectively; Stolzy 1974).

Roots and other aerobic organisms in soil consume O_2 and produce CO_2 during respiration. Thus, sustained growth requires effective gas transfer – O_2 must reach points of consumption and CO_2 must not accumulate to harmful levels. Gas transfer in soil occurs by two mechanisms: mass flow and diffusion. Mass flow, as discussed above, occurs in response to air pressure differences between the atmosphere and the matrix. Pressure differentials arise from soil water movement, surface winds, or changes in temperature or barometric pressure (Stolzy 1974; Hillel 1980). While mass flow is significant, most gas exchange occurs by diffusion (Stolzy 1974; Hillel 1980; Gliński and Stępniewski 1985; Singer and Munns 1987).

There is a strong diffusion gradient between the soil and atmosphere because, relative to atmospheric air, soil air contains a higher concentration of CO_2 and lower concentration of O_2. Gas diffusion through soil is dependent upon air-filled pores within the matrix, since diffusion rates in air are 10^4 times those in water (Stolzy 1974). However, the diffusion rate is not determined simply by the presence of air-filled pores; the diameter and tortuosity of the pores also are critical (Stolzy 1974; Singer and Munns 1987). This explains why structureless crusts are such effective aeration barriers (Hillel 1980).

If soil pores remain water-filled for long periods (as can happen in fine-textured or compacted soils with relatively small-diameter pores, or soils with elevated water tables), diffusion is inhibited. Under such conditions, oxygen is depleted rapidly from areas where respiration rates are high. Oxygen diffusion to a point of consumption is so slow in water-saturated soil that O_2 concentration can decrease from maximum to zero over a distance of 1 mm (Stolzy 1974). Under these conditions, products of anaerobic respiration (e.g., methane, nitrous oxide, hydrogen sulfide, and ammonia) may accumulate to high levels (Stolzy 1974).

Thus, the soil atmosphere is a far more dynamic, complex mixture of gases than the aboveground atmosphere. This has important consequences to root disease epidemiology.

4.2.4 Soil Temperature

Another basic component of the soil environment is temperature, which influences many physical and physiological processes central to disease development and spread. Temperature also influences physical and chemical processes within the soil matrix, such as rates of chemical reactions, evaporative water loss, and gas diffusion.

Soil is heated primarily by surface absorption of solar radiation. Heat transfer in soil follows gradients from high to low energy, generally causing conduction into the profile during the day, and out of the profile at night. The efficiency of heat conduction is influenced largely by the mineral composition of the matrix and the volume fraction of air- and water-filled pores. This is because the thermal conductivity of mineral particles is five times greater than water, and the conductivity of water is 23 times greater than air (Hillel 1980).

The amount of heat that must be conducted into soil to effect a one degree change in temperature is a function of soil heat capacity. Soil heat capacity varies in space and time, being positively related to water content and bulk density. This is because the heat capacity of soil minerals and water is much higher than air (Singer and Munns 1987). Thus, while a wet, dense soil conducts heat more efficiently than a dry, porous soil, its temperature will change more slowly.

Temperatures at the soil surface fluctuate widely in response to changes in radiant energies. However, temperatures in deeper parts of the profile fluctuate only as heat energy is transferred from or to the surface. This can be a relatively slow process and effectively insulates soil below 30 cm from diurnal temperature changes (Singer and Munns 1987).

To some extent, surface temperature can be manipulated by various practices. For example, surface temperatures can be lowered by wetting a soil (to increase its heat capacity) or covering it with a mulch (to block incident radiation). Conversely, surface temperatures can be increased by removing vegetation or allowing soil to dry (Nielsen 1974; Hillel 1980; Singer and Munns 1987). While aimed at manipulating temperature, such practices also may influence aspects of soil structure, water status and aeration, and have unexpected effects on root disease epidemiology.

4.3 Influence of Edaphic Factors on Roots and Pathogens

Not only does the physical soil environment differ uniquely from the aboveground environment, but it exerts many unique influences on the development of root diseases and the spread of soilborne pathogens. Epidemiological processes that are not even a consideration in foliar diseases (e.g., spore dispersal within a porous, solid matrix) can be significant factors in the epidemiology of root diseases.

The soil environment can (and often must) be described and considered on different scales, just as is done with the aerial environment (Zadoks and Schein 1979). However, while soils typically are classified by any of several taxonomic, usage, or capability systems (Singer and Munns 1987), these mostly describe physical properties or agronomic utility and have little meaning relative to epidemiology. Thus, for the purpose of this discussion, I will use the term "macroenvironment" to refer to a contiguous region of soil within the same soil series (a soil series is the basic unit of soil taxonomy, and consists of soils that are essentially alike in all major profile characteristics; Hausenbuiller 1972). I use the term "mesoenvironment" to refer to the soil pedon (the smallest soil volume that exhibits the normal range of horizon, texture, and structure within a soil series – usually an area no smaller than 1 m^2 with a depth no less than 1.5 m; Singer and Munns 1987).

Specific features of the macro- and mesoenvironment are determined largely by mineralogical and physical features (i.e., the size of soil particles, their chemical composition, their physical arrangement into a solid, structured matrix, and the influence of that matrix on soil water, air, and heat transfer). I use the term "microenvironment" to refer to dimensionally small soil volumes within the mesoenvironment, where physical and chemical features of a soil often are modified by the presence of roots or soil-inhabiting organisms.

In terms of root disease epidemiology, features of the macro- and mesoenvironments dominate processes such as root growth, branching, and spatial arrangement; pathogen dispersal and growth rate; and host disposition to infection. The microenvironment most strongly influences processes such as inoculum survival in soil, spore germination, and the initial infection of root tissues.

While there are many papers (see, for example, Alexander et al. 1975; Percy 1983; Sidebottom and Shew 1985; Johnson and Doyle 1986) that attempt to correlate root diseases or pathogen activity with various edaphic factors for predictive purposes, only a relatively few have attempted to link edaphic factors to epidemiology in mechanistic ways (e.g., Bloomberg 1979; Reynolds et al. 1986; Van der Werf et al. 1986). Elucidation of these linkages remains a significant research challenge. However, only when functional linkages are established will the understanding and predictive modeling of root diseases approach that of aerial diseases. In the following section, I will attempt to note some established and probable relationships between epidemiological processes and soil features, and point out areas needing research. I will emphasize the macro- and mesoenvironments, which approximate the soil environment at the whole-crop and whole-plant levels, respectively.

4.3.1 Root Growth and Spatial Arrangement

In Chapter 2, plant root systems and parameters surrounding their development and infection are delineated. As discussed below, many features of root systems that are central to disease development (i.e., growth rate, density, distribution, branching order, root hair formation, succulence, disease proneness, etc.) are very directly influenced by the soil environment.

4.3.1.1 Soil Matrix Effects

Under optimum conditions, when plants are in active growth phases, the soil volume penetrated by roots is very large and (depending upon the genetic capabilities of the plant) may extend for several meters into the profile.

Roots penetrate soil either by growing through the void space or by pushing aside soil particles. However, roots have only limited ability to push aside soil particles and enter pores with diameters smaller than the advancing root tip. In fact, root elongation decreases very abruptly in response to increasing mechanical impedance within the matrix (Taylor 1974; Russell 1977; Dexter 1987). Thus, for a root system to achieve its maximum potential, the soil must possess a large number of pores with diameters equivalent to the root tip diameter and it must be easily deformed.

Many factors can interact to cause soil conditions to deviate from ideal; some manageable and some not. For example, many soils contain subsurface zones of dense and/or indurated materials. A claypan, in which a subsurface layer has a clay content much higher than the overlying material, results in a dense zone within the soil profile. Although plastic when wet, roots may not penetrate its microporous structure easily and it presents a barrier to water infiltration. A hardpan, in which soil particles have been cemented together with precipitated carbonates of calcium or magnesium, or oxides of silicon, iron, or aluminum (Hausenbuiller 1972), is a hard, brittle subsurface layer. Such indurate zones do not soften when wetted, but may possess a macroporous structure allowing some root penetration and water infiltration.

Claypans and hardpans are inherent features of a soil series and, hence, are components of the macroenvironment. A third type of subsurface zone is not a soil characteristic per se, but is considered a feature of the macroenvironment since it often occurs at the whole-crop level. This is a subsurface zone of compacted soil, often referred to as a "tillage pan" or "plow sole". Such zones are unique to agricultural or urban soils, and result from the operation of heavy equipment on moist, compressible soil.

In addition to specific high-strength layers within the profile, the soil mass itself exhibits strength properties that fluctuate over time. This strength arises from features of the soil's texture, bulk density, chemistry, and water content. Soil strength is related positively to bulk density and negatively to water content (Taylor 1974; Russell 1977; Dexter 1988).

When roots encounter a zone of high strength within the mesoenvironment, they tend to be deflected along paths of lesser resistance. This causes root distribution to reflect the three-dimensional arrangement of soil layers and aggregates. For example, when roots encounter a hardpan or plow pan, they are deflected horizontally along the compacted zone until they encounter a crack or some path of low resistance allowing downward penetration (Whiteley and Dexter 1983; Dexter 1986). This results in an abnormal root distribution wherein a large portion of the roots is blocked into the upper parts of the profile, with only some penetrating into deeper zones. Another abnormality in root system development occurs in deep clay soils (Vertisols), which form well-developed shear planes as a result of continual soil swelling and shrinking. These shear planes often represent the only low-impedance pathways through the matrix (Taylor 1974), and cause roots to be confined largely to spatially separated soil channels.

Other abnormalities in root architecture occur commonly in high-impedance soils. For example, linear extension of roots decreases markedly in response to relatively small external pressures (resistances), resulting in overall smaller root systems (Russell 1977; Bengough and Mullins 1990). Lateral roots, which are smaller in diameter than main root axes, proliferate and penetrate smaller-than-optimum soil pores. This results in abnormally dense, branched root systems. Finally, dense formations of root hairs develop, anchoring roots against longitudinal root growth pressures (Russell 1977).

4.3.1.2 Soil Solution and Atmosphere Effects

The solid matrix is not the only factor that influences root architecture within the mesoenvironment. As stated earlier, the soil is a three-phase system, and the liquid and gaseous phases also bias root distribution.

The soil solution influences root development in three basic ways. First, roots need water to maintain cell turgor and generate the hydraulic pressures needed for extension. As soil Ψ drops to restrictive levels in parts of the mesoenvironment, root extension ceases in those areas and continues only where soil Ψ remains favorable (Russell 1977). This compensatory growth habit leads to root distributions that reflect spatial and temporal changes in soil water status within the mesoenvironment. A second way in which water affects root behavior is through its influence on the strength of the bonds between soil aggregates. The soil solution weakens interaggregate bonds, making it easier for roots to push particles aside as they penetrate the matrix. As water is lost from the matrix (through evapotranspiration) the water films surrounding particles diminish in thickness, intra- and interaggregate bonds increase in strength, and the mechanical impedance of the soil increases.

The third way in which soil water influences root growth is less direct, and actually reflects its inverse relationship with soil air. As water displaces air from soil pores, the soil environment can quickly become hypoxic, or even anoxic. Roots are very sensitive to O_2 deficits; if the O_2 concentration drops below 15% in the soil atmosphere, uptake of mineral nutrients may be impaired; if it drops below 12%, root initiation may be inhibited; if it falls below 5%, growth of existing roots may cease (Stolzy 1974). Huck (1970) showed that exposing taproots of soybean and cotton to completely anoxic conditions caused growth cessation in as little as 2–3 min, and root death within 3–5 h.

While knowledge of oxygen concentration within the soil atmosphere can be informative, gaseous concentration does not properly describe O_2 availability to roots or pathogens. This is because the ability of O_2 to diffuse from a point of relatively high concentration (i.e., the soil air) to a nearby point of consumption (i.e., a respiring root) is influenced strongly by soil temperature and the size, continuity, and tortuosity of air-filled pores. Even more limiting, however, is the diffusion rate of O_2 through water (1/10 000 the rate through the gas phase). Because roots are virtually always bathed in liquid films or surrounded by small-diameter, water-filled soil pores, the thickness and continuity of these water barriers, and rate of O_2 diffusion through them, can be significant factors influencing root respiration (Letey and Stolzy 1964).

Pore size, geometry, and water occlusion, as well as the rate of oxygen consumption, can all differ greatly from point to point within the meso-environment. This results in considerable heterogeneity in oxygen status at the microenvironment level (Letey and Stolzy 1964; Russell 1977; Stolzy and Sojka 1984) and points out why, in terms of biological activity, the oxygen diffusion rate (ODR) is often a more useful measure of soil oxygen status than gaseous concentration (Stolzy 1974; Gliński and Stępniewski 1985). Researchers have found that root growth of many plant species is retarded or prevented if ODR drops below 0.3 or $0.2\,\mu g\,O_2\,cm^{-2}$ root min^{-1}, respectively (Stolzy and Letey 1964; Gliński and Stępniewski 1985).

Low soil O_2 has many profound effects on root geometry, morphology, and physiology. For example, roots will not grow into an anoxic zone. If such a zone extends below the soil surface (such as a poorly aerated claypan or elevated water table), root growth will be restricted to the upper, aerated parts of the profile. Under chronically hypoxic conditions, roots tend to be shorter and thicker, with less branching (Gliński and Stępniewski 1985). Some plants (e.g., tomato and maize) produce shallow adventitious roots containing lysigenous aerenchyma (Jackson and Drew 1984). And if plants with established root systems are suddenly exposed to flooded, anoxic soil conditions, substantial amounts of the root system may become necrotic (Stolzy and Sojka 1984). This "root pruning" can quickly and significantly alter root: shoot ratios. Of course, the extent of root system adaptation and the significance of root necrosis vary with plant age and time of year (Russell 1977).

4.3.1.3 Epidemiological Significance of Altered Root Growth

The influence of edaphic factors on root geometry is an important but often overlooked component of root disease epidemiology. The spatial arrangement of roots within the macroenvironment, whether in horizontal or vertical aggregations, can be as critical to the onset or severity of disease as the spatial aggregation of inoculum.

For example, jarrah trees (*Eucalyptus marginata* Sm.) in south Western Australia grow predominately on deeply weathered lateritic soils (Shea et al. 1983). These soils are characterized by a relatively shallow surface horizon overlying a highly indurate (concreted) subsurface layer of sesquioxides. To obtain water during the dry season, jarrah trees have evolved a system of "vertical roots" that penetrate cracks in the concreted layer. While effective in tapping deep water supplies, this results in a root system where tree survival is totally dependent on a relatively small number of "sinker" roots. Introduction of *Phytophthora cinnamomi* into this environment has been catastrophic. Zoospores of this fungus easily disperse laterally with water at the face of the concreted soil layer and concentrate in the vertical root channels (often in association with infected roots of the highly susceptible *Banksia grandis*) (Shea et al. 1983). This leads to infection and death of the jarrah vertical roots which, because there are so few providing water to each tree, quickly leads to tree death. Clearly, the rooting pattern imposed by soil structure in this case, lends an extraordinarily high degree of significance to individual root infections.

Another example where soil structure influences disease by controlling root geometry is root rot of bean caused by *Fusarium solani* f. sp. *phaseoli* (Burke et al. 1972a,b). In the bean-growing areas of central Washington, subsurface tillage pans limit bean root growth to the upper 20–30 cm of soil. This pattern of root distribution has several consequences relating to disease: it causes root densities to be maximum where inoculum is most abundant (1×10^2 to 5.5×10^3 propagules g^{-1} soil above the plow pan, and only trace levels below; Burke et al. 1972b); it increases the likelihood of plant water stress by limiting root growth to the uppermost part of the profile; it exposes the shallow root system to large diurnal temperature changes; and it can expose large portions of the root system to transient oxygen stress after irrigation (due to slowed water infiltration through the compacted layer). Chiseling these soils to a depth of 0.5 m fractured the tillage pan and allowed roots to penetrate deeper into the profile. This treatment significantly reduced losses due to Fusarium root rot partly by allowing roots to escape inoculum, and partly by relieving root stresses (Burke et al. 1972a,b; Miller and Burke 1975).

Edaphic factors clearly can influence root architecture and associated host-pathogen interactions. While the influences may be extreme and obvious as in the samples cited above, they are often less apparent. But even subtle changes in the spatial arrangement of roots may influence the extent or nature of host-pathogen contacts and subsequent disease development. In spite of this, edaphic influences on root system development are seldom dealt with in studies of root diseases. This has led to the adoption of underlying assumptions that may not always be valid, or that may limit the utility of models to particular macroenvironments.

For example, it is frequently assumed that root growth through soil is random and can be described simply in terms of root length (or number of roots) per unit volume of soil (see reviews by Huisman 1982; Gilligan 1990), or that inoculum is randomly distributed within the mesoenvironment and can be described adequately as propagules per unit volume (see reviews by Gilligan 1983; Campbell et al. 1988). While these assumptions may be true for unstructured soils, or the uppermost layers of tilled soils, the extent of their validity is not known. Indeed, infections of root systems have seldom been tested for deviations from randomness (Gilligan 1983, 1988). One notable exception is the work of Griffin and Tomimatsu (1983), who studied infection of peanut roots by *Cylindrocladium crotalariae* in field soils. They found that infections were patchy within the root system and reflected the clumped distribution of inoculum in soils. Similarly, Friese and Koske (1991) found that spores of vesicular-arbuscular mycorrhizae occurred in clumps within the mesoenvironment, reflecting their association with roots that subsequently died and decayed.

Recognizing that roots "can reasonably be regarded not so much as growing through the soil as between the structural units of which it is composed" (Russell 1977), one can envision much nonrandomness in root/inoculum interactions within the meso- and microenvironments. In high-strength or highly structured soils, the aggregation of roots in cracks, voids, or other low-impedance pathways may lead to inoculum aggregation if the roots become infected and decay. New roots may grow through the channels left by old roots (Van Noordwijk et al. 1991), virtually assuring contact with very large numbers of propagules. Conversely, intercrop changes in soil structure (e.g., soil chiseling) may open new pathways into the profile, and allow roots of a following crop to largely bypass existing aggregations of inoculum.

Spatial patterns within root systems can influence the fundamental relationships leading to establishment of disease (Campbell and Noe 1985; Gilligan 1988) and should be accounted for in models. Failure to account for the soil environment's influence on the rate and extent of root elongation, root architecture, or spatial and temporal changes in metabolic activity may weaken certain assumptions employed in a model, or limit its utility to particular soil types or pathosystems. Detecting changes in root growth imposed by the soil environment, quantifying them, and incorporating them in studies of root disease epidemiology remain as significant research challenges.

4.3.2 Pathogen Activity

I use the term "pathogen activity" here in reference to all aspects of the life and disease cycle of soil borne pathogens. This includes their survival, germination, reproduction, dispersal, and infection of host roots. There have been many excellect books and reviews published regarding the activity of various pathogens as influenced by various soil factors (e.g., Griffin 1972, 1978; Baker and Cook 1974; Duniway 1979, 1983; Cook and Duniway 1981; Cook and Baker 1983; Weste 1983). I will not attempt a summary of these works, but rather wish to point out some of the data gaps that limit our ability to construct functional linkages between the physical soil environment and epidemiology of root diseases.

4.3.2.1 Soil Matrix Effects

As described above, soil structure can influence root architecture and, consequently, spatial patterns of inoculum within the mesoenvironment. Whether infective propagules are dispersed subsequently from their sites of formation depends largely upon propagule size, structural properties of the soil, and the energy available for dispersal.

With foliar diseases, inoculum dispersal occurs in a gaseous medium that may possess sufficient kinetic energy to transport propagules many kilometers. The soil matrix, however, with its labyrinthine pore structure, greatly restricts dispersion of root pathogens. Indeed, if propagules are isolated from soil channels, and are larger than the pores surrounding them, they are effectively trapped in the void space of what best may be described as an "impact filter". The limited ability of propagules to pass through the soil matrix led Huisman (1982) to state that, relative to root growth, inoculum movement in soil is insignificant to the establishment of host-pathogen contact.

While inoculum dispersal within the soil environment is very limited relative to the aboveground environment, it does occur. The energy for dispersal arises from both exogenous and endogenous sources. Exogenous energy is derived from forces such as tillage (which causes horizontal and vertical dispersion in the uppermost soil layer) or water flow (which contributes to horizontal dispersion at the soil surface and vertical dispersion within the matrix). Endogenous energy is derived from a pathogen's food reserves that are consumed for purposes of locomotion (i.e., motility by certain fungi, bacteria, and nematodes) or growth.

As stated previously, the kinetic energy of water in soil is small and usually ignored in calculations of energy status (Miller and Klute 1967; Hillel 1980; Singer and Munns 1987). Thus, the energy that infiltrating water imparts to sedentary propagules is small and generally inadequate to overcome the frictional and mechanical resistances of the soil matrix. With large propagules such as sclerotia, a matrix-disrupting energy (e.g., tillage; Olanya and Campbell 1988) is required for dispersal.

Smaller propagules, which can pass through existing soil pores, may receive sufficient energy from infiltrating water to effect dispersal. But even in these cases, the distances traversed appear quite limited. For example, Wilkinson et al. (1981) reported that infiltrating water carried zoospore cysts of *Phytophthora megasperma* (appx. 10 μm diameter) 15 cm into a column of sand, and 8 cm into a column of loam soil. In similar experiments, Duniway and McKeen (1987) reported that cysts of *Phytophthora cryptogea* were transported only 4 cm into columns of reconstituted soil. Even the relatively small conidia of the entomogenous fungus *Beauveria bassiana* and a prokaryotic *Streptomyces* sp. (3 × 2.5 and 1.5 μm diameter, respectively) are transported only a few centimeters with infiltrating water (Wollum and Cassel 1978; Storey et al. 1989). Propagules being transported by water eventually encounter pore necks too small to allow passage, or they are impacted on soil particles as water follows a convoluted path through the matrix.

Organisms with motile life stages are capable of greater transport with infiltrating water. For example, Wilkinson et al. (1981) reported that motile zoospores of *Phytophthora megasperma* were transported twice the distance through soil columns as nonmotile cysts. Similarly, Duniway and McKeen (1987) found that up to 40% of the motile zoospores of *P. cryptogea* added at the top of 20-cm-deep soil columns passed out the bottom with water effluent, while nonmotile cysts moved only 4 cm into the column. Also, Liddell and Parke (1989) reported that the bacterium *Pseudomonas fluorescens* was transported as deep as 14 cm into soil columns to which 54 mm of water had been applied, with depth of infiltration related to the volume of water applied. Presumably, in all these examples, motility prevented propagules from becoming lodged in small voids or remaining impacted on particle surfaces.

Although physical features of the soil matrix clearly influence propagule dispersion, our knowledge of this subject remains very incomplete. The examples described above elucidate only the influence of soil texture. Considerably less is known about the influence of soil structure.

While poorly characterized, soil structure may largely determine dispersal efficiency within the mesoenvironment. This is because even though the soil matrix functions as an impact filter, it is a filter often perforated by clefts and channels. Water initially enters soil via cracks or large interaggregate pores, in which it may have sufficient kinetic energy to dislodge and move sedentary propagules. And because cracks and large-diameter pores occur in three dimensions, they also can facilitate lateral dispersal of motile fungi (Duniway 1976) and nematodes (Jones et al. 1969). Indeed, when Jones et al. (1969) picked apart fresh soil cores, adult *Xiphinema diversicaudatum* were found only in the pore spaces between aggregates, never within aggregates. The significance of pathogen dispersal within the cracks and pores of a structured soil lies in the fact that these may be the same low-impedance pathways preferentially followed by growing roots.

Storey and Gardner (1988) and Storey et al. (1989) showed that infiltration of surface-applied spores of *Beauveria bassiana* into field soils was significantly influenced by structural features related to tillage practices. Clearly, much more research is needed before we can adequately describe the influence of the soil matrix on pathogen dispersal, or predict dispersal within natural or agricultural soils.

4.3.2.2 Soil Solution and Atmosphere Effects

The influence of soil water and atmosphere on microbial activity has been a subject of interest to plant pathologists and soil microbiologists for many years (e.g., see reviews by Griffin 1969, 1972, 1978; Cook and Papendick 1972; Baker and Cook 1974; Duniway 1979, 1983; Cook and Baker 1983; Cook and Duniway 1981; Harris 1981). However, because of the variety of experimental methods which have been employed, and real differences that exist between and within groups of microorganisms, only the broadest of generalizations can be extracted from the literature. For example, Cook and Baker (1983) noted that fungi which grow maximally at relatively high water potentials (−100 kPa to −1 MPa; e.g., *Pythium* and *Phytophthora* spp., *Rhizoctonia solani* and *Gaeumannomyces graminis* var. *tritici*) cause the greatest disease in moist soils. Conversely, fungi that grow maximally at lower water potentials (−1 to −1.5 MPa; e.g., *Fusarium* spp.) cause the greatest disease under dry soil conditions. And a few fungi (e.g., *Penicillium* and *Aspergillus* spp.) are capable of growth and survival at extremely low soil Ψ values (−30 to −40 MPa) (Griffin 1972).

One difficulty with relating much of the literature on water relations of fungi to their activity in soil is that many reports are of in vitro studies where Ψ was controlled osmotically. As described above, however, soil Ψ usually is dominated by its ψ_m component, which can have different effects on microbial activity than ψ_s. One of the most striking examples of this difference was seen in the process of zoospore release by *Phytophthora cryptogea*. In soil, zoospore release was optimum at ψ_m values of 0 to −0.1 kPa, and became almost nil when ψ_m dropped to −1 kPa. This contrasted sharply with zoospore release in various salt and sucrose solutions, which occurred at ψ_s values as low as −600 to −900 kPa (MacDonald and Duniway 1978). Thus, microbial response to ψ_s in vitro may not reliably predict responses to Ψ in vivo. Also, the responses of microorganisms to soil Ψ can be modified by concomitant effects on other organisms (e.g., antagonists) or processes (e.g., nutrient diffusion). Many of the effects of soil moisture on pathogen activity, however, are direct and influence processes such as propagule survival, germination, growth, and dispersal (see Griffin 1969, 1972, 1978; Cook and Papendick 1972; Baker and Cook 1974; Harris 1981; Cook and Baker 1983).

Although plant pathogens generally are active over the range of soil Ψ values encountered under crop cultivation (e.g., Cook and Papendick 1972; Griffin 1978; Cook and Duniway 1981), some have very specific Ψ requirements for certain aspects of their life cycle. The ψ_m requirement for zoospore release by *Phytophthora* spp. (MacDonald and Duniway 1978) is one example. Another is pathogen dispersal within the matrix. Lateral dispersal of motile pathogens actually has two requirements: (1) a soil structure possessing abundant, large-diameter pores and (2) a soil ψ_m sufficiently high that the requisite pores remain water-filled. Zoospore movement by *Phytophthora* spp., for example, requires ψ_m values >-5 kPa and an abundance of soil pores $>60\,\mu m$ diameter (Duniway 1979). If a soil does not have a sufficient volume of large-diameter pores, lateral movement will be minimal even at permissive ψ_m values. Conversely, if soil has an abundance of pores >60 μm diameter, dispersal efficiency is determined by ψ_m, since the pores that can accommodate swimming zoospores will be water-filled only at ψ_m values >−5 kPa (Duniway 1979).

Nematode migration also is influenced by pore dimension and ψ_m, but in a somewhat different manner. Unlike zoospores, larval migration by the cyst nematode (*Heterodera schachtii*) was represented as a bell-shaped curve when plotted against ψ_m; decreasing when ψ_m was low, but also decreasing when ψ_m was very high (Wallace 1958). Reduced migration in the latter case was attributed to nematodes losing frictional contact with the solid matrix. When large-diameter pores were water-filled, the lateral body motions of nematodes were not efficiently translated into forward movement. Migration was maximum in soil fractions having 30–60 μm pores and at ψ_m values bounding the deflection point of the moisture characteristic curve (Wallace 1958).

While high soil moisture generally favors microbial activity, a preponderance of water-filled pores can, as described earlier, impede gas exchange between the aboveground and below-ground atmospheres. Thus, concomitant with changes in soil water status are changes in the concentration of O_2, CO_2, ethylene, and products of anaerobic respiration. This may make it difficult to unambiguously link the activity of selected microbes to changes in soil water status independent of the soil atmosphere.

Our knowledge of the effects of soil atmosphere on root pathogens is very incomplete. In his early review, Griffin (1969) stated that "the investigation of the role of carbon dioxide in soil ecology can scarcely be claimed to have passed the introductory stages". Some 25 years later, this statement may still be valid and not limited to CO_2.

There are numerous studies that describe the effects of O_2 and CO_2 on growth of various organisms in sterile culture or in small, unstructured soil samples (e.g., Macauley and Griffin 1969; Louvet 1970). It is difficult, however, to interpret experiments that manipulate or report O_2 only as a percent of the soil atmosphere. Stolzy and coworkers (Letey and Stolzy 1964, 1967; Stolzy and Letey 1964; Stolzy and Sojka 1984) have studied soil oxygen and its effects on root growth for decades. They found that, from the standpoint of physiological function, the oxygen diffusion rate was a more relevant indicator of soil oxygen availability than O_2 percentage. One could hypothesize that the same holds true for soilborne pathogens, or that the oxygen diffusion rates conducive to pathogen activity are lower than those that limit root function ($0.2\,\mu g\,O_2\,cm^{-2}\,min^{-1}$; Stolzy and Letey 1964), but these questions remain largely unexplored and impossible to extract from existing literature.

A subject that is even less well characterized than O_2 availability is the effect of other gases on the activity of root pathogens. Rarely have experiments employing hypoxic or anoxic treatments in soil quantified concomitant changes that occur in concentrations of ethylene, methane, hydrogen sulfide, or ammonia. The potential influence of these gasses on the physiological processes being studied often is overlooked. And rarely are distinctions made between inoculum that may be aggregated in large soil pores or channels, and that which may be more isolated from major diffusion pathways.

Defining the roles and interactions of soil water and gases in root disease epidemiology is no small challenge. Experimental methods (i.e., the insertion of measuring instruments) may disrupt the system under study, or available instruments may not have the physical resolution needed to probe and measure small soil volumes. This is true especially within the micro-environment, where antagonism, predation, infection, and reproduction occur and Ψ gradients near absorbing parts of roots may be steep.

4.3.2.3 Soil Temperature Effects

Temperature influences all aspects of pathogen activity in soil through its fundamental effects on the kinetics of physical, chemical, and biochemical processes. In the case of fungi, temperature influences frequently are described in terms of direct effects on certain processes (e.g., growth rate) and the focus often is on three cardinal temperatures; the minimum, maximum, and optimum for the process under study. Much of our knowledge of temperature effects on fungi comes from experiments in which cultures grown on agar media have been exposed to a range of selected (usually constant) temperatures. This is well illustrated by the numerous reports (summarized in Zentmyer 1980) that describe the cardinal temperatures for mycelial growth, sporangium, chlamydospore and oospore formation, and spore germination by various isolates of *Phytophthora cinnamomi*.

Data of this type often are used to explain temperature effects on disease development. For example, Zentmyer (1980) noted that infection of avocado seedlings was maximum at temperatures optimal for growth of *Phytophthora cinnamomi*, but was significantly reduced at temperatures limiting mycelial growth. Likewise, root rot of cotton (caused by *Phymatotrichum omnivorum*) develops most rapidly at 27 °C, a temperature corresponding to the optimum for mycelial growth in culture (Lyda and Burnett 1971).

Although temperature effects derived from agar culture experiments can be instructive, they are not always fully predictive of temperature effects within the soil environment. This is illustrated by the rate of rhizomorph extension in *Armillaria mellea*. On malt agar, rhizomorphs grew 9.8 and 6.8 mm day^{-1} at 25 and 30 °C, respectively. In soil, however, growth was maximum at 22 °C and nil at 30 °C (Rishbeth 1968). Similarly, Taylor (1964) found that radial growth of *Cylindrocladium radicicola* on agar media was maximum at 18–25 °C and substantially reduced at 10 °C. This contrasted sharply with colonization of sterilized bean roots in soil, which was maximum at 10 °C and nil at 21–25 °C (Taylor 1964).

Temperature responses of pathogens in vivo may differ from those observed in vitro because certain processes (growth, sporulation, etc.) may be influenced simultaneously by other features of the soil environment (e.g., Ψ). For example, Chen and Griffin (1966) showed significant interactions between soil temperature and Ψ in substrate colonization by *Penicillium* and *Aspergillus* spp. Also, Sneh and McIntosh (1974) reported that hyphae of *Phytophthora cactorum* lysed more quickly in soil as temperature increased from −7 to 29 °C. At any given temperature, however, lysis was more rapid in moist (−10 kPa ψ_m) than in drier (−300 kPa ψ_m) soil. Similarly, the death rate of propagules of *Pythium ultimum* in soil increased as temperature increased from 9–27 °C, but always was higher at soil ψ_m values of −40 kPa than at −300 kPa (Lifshitz and Hancock 1984). Indeed, for certain aspects of the life cycle of pythiaceous fungi, soil Ψ may be more delimiting than temperature per se (Duniway 1979).

Other reasons for lack of agreement between in vitro and in vivo temperature effects relate to the breadth of temperature influences within the soil environment. Concomitant effects on soil physical and chemical processes, and other microorganisms, interact to affect the organism under study. These indirect effects of temperature occur mainly at the microenvironment level and our understanding of them is very incomplete. There has, however, been recent progress on this subject via research into the mechanisms of pathogen inactivation during soil solarization.

Soil solarization refers to a process whereby moist, plastic film-covered soil is heated by means of solar radiation (Katan and DeVay 1991). The use of solar energy to heat-inactivate pathogens has been described (Katan 1981) as analogous to steam treatments for eliminating pathogens from potting media (Baker and Roistacher 1957; Bollen 1985). Indeed, during solarization, temperatures in the upper 5 cm of the profile can reach or exceed the thermal death point of many pathogens (Baker and Roistacher 1957; Katan 1981; Bollen 1985; DeVay and Katan 1991). Furthermore, since exposure times are so great, thermal inactivation of propagules can occur at deeper soil depths (i.e., 15–30 cm) where temperatures may increase to only 35–40 °C (DeVay and Katan 1991).

While solar heating causes direct heat injury to propagules in the upper parts of the soil profile, solarization also has been found to reduce pathogen populations in deeper parts of the profile (>30 cm) where direct heat injury is unlikely. Research into this phenomenon has shown that modestly elevated soil temperatures can cause a general "weakening" of propagules (manifested by reduced germinability or infectivity), increased nutrient leakage from propagules (which may stimulate antagonists), and loss of fungistasis (which may cause propagules to germinate in the absence of a suitable host) (DeVay and Katan 1991). These more subtle temperature effects are attributed to changes in bacterial populations, changes in mineral nutrient availability, changes in the solubility of gases and chemicals within the soil solution, and changes in the composition of the soil atmosphere (Katan 1981; DeVay and Katan 1991).

While a good deal of research into temperature effects has been directed recently toward pathogen inactivation at high temperatures, low temperatures also influence pathogen activity. Indeed, the inability of *Phymatotrichum omnivorum* to tolerate prolonged exposure to temperatures <15 °C is considered a major factor defining its geographic distribution on the North American continent (Percy 1983). More typically, the activity of soil fungi relates to diurnal or even seasonal fluctuations in temperature. For example, seasonal fluctuations in the numbers of viable propagules of *Phytophthora cinnamomi* in Australian forest soils (high numbers in spring, summer,and autumn; low numbers in winter) have been attributed to sensitivity of the fungus to soil temperatures <10 °C (Weste and Ruppin 1977). And at very low temperatures (≤0 °C), chlamydospores of *P. cinnamomi* are killed in sandy loam soil at a rate correlated with the number of degrees below freezing (Benson 1982).

The response of soilborne pathogens to changing temperatures often may be an important factor in root disease epidemiology. However, to make good use of soil temperature data in epidemiological studies, it is critical to know whether pathogen activity is related to daily means, high or low extremes, the rates of change, the duration, or the cumulative effects of temperature. Attempts to predict fungal response to different temperatures have included calculations of temperature coefficients (Q_{10}, the ratio of activity at X °C to activity at X − 10 °C) (Yarwood 1975; Benson 1978) and thermal death curves (Pullman et al. 1981). While providing information useful for certain purposes (e.g., estimating heat therapy exposures), these calculations have limited application to root disease epidemiology since they are derived from fungal responses at constant temperatures.

A more meaningful expression of pathogen response to temperature might be to quantify response in terms of thermal units. Mihail and Alcorn (1984) employed this concept in evaluating the responses of *Sclerotium rolfsii* and *Macrophomina phaseolina* to solarization. They expressed treatment effects in terms of degree hours – the cumulative time at which soil temperature exceeded 45 or 50 °C. This approach was analogous to the use of "degree days" for expressing physiological time in crop and insect growth models, but employed a time factor (hours) more relevant to microbial response. While potentially useful, "degree hours" are referenced to specific threshold temperatures. Thermal death curves and Q_{10} values, however, clearly illustrate that fungi respond differentially to the whole temperature spectrum (Yarwood 1975; Pullman et al. 1881). Thus, rather than focusing on certain selected temperatures, better predictions of pathogen activity might be obtained if soil temperature and time were more thoroughly integrated in thermal unit calculations.

4.3.3 Host Disposition

In preceding sections, edaphic factors are discussed as they influence root growth and pathogen activity. Clearly, from the perspective of each organism inhabiting the soil environment, there is an optimum range for every environmental component. Prolonged exposure to environmental factors (temperature, oxygen, water, salts, etc.) outside the optimum range can be moderately to severely disruptive. Plants express the result of severe disruption in the form of abiotic disease. But environmental stresses of lesser magnitude or duration also are significant, since they may alter host disposition to disease (Colhoun 1973; Schoeneweiss 1975). While knowledge of this subject is limited to a relatively few examples, it is clear that even moderate root stress (that from which a plant could recover with minimal injury) significantly influences the establishment or severity of some root diseases. Distinguishing whether the soil environment is acting negatively on a host or positively on a pathogen may be difficult to discern, but absolutely critical to understanding disease epidemiology.

4.3.3.1 Soil Solution and Atmosphere

As illustrated by the success of hydroponic plant culture, saturated conditions per se have little harmful effect on root physiology. And, as described in preceding sections, high soil moisture can stimulate the growth, inoculum production, and dissemination of many pathogens. In fact, the activity of *Phytophthora* spp. is so generally favored by high soil moisture (Duniway 1979) that the association between Phytophthora root rots and wet soil conditions has long been attributed to the influence of water on the fungus.

Indeed, there are examples of Phytophthora root rots where the promotive effects of soil moisture relate most closely to enhanced production and dispersal of inoculum (e.g., Kenerley et al. 1984). But saturated soil conditions also can influence disease by inducing O_2 stress in roots. An early example of this was presented by Stolzy et al. (1965), who reported that Phytophthora root rot severity increased in citrus seedlings as the duration of flooding periods increased or as O_2 concentration within the root zone decreased. Other researchers (e.g., Kuan and Erwin 1980; Blaker and MacDonald 1981; Wilcox and Mircetich 1985) showed that episodes of soil flooding actually increased root susceptibility to infection and colonization by *Phytophthora* spp. While this effect was attributed to oxygen stress, none of these reports included measurements of root zone O_2. Heritage and Duniway (1983), however, did measure and control O_2 concentration in the gasses used to aerate hydroponically grown safflower plants, and found that the severity of *P. cryptogea* infections increased at O_2 concentrations below 0.5%. Similarly, exposing roots of shortleaf and loblolly pines (Fraedrich and Tainter 1989) or cork oak (Jacobs 1991) to low O_2 concentrations increased their susceptibility to infection by *P. cinnamomi*. In the case of cork oak, this was attributed partly to reduced accumulation of lignin and suberin in oxygen-stressed roots (Jacobs 1991).

While much recent research has focused on Phytophthora root rots, oxygen stress can predispose roots to other pathogens as well. For example, Miller and Burke (1977) reported that Fusarium root rot of bean was most severe in field soils where oxygen diffusion rates dropped below $0.2\,\mu g\,O_2\,cm^{-2}\,min^{-1}$ for up to 24 h after irrigation. This work is particularly significant because it illustrates how the description of soil oxygen status only in terms of O_2 percentage can lead to predictive errors. Analysis of gas samples extracted from field soils indicated that the percent O_2 in the soil atmosphere was not substantially different between heavily and normally irrigated treatments. And, based on laboratory experiments, the O_2 concentration in field soils was not considered low enough to affect disease. However, oxygen diffusion to roots was inhibited by wet soil conditions, which led to increased root infection. Because ODR reflects the interaction of several components of the soil environment, increased emphasis on this parameter could be very informative in epidemiological studies.

In contrast to the indirect stresses caused by high soil moisture (via O_2 stress), low soil moisture causes direct physiological stress that can predispose roots to infection. There are numerous examples illustrating the pathological consequences of water stress (e.g., Ghaffar and Erwin 1969; Cook and Papendick 1972; Duniway 1977; Blaker and MacDonald 1981; Ristaino and Duniway 1989). And, as noted in a previous section, certain physical features of the soil environment can impede water entry into the profile or limit root exploration of the soil profile, thereby contributing to the occurrence or severity of water stress.

While the foregoing discussion mainly concerns the availability of soil water and its interrelation with the soil atmosphere, the chemical composition of the soil solution also can influence host disposition to disease. As noted earlier, the specific composition of the soil solution is determined by the mineral and organic composition of the soil matrix, and complex chemical reactions that occur at those surfaces as part of mineral weathering. In agricultural soils, the chemistry of the soil solution also is influenced by nutrient amendments.

The stimulatory or suppressive roles of specific macro- and micronutrients in certain diseases have been reviewed recently (Engelhard 1989). A special case of mineral ion excess is salinity stress, which results (usually) from the accumulation of mineral ions carried into soil via irrigation water. Aside from being an independent cause of predisposing plant stress (e.g., MacDonald 1982, 1984; Blaker and MacDonald 1986; Swiecki and MacDonald 1988, 1991; Bouchibi et al. 1990; Snapp et al. 1991; Sulistyowati and Keane 1992), soil salinity also interacts significantly with other features of the physical environment. For example, the accumulation of Na^+ in soil, as noted earlier, leads to dispersion of clay floccules and disintegration of soil structure (Hillel 1980). Salt accumulations also can increase the significance of the ψ_s component of soil Ψ – particularly as water is lost from the soil by evapotranspiration and salts are concentrated into ever smaller water volumes. Thus, in addition to severe ionic imbalances, salinity can aggravate both water and oxygen stresses within roots.

4.3.3.2 Soil Temperature Effects

There are mumerous reports (summarized in Baker and Cook 1974) indicating that roots chronically exposed to temperatures above or below their growth optimum are more prone to infection if the temperatures also do not inhibit pathogen activity. This effect was conceptualized by Leach (1947) as a "relative growth advantage" of the pathogen over the host. While having some merit, this concept must be viewed as an oversimplification since reduced growth implies physiological stress and all the attendant disturbances of host function. Indeed, in many of the examples studied (e.g., Hayman 1969), moderately high or low soil temperatures have been found to increase root exudation (via increased membrane permeability), which subsequently stimulates microbial activity.

Beyond temperatures that lie on either side of the physiological optimum are temperatures that approach the far limits of a plant's tolerance and cause severe stress or physical injury. Roots exposed to freezing temperatures may be injured directly by ice formation within tissues or indirectly through frost heaving of soil and resultant root tearing (e.g., Cephalosporium stripe of wheat; Pool and Sharp 1969). Additionally, roots in the uppermost part of the profile may be exposed to periods of heat stress if the soil surface is fully exposed to solar radiation. Roots exposed to heat stress may be predisposed to fungal infection (e.g., MacDonald 1991), and edaphic factors that limit root growth to the upper parts of the profile increase the likelihood of such stress.

4.3.3.3 Mechanism of Predisposition

While there are many individual examples of predisposition due to edaphic factors, there are only a few examples where information has been developed to a point where it is useful for prediction or management of root diseases (e.g., Heritage and Harrigan 1984; Ristaino and Duniway 1989). This is unfortunate, because stress induction is one area where the physical environment can play a major role in disease development; events occur on a large enough scale that they can be reliably quantified and monitored over time; and intervention is sometimes possible.

There still are significant gaps in our understanding of predisposition. For example, we know very little about interactions between multiple stresses, or whether the physiological hardening that occurs with repetitive stress alters predisposition thresholds. Also, very little is known about the mechanisms involved in predisposition – although it is clear that there are physiological changes within root tissues (Duniway 1977; Kuan and Erwin 1980; Blaker and MacDonald 1981, 1986; MacDonald 1982, 1984; Heritage and Duniway 1985; Swiecki and MacDonald 1988; Sulistyowati and Keane 1992). However, too little is known about normal resistance reactions to interpret these data mechanistically.

Greater knowledge of plant signals (e.g., stress hormones; Bostock et al. 1990) associated with predisposition could enable early detection of stress onset and allow remediation before predisposing thresholds are crossed. This could be particularly beneficial for irrigated crops, where soil moisture and salinity can be manipulated through various means. In addition, greater knowledge of stress/disease interactions would aid efforts to detect and deploy stable host resistance genes (Heritage and Harrigan 1984).

4.4 Conclusions

Knowledge of the physical soil environment and its influence on roots and pathogens has increased greatly over the past two decades. In spite of substantial progress, however, it is clear that our knowledge of permissive or restrictive environmental parameters is far less developed for root diseases than for foliar/shoot diseases. This is evidenced by the comparatively small number of simulation models that have been attempted for root diseases (fewer than 25% of all published models; see Campbell and Madden 1990). The dearth of even simple, descriptive models of important root diseases stands out as a major constraint to furthering our knowledge of the soil environment and its influence on root disease epidemiology. It is only through the conceptualization, testing, and refinement of models that researchers will be able to identify and delineate the key environmental parameters affecting individual diseases.

One reason for the relatively few attempts at modeling root diseases undoubtedly stems from the immense complexity of the soil environment. As a three-phase system, measuring changes in any single component (e.g., soil Ψ) fails to account for concomitant changes in other components (e.g., the soil atmosphere), or the subparts of those components (e.g., O_2, CO_2, ethylene, oxygen diffusion rate, redox potential, etc.). The opacity, heterogeneity, and mutability of the soil matrix, combined with the interrelatedness of soil features, generally complicates efforts to measure and account for environmental factors within a disease-relevant time frame. Furthermore, at the microenvironment level, there are technical difficulties associated with the physical dimensions of sensing equipment and the spatial dynamics of these "spheres of influence" within the mesoenvironment.

While thorough characterization of the soil environment may seem impracticable, it also may not always be necessary. This is because, for any given pathosystem, certain aspects of the soil environment may have over-riding influences on root or pathogen activity. Such bounding assumptions were employed by Bloomberg (1979), who modeled damping-off of Douglas fir seedlings (caused by *Fusarium oxysporum*) using soil temperature as the determinate environmental factor. Similarly, Reynolds et al. (1986) assumed that soil temperature and ψ_m were the principal environmental variables affecting spread of Phytophthora root rot in Fraser fir nursery beds, and van der Werf et al. (1986) made similar assumptions when simulating cyst nematode migration to sugarbeet seedlings.

Whether such initial assumptions are valid or underrepresent environmental interactions can only be determined through model validation studies. But validation studies are another area where significant differences exist between aerial and soilborne diseases. Environmental variables in the aerial environment (e.g., temperature, humidity, wetness periods, etc.) have similar effects on given pathosystems across broad geographic regions. In the soil environment, however, using the term "macroenvironment" to refer to individual soil series (see above) clearly connotes major environmental differences over relatively small geographic areas. Hence, bounding assumptions employed in a model developed in one macroenvironment (i.e., soil series) may not be valid in another. A good example of this is revealed in movement of zoospores and nematodes within the soil matrix. In soils with numerous, large-diameter pores, migration efficiency will be regulated largely by soil ψ_m (Wallace 1985; Duniway 1976). However, in soils with limiting pore structure, ψ_m will not predict migration.

Taken a step farther, it also should be apparent that even within a given soil series, changes in the soil environment occur through the very practice of agriculture. Some changes may be short-term, including fluctuations in water status (via irrigation), composition of the soil solution (via fertilizer amendments), strutural properties of the soil surface (via tillage), and soil temperature (via crop canopy cover and irrigation). Other changes may be more long-term, including changes in soil organic matter content (via crop selection and residue management), soil structure (via crop rooting habit and vehicular traffic), and soil salinity (via irrigation), to name but a few. Indeed, Boag (1985) noted that nematode populations decreased sharply in soil compressed under the wheels of a tractor. Such management-induced changes in the soil environment may be detected by spatial analysis of disease within a crop (Campbell and Noe 1985; Gilligan 1988). But, rather than serving only to complicate modeling efforts, the spatial and temporal heterogeneity of the soil environment actually represents a research tool for elucidating interactions between environmental factors, as well as assessing and predicting the impacts of certain agricultural practices.

While technical difficulties associated with quantifying aspects of the soil environment and their interactions are not trivial, there still is much that can be accomplished with current technology. Many of the classical methods developed for analysis of the soil environment are summarized in a variety of excellent reference books (e.g., Böhm 1979; Page et al. 1982; Gliński and

Stępniewski 1985; Klute 1986; Hanks and Brown 1987; Singleton et al. 1992). And many newer, easier, more reliable, or more precise methods are being developed all the time. The challenge to plant pathologists lies in the appropriate and critical utilization of these methods, and their interpretation, to reveal functional linkages between the soil environment and root disease epidemiology.

References

Alexander SA, Skelly JM, Morris CI (1975) Edaphic factors associated with the incidence and severity of disease caused by *Fomes annosus* in loblolly pine plantations in Virginia. Phytopathology 65:585–591

Baker KF, Cook RJ (1974) Biological control of plant pathogens. WH Freeman, San Francisco

Baker KF, Roistacher CN (1957) Heat treatment of soil. In: Baker KF (ed) The U.C. system for producing healthy container-grown plants. UC Manual 23. Univ California Press, Berkeley, pp 123–137

Barber SA (1984) Soil nutrient bioavailability: a mechanistic approach. John Wiley & Sons, New York

Benson DM (1978) Thermal inactivation of *Phytophthora cinnamomi* for control of Fraser fir root rot. Phytopathology 68:1373–1376

Benson DM (1982) Cold inactivation of *Phytophthora cinnamomi*. Phytopathology 72:560–563

Bengough AG, Mullins CE (1990) Mechanical impedance to root growth: a review of experimental techniques and root growth responses. J Soil Sci 41:341–358

Blaker NS, MacDonald JD (1981) Predisposing effects of soil moisture on the susceptibility of rhododendron to Phytophthora root and crown rot. Phytopathology 71:831–834

Blaker NS, MacDonald JD (1986) The role of salinity in the development of Phytophthora root rot of citrus. Phytopathology 76:970–975

Bloomberg WJ (1979) A model of damping-off and root rot of Douglas-fir seedlings caused by *Fusarium oxysporum*. Phytopathology 69:74–81

Boag B (1985) Effect of soil compaction on migratory plant-parasitic nematodes. Crop Res 25:63–67

Böhm W (1979) Methods of studying root systems. Ecological studies, vol 33. Springer, Berlin Heidelberg New York

Bohn HL, McNeal BL, O'Connor GA (1985) Soil chemistry, 2nd edn. John Wiley & Sons, New York

Bollen GJ (1985) Lethal temperatures of soil fungi. In: Parker CA, Rovira AD, Moore KJ, Wong PTW (eds) Ecology and management of soilborne plant pathogens. Am Phytopathol Soc, St Paul, pp 191–193

Bostock RM, MacDonald JD, Duniway JM, Stites J (1990) Abscisic acid as a determinant of host susceptibility in salinity predisposition to Phytophthora root rot in chrysanthemum and tomato. Phytopathology 80:1023 (Abstr)

Bowen GD, Rovira AD (1976) Microbial colonization of plant roots. Annu Rev Phytopathol 14:121–144

Bouchibi N, van Bruggen AHC, MacDonald JD (1990) Effect of ion concentration and sodium: calcium ratio of a nutrient solution on Phytophthora root rot of tomato and zoospore motility and viability of *Phytophthora parasitica*. Phytopathology 80:1323–1329

Burke DW, Miller DE, Holmes LD, Barker AW (1972a) Counteracting bean root rot by loosening the soil. Phytopathology 62:306–309
Burke DW, Holmes LD, Barker AW (1972b) Distribution of *Fusarium solani* f. sp. *phaseoli* and bean roots in relation to tillage and soil compaction. Phytopathology 62:550–554
Campbell CL, Madden LV (1990) Introduction to plant disease epidemiology. John Wiley & Sons, New York
Campbell CL, Noe JP (1985) The spatial analysis of soilborne pathogens and root diseases. Annu Rev Phytopathol 23:129–148
Campbell CL, Reynolds KM, Madden LV (1988) Modeling epidemics of root diseases and development of simulators. In: Kranz J, Rotem J (eds) Experimental techniques in plant disease epidemiology. Springer, Berlin Heidelberg New York, pp 253–265
Chapman HD (1970) Chemical factors of the soil as they affect microorganisms. In: Baker KF, Snyder WC (eds) Ecology of soil-borne plant pathogens. Univ California Press, Berkeley, pp 120–141
Chen AW-C, Griffin DM (1966) Soil physical factors and the ecology of fungi. VI. Interaction between temperature and soil moisture. Trans Br Mycol Soc 49:551–561
Cook RJ, Baker KF (1983) The nature and practice of biological control of plant pathogens. Am Phytopathol Soc, St Paul
Cook RJ, Duniway JM (1981) Water relations in the life-cycles of soilborne plant pathogens. In: Parr JF, Gardner WR, Elliott LF (eds) Water potential relations in soil microbiology. SSSA Spec Publ no 9, Soil Sci Soc Am, Madison, pp 119–139
Cook RJ, Papendick RI (1972) Influence of water potential of soils and plants on root disease. Annu Rev Phytopathol 10:349–374
Colhoun J (1973) Effects of environmental factors on plant disease. Annu Rev Phytopathol 11:343–364
Day PR, Bolt GH, Anderson DM (1967) Nature of soil water. In: Hagen RM, Haise HR, Edminster TW (eds) Irrigation of agricultural lands. Agronomy Monogr no 11, Am Soc Agron, Madison, pp 193–208
DeVay JE, Katan J (1991) Mechanisms of pathogen control in solarized soils. In: Katan J, DeVay JE (eds) Soil solarization. CRC Press, Boca Raton, pp 87–101
Dexter AR (1986) Model experiments on the behaviour of roots at the interface between a tilled seed-bed and a compacted sub-soil. II. Entry of pea and wheat roots into subsoil cracks. Plant Soil 95:135–147
Dexter AR (1987) Mechanics of root growth. Plant Soil 98:303–312
Dexter AR (1988) Advances in characterization of soil structure. Soil Tillage Res 11:199–238
Duniway JM (1976) Movement of zoospores of *Phytophthora cryptogea* in soils of various textures and matric potentials. Phytopathology 66:877–882
Duniway JM (1977) Predisposing effect of water stress on the severity of Phytophthora root rot in safflower. Phytopathology 67:884–889
Duniway JM (1979) Water relations of water molds. Annu Rev Phytopathol 17:431–460
Duniway JM (1983) Role of physical factors in the development of *Phytophthora* diseases. In: Erwin D, Bartnicki-Garcia S, Tsao PH (eds) *Phytophthora*: its biology, taxonomy, ecology, and pathology. Am Phytopathol Soc, St Paul, pp 175–187
Duniway JM, McKeen CD (1987) Dispersal of *Phytophthora cryptogea* zoospores in soils and glass microbeads by water flow. Phytopathology 77:1744 (Abstr)
Engelhard AW (ed) (1989) Soilborne plant pathogens: management of diseases with macro- and microelements. Am Phytopathol Soc, St Paul

Foster RC (1986) The ultrastructure of the rhizoplane and rhizosphere. Annu Rev Phytopathol 24:211–234
Foster RC, Rovira AD, Cock TW (1983) Ultrastructure of the root-soil interface. Am Phytopathol Soc, St Paul
Fraedrich SW, Tainter FH (1989) Effect of dissolved oxygen concentration on the relative susceptibility of shortleaf and loblolly pine root tips to *Phytophthora cinnamomi*. Phytopathology 79:1114–1118
Friese CF, Koske RE (1991) The spatial dispersion of spores of vesicular-arbuscular mycorrhizal fungi in a sand dune: microscale patterns associated with the root architecture of American beachgrass. Mycological Res 95:952–957
Fry WE (1982) Princples of plant disease management. Academic Press, New York
Garrett SD (1970) Pathogenic root-infecting fungi. Cambridge Univ Press, London
Ghaffar A, Erwin DC (1969) Effect of soil water stress on root rot of cotton caused by *Macrophomina phaseoli*. Phytopathology 59:795–797
Gilligan CA (1983) Modeling of soilborne pathogens. Annu Rev Phytopathol 21: 45–64
Gilligan CA (1988) Analysis of the spatial pattern of soilborne pathogens. In: Kranz J, Rotem J (eds) Techniques in plant disease epidemiology. Springer, Berlin Heidelberg New York, pp 85–98
Gilligan CA (1990) Mathmatical modeling and analysis of soilborne pathogens. In: Kranz J (ed) Epidemics of plant diseases: mathematical analysis and modeling. Springer, Berlin Heidelberg New York, pp 98–142
Gliński J, Stępniewski W (1985) Soil aeration and its role for plants. CRC Press, Boca Raton
Griffin DM (1969) Soil water in the ecology of fungi. Annu Rev Phytopathol 7:289–310
Griffin DM (1972) Ecology of soil fungi. Syracuse Univ Press, Syracuse
Griffin DM (1978) Effect of soil moisture on survival and spread of pathogens. In: Kozlowski TT (ed) Water deficits and plant growth: water and plant disease, vol 5. Academic Press, New York, pp 175–197
Griffin GJ, Tomimatsu GS (1983) Root infection pattern, infection efficiency, and infection density-disease incidence relationships of *Cylindrocladium crotalariae* on peanut in field soil. Can J Plant Pathol 5:81–88
Hanks RJ, Brown RW (eds) (1987) Proceedings of the international conference on measurement of soil and plant water status: soils, vol 1. Utah State Univ, Logan
Harris RF (1981) Effect of water potential on microbial growth and activity. In: Water potential relations in soil microbiology. SSSA Spec Publ no 9, Soil Sci Soc Am, Madison, pp 23–94
Hausenbuiller RL (1972) Soil science principles and practices. Wm C Brown, Dubuque
Hayman DS (1969) The influence of temperature on the exudation of nutrients from cotton seeds and on preemergence damping-off by *Rhizoctonia solani*. Can J Bot 47:1663–1669
Heritage AD, Duniway JM (1983) Influence of depleted oxygen supply on Phytophthora root rot of safflower in nutrient solution. In: Parker CA, Rovira AD, Moore KJ, Wong PTW (eds) Ecology and management of soilborne plant pathogens. Am Phytopathol Soc, St Paul, pp 199–202
Heritage AD, Harrigan EKS (1984) Environmental factors influencing safflower screening for resistance to *Phytophthora cryptogea*. Plant Dis 68:767–769
Hillel D (1980) Fundamentals of soil physics. Academic Press, New York
Holmes JW, Taylor SA, Richards SJ (1967) Measurement of soil water. In: Hagen RM, Haise HR, Edminster TW (eds) Irrigation of agricultural lands. Agronomy Monogr 11. Am Soc Agron, Madison, pp 275–303

Huck MG (1970) Variation in taproot elongation as influenced by composition of the soil air. Agron J 62:815–818

Huisman OC (1982) Interrelations of root growth dynamics to epidemiology of root-invading fungi. Annu Rev Phytopathol 20:303–327

Jackson MB, Drew MC (1984) Effects of flooding on growth and metabolism of herbaceous plants. In: Kozlowski TT (ed) Flooding and plant growth. Academic Press, New York, pp 47–128

Jacobs KA (1991) Oak root pathology: effect of oxygen stress on root growth and infection by *Phytophthora cinnamomi* and identification of *Armillaria* spp. in California. PhD Thesis, Univ California, Davis

Johnson LF, Doyle JH (1986) Relationships of seedling disease of cotton to characteristics of loessial soils in Tennessee. Phytopathology 76:286–290

Jones FGW, Larbey DW, Parrott DM (1969) The influence of soil structure and moisture on nematodes, especially *Xiphinema*, *Longidorus*, *Trichodorus* and *Heterodera* spp. Soil Biol Biochem 1:153–165

Katan J (1981) Solar heating (solarization) of soil for control of soilborne pests. Annu Rev Phytopathol 19:211–236

Katan J, DeVay JE (eds) (1991) Soil solarization. CRC Press, Boca Raton

Kenerley CM, Papke K, Bruck RI (1984) Effect of flooding on development of Phytophthora root rot in Fraser fir seedlings. Phytopathology 74:401–404

Klute A (ed) (1986) Methods of soil analysis. Part I. Physical and mineralogical methods, 2nd edn. Am Soc Agron Soil Sci Soc Am, Madison

Kuan TL, Erwin DC (1980) Predisposition effect of water saturation of soil on Phytophthora root rot of alfalfa. Phytopathology 70:981–986

Leach LD (1947) Growth rates of host and pathogen as factors determining the severity of preemergence damping-off. J Agric Res 75:161–179

Letey J, Stolzy LH (1964) Measurement of oxygen diffusion rates with the platinum microelectrode. I. Theory and equipment. Hilgardia 35:545–554

Letey J, Stolzy LH (1967) Limiting distances between root and gas phase for adequate oxygen supply. Soil Sci 103:404–409

Liddell CM, Parke JL (1989) Enhanced colonization of pea taproots by a fluorescent pseudomonad biocontrol agent by water infiltration into soil. Phytopathology 79:1327–1332

Lifshitz R, Hancock JG (1984) Environmental influences on the passive survival of *Pythium ultimum* in soil. Phytopathology 74:128–132

Lockwood JL (1964) Soil fungistasis. Annu Rev Phytopathol 2:341–362

Lockwood JL (1986) Soilborne plant pathogens: concepts and connections. Phytopathology 76:20–27

Louvet J (1970) Effect of aeration and of concentration of carbon dioxide on the activity of plant pathogenic fungi in the soil. In: Toussoun TA, Bega RV, Nelson PE (eds) Root diseases and soil-borne pathogens. Univ California Press, Berkeley, pp 89–91

Lyda SD, Burnett E (1971) Influence of temperature on *Phymatotrichum* sclerotial formation and disease development. Phytopathology 61:728–730

MacDonald JD (1982) Effect of salinity stress on the development of Phytophthora root rot of chrysanthemum. Phytopathology 72:214–219

MacDonald JD (1984) Salinity effects on the susceptibility of chrysanthemum roots to *Phytophthora cryptogea*. Phytopathology 74:621–624

MacDonald JD (1991) Heat stress enhances Phytophthora root rot severity in container-grown chrysanthemums. J Am Soc Hortic Sci 116:36–41

MacDonald JD, Duniway JM (1978) Influence of the matric and osmotic components of water potential on zoospore discharge in *Phytophthora*. Phytopathology 68:751–757

Macauley BJ, Griffin DM (1969) Effects of carbon dioxide and oxygen on the activity of some soil fungi. Trans Br Mycol Soc 53:53–62
Mihail JD, Alcorn SM (1984) Effects of soil solarization on *Macrophomina phaseolina* and *Sclerotium rolfsii*. Plant Dis 68:156–159
Miller EE, Klute A (1967) The dynamics of soil water: mechanical forces. In: Hagen RM, Haise HR, Edminster TW (eds) Irrigation of agricultural lands. Agronomy Monogr no 11. Am Soc Agron, Madison, pp 209–244
Miller DE, Burke DW (1975) Effect of soil aeration on Fusarium root rot of beans. Phytopathology 65:519–523
Miller DE, Burke DW (1977) Effect of temporary excessive wetting on soil aeration and Fusarium root rot of beans. Plant Dis Rep 61:175–179
Miller RM, Jastrow JM (1990) Hierarchy of root and mycorrhizal fungal interactions with soil aggregation. Soil Biol Biochem 22:579–584
Nielsen KF (1974) Roots and root temperatures. In: Carson EW (ed) The plant root and its environment. Univ Press Virginia, Charlottsville, pp 293–333
Oades JM (1984) Soil organic matter and structural stability: mechanisms and implications for management. Plant Soil 76:319–337
Olanya OM, Campbell CL (1988) Effects of tillage on the spatial pattern of microsclerotia of *Macrophomina phaseolina*. Phytopathology 78:217–221
Page AL, Miller RH, Keeney DR (1982) Methods of soil analysis. Part 2. Chemical and microbiological properties, 2nd edn. Am Soc Agron Soil Sci Soc Am, Madison
Papendick RI, Campbell GS (1981) Theory amd measurement of water potential. In: Parr JF, Gardner WR, Elliott LF (eds) Water potential relations in soil microbiology. SSSA Spec Publ 9, Soil Sci Soc Am, Madison, pp 1–22
Percy RG (1983) Potential range of *Phymatotrichum omnivorum* as determined by edaphic factors. Plant Dis 67:981–983
Pool RAF, Sharp EL (1969) Some environmental and cultural factors affecting Cephalosporium stripe of winter wheat. Plant Dis Rep 53:898–902
Pullman GS, DeVay JE, Garber RH (1981) Soil solarization and thermal death: a logarithmic relationship between time and temperature for four soilborne plant pathogens. Phytopathology 71:959–964
Raney WA (1970) Physical factors of the soil as they affect soil microorganisms. In: Baker KF, Snyder WC (eds) Ecology of soil-borne plant pathogens: prelude to biological control. Univ California Press, Berkeley, pp 115–119
Reynolds KM, Gold HJ, Bruck RI, Benson DM, Campbell CL (1986) Simulation of the spread of *Phytophthora cinnamomi* causing a root rot of Fraser fir in nursery beds. Phytopathology 76:1190–1201
Rishbeth J (1968) The growth rate of *Armillaria mellea*. Trans Br Mycol Soc 51:575–586
Ristaino JB, Duniway JM (1989) Effect of preinoculation and postinoculation water stress on the severity of Phytophthora root rot in processing tomatoes. Plant Dis 73:349–352
Russell RS (1977) Plant root systems: their function and interaction with the soil. McGraw-Hill, London
Schoeneweiss DF (1975) Predisposition, stress, and plant disease. Annu Rev Phytopathol 13:193–211
Schmitthenner AF, Canaday CH (1983) Role of chemical factors in development of *Phytophthora* diseases. In: Erwin DC, Bartnicki-Garcia S, Tsao PH (eds) *Phytophthora*: its biology, ecology and pathology. Am Phytopathol Soc, St Paul, pp 189–196
Shea SR, Shearer BL, Tippett JT, Deegan PM (1983) Distribution, reproduction, and movement of *Phytophthora cinnamomi* on sites highly conducive to jarrah dieback in south western Australia. Plant Dis 67:970–973

Sidebottom JR, Shew HD (1985) Effect of soil type and soil matric potential on infection of tobacco by *Phytophthora parasitica* var. *nicotianae*. Phytopathology 75:1439–1443
Singer MJ, Munns DN (1987) Soils: an introduction. MacMillan, New York
Singleton LL, Mihail JD, Rush CM (eds) (1992) Methods for research on soilborne phytopathogenic fungi. Am Phytopathol Soc, St Paul
Snapp SS, Shennan C, van Bruggen AHC (1991) Effects of salinity on severity of infection by *Phytophthora parasitica* Dast., ion concentrations and growth of tomato, *Lycopersicon esculentum* Mill. New Phytol 119:275–284
Sneh B, McIntosh DL (1974) Studies on the behavior and survival of *Phytophthora cactorum* in soil. Can J Bot 52:795–802
Stolzy LH (1974) Soil atmosphere. In: Carson EW (ed) The plant root and its environment. Univ Press Virginia, Charlottesville, pp 335–361
Stolzy LH, Letey J (1964) Measurement of oxygen diffusion rates with the platinum microelectrode. III. Correlation of plant response to soil oxygen diffusion rates. Hilgardia 35:567–576
Stolzy LH, Letey J, Klotz LJ, Labanauskas CK (1965) Water and aeration as factors in root decay of *Citrus sinensis*. Phytopathology 55:270–275
Stolzy LH, Sojka RE (1984) Effects of flooding on plant disease. In: Kozlowski TT (ed) Flooding and plant growth. Academic Press, New York, pp 221–264
Storey LH, Gardner WA (1988) Movement of an aqueous spray of *Beauvaria bassiana* into the profile of four Georgia soils. Environ Entomol 17:135–139
Storey GK, Gardner WA, Tollner EW (1989) Penetration and persistance of commercially formulated *Beauvaria bassiana* conidia in soil of two tillage systems. Environ Entomol 18:835–839
Sulistyowati L, Keane PJ (1992) Effect of soil salinity and water content on stem rot caused by *Phytophthora citrophthora* and accumulation of phytoalexin in citrus rootstocks. Phytopathology 82:771–777
Swiecki TJ, MacDonald JD (1988) Histology of chrysanthemum roots exposed to salinity stress and *Phytophthora cryptogea*. Can J Bot 66:280–288
Swiecki TJ, MacDonald JD (1991) Soil salinity enhances Phytophthora root rot of tomato but hinders asexual reproduction by *Phytophthora parasitica*. J Am Soc Hort Sci 116:471–477
Taylor GS (1964) *Fusarium oxysporum* and *Cylindrocarpon radicicola* in relation to their association with plant roots. Trans Br Mycol Soc 47:381–392
Taylor HM (1974) Root behavior as affected by soil structure and strength. In: Carson EW (ed) The plant root and its environment. Univ Press Virginia, Charlottesville, pp 271–291
Tisdale JM, Oades JM (1982) Organic matter and the water-stable aggregates in soils. J Soil Sci 33:141–163
Van der Werf W, Rabbinge R, Heijbroek W (1986) Spring invasion of *Heterodera schachtii* in sugarbeet; a simulation study. Neth J Plant Pathol 92:33–42
Van Noordwijk M, Widianto, Heinen M, Hairiah K (1991) Old tree root channels in acid soils in the humid tropics: important for crop root penetration, water infiltration and nitrogen management. Plant Soil 134:37–44
Wallace HR (1958) Movement of eelworms. I. The influence of pore size and moisture content of the soil on the migration of larvae of the beet eelworm, *Heterodera schachtii* Schmidt. Ann Appl Biol 46:74–85
Weller DM (1988) Biological control of soilborne plant pathogens in the rhizosphere with bacteria. Annu Rev Phytopathol 26:379–407
Weste G (1983) Population dynamics and survival of *Phytophthora*. In: Erwin D, Bartnicki-Garcia S, Tsao PH (eds) *Phytophthora*: its biology, taxonomy, ecology, and pathology. Am Phytopathol Soc, St Paul, pp 237–257

Weste G, Ruppin P (1977) *Phytophthora cinnamomi*: population densities in forest soils. Aust J Bot 25:461–475
Whitely GM, Dexter AR (1983) Behaviour of roots in cracks between soil peds. Plant Soil 74:153–162
Wilcox WF, Mircetich SM (1985) Effects of flooding duration on the development of Phytophthora root and crown rots of cherry. Phytopathology 75:1451–1455
Wilkinson HT, Miller RD, Millar RL (1981) Infiltration of fungal and bacterial propagules into soil. Soil Sci Soc Am J 45:1034–1039
Wollum AG, Cassel DK (1978) Transport of microorganisms in sand columns. Soil Sci Soc Am J 42:72–76
Yarwood CE (1975) Temperature coefficients in plant pathology. Phytopathology 65:1198–1201
Zadoks JC, Schein RD (1979) Epidemiology and plant disease management. Oxford Univ Press, New York
Zentmyer GA (1980) *Phytophthora cinnamomi* and the diseases it causes. Monograph no 10. Am Phytopathol Soc, St Paul

5 Estimating Disease Severity and Incidence

C.L. Campbell and D.A. Neher

5.1 Role of Disease Assessment in Root Disease Epidemiology

Disease assessment is one of the most challenging tasks in working with plant diseases. A diseased plant or group of diseased plants is often recognized easily once symptoms or signs become visible; however, it is the quantification of the disease that presents the challenge. The assessment of disease incidence (i.e., the number or proportion of diseased plants in a population) is an apparently simple counting task, but is subject to the usual limitations of interpretation related to sample size. The accurate and precise estimation or measurement of disease severity (i.e., the area or proportion of plant tissue that is symptomatic) can be a formidable task because of visual and measurement errors and the need for samples to be representative of the area considered and to be of adequate number.

Many factors influence the quality of disease assessment data. Our objectives in this chapter are to examine these factors through (1) the identification of the needs and goals for disease assessment; (2) consideration of the types of symptoms and signs that may be assessed along with the challenges associated with such assessments; and (3) analysis of the methods and techniques available to quantify root diseases. We will also address some of the challenges in disease assessment that are unique to root disease epidemiology.

5.1.1 The Need for Disease Assessment

Disease assessment, the cornerstone of epidemic analysis, is the most important task in the study of plant disease epidemics. It is the starting point in the characterization of any epidemic and represents the foundation for all subsequent analyses and interpretations. In essence, assessments of disease intensity are the currency by which epidemics are characterized and compared.

Assessments of disease are vital to our interpretation as to whether disease management practices are successful. Relative magnitudes of success in disease management are judged on a comparative basis by growers and

C.L. Campbell/D.M. Benson (Eds.)
Epidemiology and Management of Root Diseases

scientists through the process of disease assessment. The ultimate extension of such assessments is the translation of impressions of how much disease is present into mental or mathematical estimates of crop losses due to disease.

In many cases, assessment of disease during the course of root epidemics is the most costly part of a study. The cost is apportioned among the time needed to plan and carry out the assessment, the equipment required, and the degree of crop destruction brought about by the assessment. Quite often, the time component provides the largest real cost factor in an assessment of root disease intensity.

5.1.2 Goals of Disease Assessment

The goals of an assessment depend on a clear conception and statement of the objectives of the particular study. A survey of fields of crop x in region y to determine the prevalence of root disease z will have different assessment goals than a study to determine how moisture stress and soil physical factors affect the spatiotemporal progress of disease z on crop x in one or two fields. Not only will the goals differ, but the data quality objectives will undoubtedly be different for two so different studies. Thus, the specific goals for disease assessment must be formulated for each specific study.

The general goal of disease assessment, however, will remain constant for all types of studies. This goal is to provide *reliable* estimates of the *amount of disease* in an *area* (plot, field, farm, county, region, etc.) based upon the evaluation of specific symptoms and signs, which are known to be *characteristic for a disease*, at the lowest reasonable *cost* with *known confidence*. Few actual disease assessments fulfill this goal completely. An examination of the key components of this general goal is instructive in developing and evaluating disease assessment schemes.

Disease assessment must be *reliable*. The estimates or measurements of disease should be accurate, precise, and reproducible. *Accuracy* refers to the closeness of the sample mean from the assessment to the true population mean and is measured as bias, where bias = $|\bar{x} - u|$ with $\bar{x}$ = the sample mean and u = the population mean. Bias is difficult to measure in most actual field studies, because the required knowledge of the population mean is usually not available. Bias can, however, be measured in a preliminary study where a census of plants is taken in an area and measurements of the amount of disease are made. A sample mean can then be obtained by assessing disease in the manner that would be used under actual field conditions on a set of plants chosen randomly. The true mean and the sample mean can then be compared to estimate the bias.

Precision is a measure of the relative closeness or scatter of estimates obtained about the mean and is estimated by sample variance, s^2. The greater the sample variance, the lower the degree of precision in estimating $\bar{x}$. A low degree of precision (= high s^2) can be obtained due to errors in the assessment or because of actual variation in the relative amounts of disease on individual plants. The degree of spatial aggregation will have a significant influence on the magnitude of s^2 obtained for root diseases (see Chap. 7).

Reproducibility, a measure of the ability of an evaluator to obtain the same estimate of disease at a second, third, fourth, etc. assessment of the same diseased plant, plot, or other area within a short time interval, can be estimated by the degree of correlation between values from sets of assessments for the same plants, plots, etc. made at different times within a short time interval by the same evaluator. If the values for the amount of disease in populations of specific plants or plots from two assessment times are compared by a correlation analysis, a high degree of correlation (i.e., a coefficient $r \geq 0.80$) suggests a high degree of reproducibility for the assessment.

A slightly different approach to measuring reliability was taken by Shokes et al. (1987) in the assessment of the foliar disease of late leaf spot of peanut; however, the approach should be equally applicable to root diseases. The goal was to provide a measure of agreement among several disease evaluators or judges (i.e., inter-rater reliability) to indicate how reliable estimates of disease were. Each judge estimated the amount of disease present in each of several plots and variance components associated with the true variation among plots (σ_T^2), the variation among evaluators (σ_j^2), and the interaction among evaluators and plots or error (σ_e^2) were calculated. Variance components were used to calculate a coefficient of reliability (ρ) such that

$$\rho = \sigma_T^2/(\sigma_T^2 + \sigma_j^2 + \sigma_e^2).$$

The greater the magnitude of ρ, the greater the degree of inter-rater reliability. Such a measure will be quite useful if it is necessary to have more than one disease evaluator, particularly during evaluator training.

The way in which the *amount of disease* is estimated or measured will depend on the type of disease and purpose of the study. *Disease intensity* refers to the quantity of disease present in an area (Seem 1984; Kranz 1988; Campbell and Madden 1990). Within disease intensity, the measures of incidence and disease severity can be distinguished. *Disease incidence* refers to the number of plant units (e.g., plants, primary roots, secondary roots) that are visibly diseased (James 1974; Kranz 1988; Campbell and Madden 1990). *Disease severity* is the area or volume of plant tissue that is visibly diseased (Kranz 1988; Campbell and Madden 1990) relative to the total area or volume. *Disease prevalence* can be used as an extension of disease incidence when it is taken to be the percentage or proportion of fields in a region in which the disease is observed (Zadoks and Schein 1979).

Area is an important component of disease assessment because it indicates the spatial or geographic scale at which the assessment was made. Whereas it may be possible to examine the co-occurrence of events at a regional scale, at the scale of a rhizosphere, it is possible to examine direct causes, e.g., different symptoms due to different pathogens. The complexity of the logistics associated with the assessment of disease, as well as the relative accuracy and precision of the values obtained, will vary as the spatial scale increases from that of a single plant or rhizosphere to a county or region. Disease assessment at a regional scale will involve more evaluators and a longer period of time will be required for an assessment. Changes that occur during the assessment period will be more difficult to take into account for a regional assessment than those that occur in a relatively shorter assessment period for a single field. Techniques available to estimate or measure disease intensity and the general sampling approach will also vary with the scale of interest. Because of differences in the relative precision of the techniques used, estimates of disease from different spatial scales may not be comparable directly. Expansion factors for data obtained in small areas to provide estimates of disease for larger areas (e.g., regions) should be considered prior to developing the sampling plan.

With root diseases there are often several types of symptoms and signs associated with a single disease. There may be both root and shoot symptoms (see Sect. 5.3) that are *characteristic for a disease*. For example, a plant with root or crown rot caused by a *Phytophthora* spp. may first exhibit lesions on small roots or near the crown; however, the first visible (i.e., aboveground) symptom may be wilting of the shoots and leaves. Care must be taken to insure that the wilting is indeed due to the root disease and not purely to drought stress or plant injury. Stunting of plants is another symptom that may be characteristic of a root disease, but could also be due to other causes.

Cost is a factor of primary importance in planning the assessment of disease. Cost of effort must be balanced against the need for a certain amount of information. In experimental plots associated with specific research projects, the expenditure of large amounts of time and effort may be warranted to fulfill the objectives of the study. The intensive efforts required make the estimates of disease intensity high-priced, but such cost is expected with research. In survey assessments, however, the need may be more to cover a given area within a specified amount of time. In this case, as indeed in all cases, the cost of the estimates must be balanced against the accuracy and precision of the estimates obtained. More accurate and precise estimates are usually more costly than estimates with a lower accuracy and precision.

Data quality objectives and statistical confidence are issues that require much more attention in disease assessment than they have received previously. Both are attributes of the *known confidence* that must be associated with assessments of disease. A data quality objective is a statement of the magnitude of change that must be detectable for a disease over a specific time interval. Certain levels of statistical confidence are required for the establishment of the data quality objective. For example, based upon a knowledge of yield losses or a set of management decision rules, an investigator may decide that a 5% change in the level of disease is the smallest interval to be detected. The time interval of interest is a 2-week period. Thus, the first part of the data quality objective would be: *to detect a 5% change in disease within a 2-week period*. Statistical confidence in reaching the data quality objective comes from assigning specific, acceptable levels of α (probability of making a type I error, i.e., falsely rejecting the null hypothesis or identifying differences that actually do not exist) and β (probability of making a type II error, i.e., falsely accepting the null hypothesis or failing to identify differences that actually do exist). It may be reasonable to set α at 0.05, so that the probability of not making a type I error is 0.95 and to set β at 0.10 so that the probability of not making a type II error is 0.90.

The issue of degree of statistical confidence in the data will have an impact on assessment costs and is important in designing the assessment scheme. To gain greater confidence in assessment data, evaluators' errors can be reduced and sample size (n) can be increased. Also, if the level of differences to be detected is, for example, 2%, the use of a disease assessment or rating scale with a resolution of only 5% is not appropriate. Similarly, a five-point scale (0 = healthy, 1–25, 26–50, 51–75, 76–100%) would not be appropriate to detect a 2% change. Thus, data quality objectives need to be considered in relation to the overall purpose of the assessment and the resources available to obtain the assessment data needed.

5.2 Signs and Symptoms

Signs are the parts of a pathogen or its products seen on the diseased host plant. *Symptoms* are the internal or external reactions or alterations of a plant due to disease (Lucas et al. 1992). Most assessments of root diseases rely on root and shoot symptoms; however, in those instances where signs are present and readily observed, they can serve as confirmatory evidence of the presence of a specific pathogen. Signs associated with root diseases most often include sclerotia of fungi such as *Sclerotium rolfsii* and *Sclerotinia* spp., the mycelium of certain fungi such as *Rhizoctonia solani*, or basidiocarps of certain wood-decaying fungi such as *Armillaria mellea*. For root diseases caused by nematodes such as *Heterodera* spp., the presence of visible cysts is a characteristic sign.

Symptoms associated with root diseases range from minute lesions on fine rootlets to wilting of the entire plant. The dilemma in assessment of root diseases is that the symptoms on roots will often precede symptoms visible on aboveground portions of plants, but shoot symptoms may be much easier to evaluate than root symptoms. This dilemma is discussed further in Section 5.3. The point here is to raise the issue of what symptoms are to be evaluated to fulfill the general goals of assessment identified previously.

5.3 Root and Shoot Symptoms

Concepts applied to assessment of foliar diseases can be utilized for root diseases (Kranz 1988). However, symptom assessment of root diseases present a challenge not equaled by foliar diseases. Direct assessment of roots requires destructive sampling which prevents repeated assessment on individual roots or plants through time and may influence epidemic development in plant populations (Table 5.1). The "picture" of the epidemic may also change because variation over time (= different sample units at each time) may not be at all constant, especially with diseases characterized by aggregation and/or clustering of diseased plants. If the assessment is conducted only at harvest, roots may have been sloughed prior to sampling and, thus, an underestimate of disease intensity may be obtained (Berger 1980).

It is tempting to rate only the shoot symptoms associated with many root diseases because of the problems associated with actual root assessments. However, shoot symptoms may differ quantitatively and qualitatively

Table 5.1. Advantages and disadvantages of destructive and nondestructive sampling for the assessment of root diseases

Sample type	Advantage	Disadvantage
Destructive	– Actual root evaluated, not a surrogate – Allows determination of causal agent – Assessment of reduction in root volume is possible when compared to healthy plants – Sampling units independent or nearly so	– Sample unit is destroyed – Adjacent plant roots and root environment are disturbed – Usually very time-consuming and costly – Disease progress cannot be observed temporally on same plants
Nondestructive	– Plants remain intact and can be evaluated repeatedly – Root environment remains undisturbed – Sample units independent only if new sample of plants evaluated each time	– Visible, above ground symptoms may have a range of temporal and physiological relationships to root symptoms – Assessment values may be autocorrelated if the same plants are evaluated repeatedly – No estimate of reduction of size in root system is obtained – No estimate of severity of symptoms on roots

from root symptoms and often develop at variable times after root infections occur. Physiological, environmental, and temporal relationships between development of root and shoot symptoms should be understood before a specific symptom type is selected for monitoring epidemics and evaluating management procedures.

5.3.1 Physiological Relationships

In some cases, the actual effects of root disease are greater than the visual estimates of damage to roots would indicate. For example, when root diseases cause plants to wilt in a relatively wet soil under mild transpirational stress, large resistances to water uptake by the plant may have developed. At all stages of the development of Phytophthora root rot in safflower, resistance to water flow through healthy and infected safflower roots was much greater than would be expected due to differences in the size of root systems (Duniway 1977). Because the reduction in root surface area only accounted for a portion of the resistance to water flow induced by infection, a large portion of that resistance to flow in infected roots could be attributed to the xylem (Duniway 1977). In another system, Cylindrocladium black rot of peanut, stem and leaf wilting may result from reduced water flow caused by an accumulation of gums in vessels of fibrous and tap roots (Harris and Beute 1981).

In order to ascertain the physiological effects of disease on plants, the position and depth of lesions on roots must be evaluated (Hornby and Fitt 1981). Fitter (1982, 1987) defined a quantitative morphometric root analysis system which classified root systems by root order (first, second, or third) and type (lateral, tap root) (Fig. 5.1). The type and order of root segment may influence the susceptibility of root tissue as shown for infections of tobacco roots by *Phytophthora parasitica* var. *nicotianae* (English and Mitchell 1989). Also, infections and/or lesions on third-order (tap) roots would likely reduce water flow to stems more than lesions on first- or second-order (lateral) roots (Fig. 5.1; Bowers et al. 1991). This was apparent for Phytophthora root rot on processing tomato (Neher 1990) and tobacco black shank (Jones 1990) where symptoms on shoots were visible when the tap root had lesions, but not when lesions were restricted to lateral roots.

Some root diseases may elevate or alter levels of plant hormones that at least partially account for symptoms visible on roots or shoots. For example, increased ethylene production occurs with Verticillium wilt and plays an interactive role in the chlorosis and wilting of tomato leaves, as well as in the defoliation of cotton plants (Pegg 1981). Plant hormone shifts can alter patterns of photosynthate partitioning as illustrated by corky root of lettuce (O'Brien and van Bruggen 1992b) caused by *Rhizomonas suberifaciens* (van Bruggen et al. 1990). Diseased roots were significantly greater in diameter, had greater biomass, and much lower water content than apparently healthy

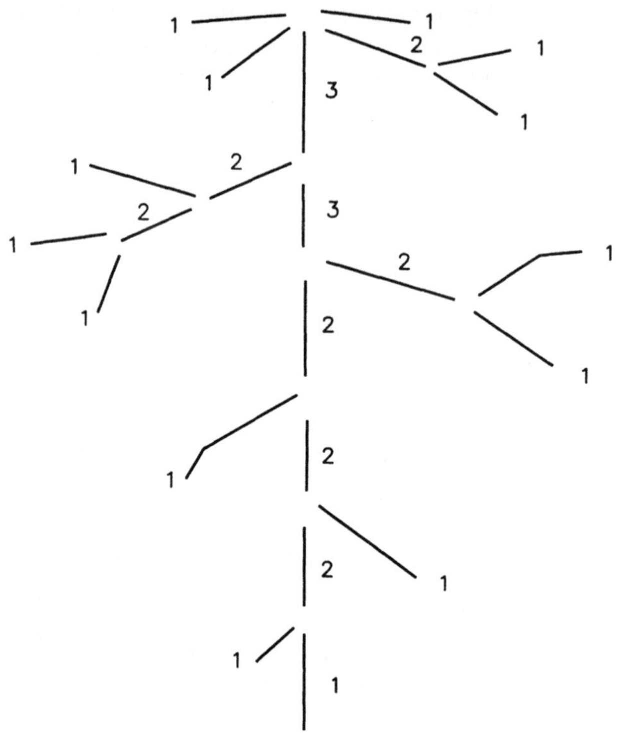

Fig. 5.1. Schematic representation of a pepper root system broken into root segments and described by the classification scheme of the morphometric root analysis system; *1*, *2*, *3* = root orders. (Bowers et al. 1991)

roots; more photosynthate was partitioned to roots relative to shoots, thus resulting in "corkiness" of the tap root (O'Brien and van Bruggen 1992b).

Mineral deficiencies and toxicities can confuse the interpretation of disease symptoms. Mineral nutrition may accentuate disease symptoms (Huber and Watson 1974) or lead to an incorrect diagnosis of the cause of a plant's poor health, especially by an inexperienced observer. For example, nitrogen-deficient hop plants may not show symptoms even though *Verticillium* spp. can be isolated from roots (Huber and Watson 1974). The type and abundance of nutrients in soil can also influence the severity of disease. For example, different forms of nitrogen, i.e., nitrate or ammonium, may differentially increase or decrease the severity of disease symptoms. No single form of nitrogen influences symptom development the same way for all diseases; therefore, each disease must be considered individually. For example, cortical and root diseases caused by species of *Fusarium*, *Rhizoctonia*, *Aphanomyces*, *Cercosporella*, *Poria*, and *Armillaria* may be reduced by applications of NO_3-N and increased by applications of NH_4-N, whereas diseases caused by species of *Ophiobolus*, *Diplodia*, *Pythium*, and *Strep-*

tomyces respond in an opposite manner (Huber and Watson 1974). The nitrogen form can affect the composition of root exudates which in turn may influence pathogenesis (Schroth and Hildebrand 1964; Huber and Watson 1974). In addition, nitrogen may influence microbial interactions, for example, suppression of pea root rot in soils amended with NO_3-N was correlated with an increased abundance of bacteria and actinomycetes antagonistic to the pathogen, *Aphanomyces euteiches*; the relationship was not observed in soils amended with NH_4-N (Carley 1969).

5.3.2 Environmental Effects

Temperature, moisture, and salinity stresses may influence the expression of root and shoot symptoms, as well as the activity of other root-infecting organisms. The effects of temperature may differ for each the pathogen, host, and disease. The most severe symptoms on processing tomato roots caused by *Phytophthora parasitica* occurred at temperatures that surpassed those favorable for either tomato growth or *P. parasitica* alone (Neher 1990; Fig. 5.2). A different relationship was observed, however, for Phytophthora root rot on avocado. The growth curve of *Phytophthora cinnamomi* in response to temperature was similar to that of avocado, except at 33 °C where the host grew well, but the pathogen did not (Zentmyer 1981). Disease development was greatest at 15–27 °C, but the host outgrew the pathogen at 33 °C (Zentmyer 1981). In the case of black root rot of tobacco, the temperatures optimum for the pathogen, *Thielaviopsis basicola* (= *Chalara elegans*), and its tobacco host were similar (28–30 °C) (Johnson and Hartman 1919, cited by Zentmyer 1981). However, the disease symptoms caused *T. basicola* on tobacco were most severe at 17–23 °C, and little damage occurred at 28–30 °C (Zentmyer 1981). Apparently, *T. basicola* was unable to cause disease in a vigorously growing plant at the higher temperatures. In contrast, disease development in corn and wheat, infected with *Gibberella* spp., was most severe at temperatures unfavorable for the host (Dickson 1923, cited by Zentmyer 1981).

The development of wilt symptoms may be influenced by high transpirational demand as in Phytophthora root rot in processing tomato, caused by *P. parasitica* (Neher 1990) and black shank of tobacco, caused by *P. parasitica* var. *nicotianae* (Jacobi et al. 1983). For processing tomato, soil temperatures influenced the severity of root symptoms that developed before shoot symptoms were apparent (Neher 1990; Fig. 5.3). Infections by *P. parasitica* on tomato and parsley and *P. cinnamomi* on avocado occurred at low soil temperatures, but symptoms were delayed until soil temperatures exceeded a threshold level during an early vegetative phase of the host (Hine and Aragaki 1963; Zentmyer 1981; Neher 1990). Black shank symptoms developed rapidly on tobacco after periods of drought and high transpiration. These stressful environmental conditions coincided with physio-

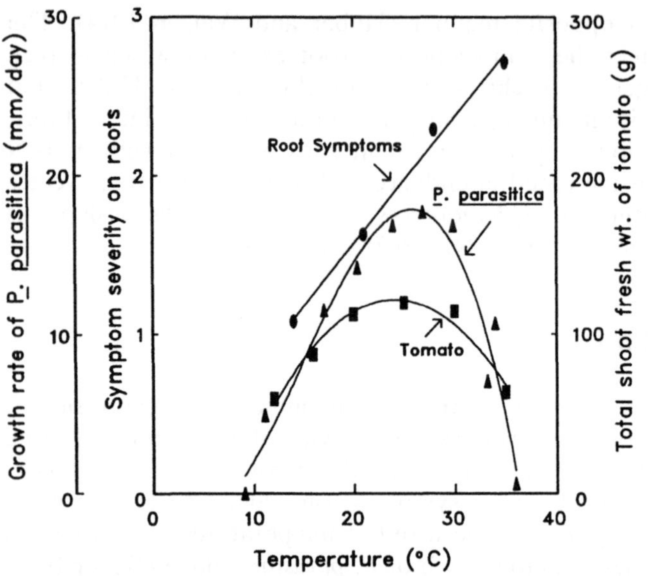

Fig. 5.2. Effect of soil temperature on disease development (*ovals*) and tomato growth (*squares*) and substrate temperature on in vitro vegetative growth of *Phytophthora parasitica* (*triangles*). Disease severity values represent the mean root symptom severity across replicate experiments of a greenhouse study. Tomato growth represents shoot fresh weight of 28-day-old tomatoes averaged for the seasons comparable to the periods that the different replicates of the greenhouse experiments were conducted (data shown from Harssema 1977). Growth rate (mm/day) of *P. parasitica* was measured on lima bean agar (C.D. McKeen and J.M. Duniway, unpubl. data). Curves illustrating the effect of temperature (T) were fitted using regression analysis: tomato growth = 141.70 + 21.87 (T) − 0.45 (T^2), R^2 = 0.96; disease symptoms = −1.50 + 7.93 (T), R^2 = 0.99; *P. parasitica* growth = 2.38 − 10.86 (T) + 1.64 (T^2) − 0.04 (T^2), R^2 = 0.94. (Neher 1990)

logical maturation of plants (Jacobi et al. 1983). In contrast, periods of high air temperatures (≥28 °C) inhibited the development of foliar symptoms of cotton caused by Verticillium wilt by decreasing the susceptibility of the host (Pullman and DeVay 1982). For research studies, wilt symptoms on shoots can be quantified as leaf water potential using a portable pressure chamber (Kramer 1983).

Soil moisture is also known to influence the development of root and shoot symptoms. Symptoms of Fusarium foot rot on mature wheat were more severe in plants experiencing water stress than in plants not experiencing water stress (Papendick and Cook 1974). Papendick and Cook (1974) suggested that water stress predisposed infected plants to rapid development of foot rot rather than the disease causing the measured water stress. Water stress also predisposes other hosts to pathogens, for example, *Fusarium solani* f. sp. *phaseoli* on beans, *Heterobasidion annosum* on pine seedlings,

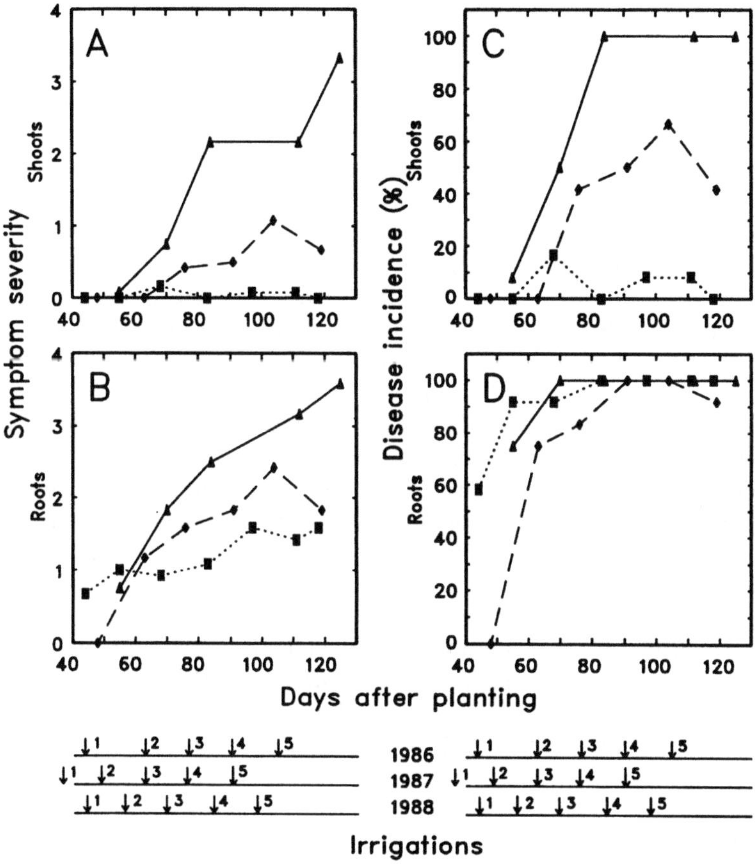

Fig. 5.3A–D. Mean severities of symptoms on shoots (**A**) or roots (**B**) and the incidence of processing tomato plants with shoot (**C**) or root (**D**) symptoms of Phytophthora root rot plotted as a function of days after planting in 1986 (*triangles*) (data for 1986 from Ristaino et al. 1989), 1987 (*squares*), and 1988 (*diamonds*). Symptom severity was based on a scale of 0–4 where 0 = healthy and 4 = extreme symptoms. Values represent means of 12 plants destructively sampled in four replicate plots. *Arrows* represent the dates of irrigations. The first irrigation was always 2 days after inoculation. (Neher 1990)

and *Macrophomina phaseolina* on sorghum and cotton (Cook and Papendick 1972). In contrast, wet soils favor development of Phytophthora root rots by creating environmental conditions favorable to growth and dispersal of the pathogen (Duniway and Gordon 1986).

If some factor of the environment is unfavorable to the host, various physiological processes, including disease resistance, are commonly impaired (Cook and Papendick 1972). For example, flooding, drought, and salinity stresses may predispose various hosts to Phytophthora root rot (Duniway and Gordon 1986). The mechanisms of predisposition are not known clearly but low oxygen levels in saturated soils may make roots leak unusually large quantities of root exudates and, thus become more attractive to zoospores (Kuan and Erwin 1980). In addition, a lack of regeneration or growth of new roots under drought or saline conditions in soil may impair a plant's capacity to tolerate root rot (Duniway and Gordon 1986).

Plants can be predisposed to root diseases caused by fungi and bacteria by other root-infecting organisms, such as nematodes. Root cracking, galling, and giant cell formation caused by *Meloidogyne incognita* may predispose cotton to other root diseases, e.g., Fusarium wilt (Shepherd and Huck 1989). Nematodes may also interact synergistically with other disease organisms to increase the relative incidence and severity of symptoms (Powell 1971; Shepherd and Huck 1989). Inoculation of tobacco with *Pratylenchus brachyurus* simultaneously or 1 week before *Phytophthora parasitica* var. *nicotianae* increased the severity of black shank symptoms in comparison to those observed when *P. parasitica* var. *nicotianae* was added alone (Inagaki and Powell 1969). This suggests that black shank symptoms developed faster in plants that were wounded mechanically by the nematodes. Peanuts were predisposed to Cylindrocladium black rot mostly by the nematodes *Meloidogyne hapla* and less by *Macroposthonia ornata* (Diomande and Beute 1981). Severity of foliar symptoms and yield loss were greater for potatoes with early dying disease when *Verticillium dahliae* was present in combination with *Pratylenchus penetrans* than when *V. dahliae* or *P. penetrans* was present alone (MacGuidwin and Rouse 1990).

5.3.3 Temporal Aspects

A time lag often occurs between infection by the pathogen and symptom development in roots, and between symptom expression on roots and shoots. For example, colonization of cotton roots by *Verticillium dahliae* lagged increases in inoculum density by 3 to 4 weeks (Huisman 1988). Infection by *Phytophthora parasitica* var. *nicotiane* may occur on tobacco roots soon after transplanting, but shoot symptoms of black shank may not be apparent until 5–6 weeks after infection (Jacobi et al. 1983). Visible symptoms of Phytophthora root rot were not apparent on tomato roots until approximately 9 days after infestation of soil under field conditions; shoot symptoms were not apparent until at least 14 days after root symptoms were apparent (Neher and Duniway 1991).

The expression of symptoms may be linked to host plant phenology. The expression of symptoms of Phymatotrichum root rot on cotton is closely associated with cortical senescence in roots. Disease symptoms are restricted initially to the roots, and foliar symptoms are not visible until the root cortex is sloughed, at least 27 days after emergence. This suggests that the fungus was unable to grow beyond cortical tissue without an external energy source. Regardless of plant age (2–22 days) when exposed to sclerotial inoculum of *P. omnivorum* in containers, the root cortex sloughed 18–24 days and plants died 27–50 days after seedling emergence (Rush et al. 1984).

For many plants, infections must occur within a given temporal "window of susceptibility" for symptoms to develop and become severe. Susceptibility may coincide with the period of maximum growth of a plant part, for example, damping-off caused by *Pythium ultimum* killed snapdragon seedings less than 15 days old but not seedlings greater than 20 days old (Populer 1978). Similarly, seedlings of beans, sugar beets, and many other plants are initially susceptible to damping-off caused by *Rhizoctonia solani* but become resistant about 21 days after emergence due to physiological changes (Bateman and Lumsden 1965). Susceptibility may change with age of an infected plant part, for example, the severity of Fusarium wilt on peas decreases with plant age due to restricted fungal growth in the xylem vessels rather than to decreased infections of the roots (Populer 1978). Susceptibility of potato tubers to *Streptomyces scabies* was restricted to a period of about 6 days, when the tuber had matured to the stage equivalent to development of three or four internodes (Lapwood and Adams 1973). Susceptibility may be bimodal, for example, potato tubers are susceptible to infection by *Fusarium caeruleum* during the first 2 months of tuber growth in the field and when tubers are in winter storage, but are relatively resistant shortly after the haulm is removed or senesces (Boyd 1967).

Crop yield losses may be related directly to the time that foliar symptoms appear relative to host plant phenology. When foliar symptoms appear late in the season, cotton lint reductions due to Verticillium wilt are small (Pullman and DeVay 1982). The effects of Verticillium wilt on yield are less severe when foliar symptoms develop during fruit development, because the plant has already slowed vegetative growth and partitioned more photosynthate to fruit for development (Pullman and DeVay 1982). For diseases that are not primarily systemic, such as Phytophthora root rots, early times of infection may indirectly reduce yield loss by lengthening the period over which symptoms increase in severity and alter host phenology to the extent that it results in yield reductions (Ristaino et al. 1989; Neher et al. 1993). For proper comparison, yield should be assessed for the same plants or plots on which disease severity was assessed (Kranz 1988).

Shoot symptoms should not be rated in isolation of root symptoms for root diseases, especially when interactions with plant nutrition and other microorganisms in soil may be involved. After confirmation of the cause of disease by examination of the roots or appropriate isolations, shoot symptoms can be relied on more exclusively for systemic diseases such as Verticillium wilts than for nonsystemic diseases such as Phytophthora root rots. In some cases where disease affects water transport in plants, the position and depth of lesions on roots may be a useful measure. Diameter and biomass of roots or maturation of leaves may be an important parameter in assessment of root diseases which alter plant hormones. The time of symptom ratings should correspond to periods of host susceptibility or rapid epidemic development. Interpretation of disease symptoms should consider the environmental variables that may influence symptom expression and the growth of the host or pathogen such as temperature and humidity of the air and temperature and moisture of the soil.

5.4 Disease Assessment

Ideally, disease assessment methods should be (1) accurate, precise, and reproducible by many observers; (2) applicable over a range of conditions; (3) economical; and (4) simple (Kranz 1988; Berger 1988; Campbell and Madden 1990). Assessment methods and assessors can be evaluated in relation to a set of known or measured standards for accuracy and precision by linear regression (Amanat 1977; Kranz 1988; Campbell and Madden 1990). Reproducibility can be evaluated by correlation analysis. Inter-rater reliability can be evaluated by comparing the relative assessment of plant disease over a range of conditions by several observers (Shokes et al. 1987).

The purpose and scale of assessment must be considered when choosing an assessment method. Disease assessments may be conducted for a variety of purposes, including estimation of disease effects on yield, plant disease surveys, screening of fungicides, varietal trials, surveys to evaluate disease control methods, and forecasting (Large 1966). Methods for rating individual plants and plant populations may be different than for assessing disease for a large geographic region. Diagrammatic keys and scales (e.g., Horsfall and Barratt 1945) may be appropriate for rating individual plants and populations of plants, but remote sensing in surveys may be more appropriate for regional or continental estimations or comparisons.

It is also important to consider the plant organ(s) physiologically linked to symptom expression when selecting an assessment method. If fleshy or storage roots and fibrous roots are present or if woody and vegetative roots are present, such as with most trees, it may be necessary to evaluate disease on both types of roots. The focus of disease assessment may differ for plants for which the root (e.g., sweet potatoes, sugar beets, carrots) or the above-ground plant parts (e.g., tomatoes, oranges, corn, tobacco) are harvested as the yield component. For example, severity of soil rot or pox (*Streptomyces ipomoea*) symptoms on the fleshy storage roots of sweet potatoes has more effect on yield than symptoms on fibrous roots or shoots (Clark and Moyer 1988; Ristaino and Averre 1992). Cavity spot (*Pythium violae*) (Vivoda et al. 1991) and forked root (*P. irregulare* and *P. ultimum*) (Liddell et al. 1989) diseases of carrot affect primarily the roots and, thus, the economic yield. For *Phytophthora parasitica* var. *nicotianae* on tobacco, wilt symptoms on shoots directly affect the economically important leaves (Shew and Lucas

1991). In contrast, when fruit are harvested from above-ground plant parts, the severity of symptoms of shoots may influence yield more than the severity of root symptoms, for example Phytophthora root rot in processing tomatoes (Neher et al. 1993).

5.4.1 Illusions and Hazards of Disease Assessment

Errors in disease assessment may arise from perceptions of individual raters, from factors that influence the object rated, such as three-dimentional complexity of the sampling unit, size and shape of lesions, color of lesions in sampling unit and light conditions, and from the biological characteristic of a diseased plant such as root excision or defoliation due to disease. Errors in assessments are often cryptic and can lead to false impressions concerning the success or failure of management practices. Although not all of the illusions or hazards associated with disease assessments can be resolved simultaneously, awareness of potential problems can aid in improving the quality of assessments.

One concern is the accuracy of presenting maximum disease as 100%. Several authors (Kranz 1977, 1988; Sherwood et al. 1983) have discussed the problems with overestimation of disease. This illusion of disease assessment is compounded by the tendency to equate 100% with maximum disease severity. Some disease assessment scales such as the modified Cobb scale (Large 1966) have attempted to account for the fact that maximum disease severity may be less than 100%. In the modified Cobb scale for rust disease, 37% was chosen arbitrarily as the maximum proportion of leaf area that could be covered with rust pustules and this was labeled as 100% (Large 1966). Kranz (1977) also has suggested 37% as the maximum possible severity for many foliar diseases. Similar maximum limits have not been established with root diseases. There are also analytical problems in estimating the rate of disease progress when maximum disease is assumed to be 100% and it is actually lower (Analytis 1973, 1979; Park and Lim 1985; Neher and Campbell 1992). Incidence-severity relationships when incidence is 100% also cannot be assumed to be constant for all root diseases. Disease expressed as 100% incidence may be representative of severity for a systemic disease such as Verticillium wilt of cotton; however, 100% incidence may correspond to a severity rating of much less than 100% for a nonsystemic disease such as Phytophthora root rot on citrus.

Another hazard associated with assessment of symptoms of root diseases is that one of the main effects of root rots is the actual rotting of root tissues in soil. At first examination this statement may seem trivial; however, when a plant is removed from the soil, the symptom to be evaluated actually may be the amount of root tissue that is not present when it should be. The roots lost due to disease and subsequent decay in addition to normal turnover result in a smaller root system. The role of rotted or missing roots in disease assessment is analogous to that of defoliation in the assessment of foliar diseases.

Another issue which poses difficulties in assessment of root diseases is the growth of the host during the period of time over which assessments are conducted. This difficulty is analogous to that of assessment of foliar diseases on a host that continues to add significant amounts of foliage during the course of an epidemic (Kranz 1988; Campbell and Madden 1990). What is needed is a measure or estimate of host growth or amount of host tissue present. If roots are removed from soil, root systems can be measured as discussed by Böhm (1979) and measures of root length can be made by direct measurements or by line-intersection, photoelectric, or digitizing methods (Newman 1966; Marsh 1971; Rowse and Phillips 1974; Tennant 1975; Voorhees et al. 1980; Campbell and Madden 1990). In addition, morphometric measurements of root systems provide information about root tissue type, and thus root function, which may relate to tissue susceptibility (Fitter 1982, 1987; English and Mitchell 1989).

5.4.2 Methodology

It is easier and less error-prone to measure incidence (Kranz 1988), therefore, it would be desirable to quantify incidence and indirectly relate it to disease severity. Incidence may be as informative as severity for some diseases, for example, those that (1) are systemic, e.g., fungal wilts, damping-off, and viral diseases (Seem 1984; Kranz 1988); (2) have little variation among severity levels within a sampling unit; or (3) the severity levels are low, i.e., in the early stages of epidemic development (Seem 1984). However, for other diseases such as Phytophthora root rot on citrus, incidence may be less informative and less accurate than disease severity (Horsfall and Cowling 1978; Seem 1984) in reference to the effects of disease on plant growth and yield. If relationships between disease incidence and severity are proven to be highly statistically significant by techniques such as correlation, linear regression, and nonlinear regression, it may be possible to quantify disease incidence and relate it to disease severity (Seem 1984). However, a variety of factors complicate the ability to quantify incidence with severity including differential tissue susceptibility, crop age or growth, and year-to-year and site-to-site variations in weather and environment (Seem 1984; Kranz 1988).

There are essentially two ways to measure disease intensity of root diseases: (1) destructively sampling "sacrificial" plants, or (2) assessing the intensity of root disease according to features of above-ground parts (Kranz 1988), i.e., visible symptoms or an expression of growth or stress. Both approaches have significant advantages and disadvantages (Table 1). In the first approach, it is essential to excavate the root systems of diseased plants several times during a growing season to assess the extent of diseased roots accurately (Hornby and Fitt 1981). Because it is nearly impossible to excavate an entire root system, it may be necessary to dig plants with symptomless roots for comparison to root systems with disease symptoms.

The comparative excavation approach can be implemented in research studies by establishing experimental plots with paired rows of the crop. One row can be assessed for above-ground symptoms with nondestructive techniques and the other row for above- and below-ground symptoms by destructive sampling. In the row for destructive sampling, a randomly selected plant or quadrat of plants, which represent a small fraction of the total plants in a row, is removed at repeated intervals through the growing season to assess both shoot and root symptoms. Ristaino et al. (1989) and Neher and Duniway (1991) used this approach to characterize epidemics of *Phytophthora parasitica* on tomato. Campbell et al. (1980a,b) used a similar approach of removing only a small fraction of the entire host population to characterize epidemics of bean hypocotyl rot caused by *Rhizoctonia solani*. The level of impact that plant removal has on epidemic progress must be assessed. For example, above-ground symptoms can be compared between paired rows of destructively and nondestructively sampled plants to determine the effect of the destructive sampling on epidemic development.

Caution must be taken in inferring the extent of root disease by observing only foliar symptoms such as yellowing, epinasty, chlorosis, and wilting (Hornby and Fitt 1981). However, it is possible to indirectly assess symptoms of root disease by observing above-ground symptoms for plants with predictable patterns of symptom development on above-ground plant parts. For example, Phymatotrichum root rot of cotton initially causes the rapid wilting of upper leaves followed by wilting of lower leaves and plant death (Jeger and Lyda 1986). Also, statistically significant relationships have been established between incidence of above-ground symptoms of Cylindrocladium black rot of peanut and yield (Pataky et al. 1983). Disease incidence is rated on shoots throughout the season and roots are dug and evaluated at the end of the season (Pataky et al. 1984). Correlation of symptom severity on roots with above-ground plant symptoms may not be reliable until repeated research observations have been conducted and the associations confirmed (Kranz 1988).

An alternative to comparing root symptoms to shoot symptoms is to correlate root symptoms with above-ground plant growth, mass, color, leaf numbers, and respiration (Bald 1969). Assessment of host growth using leaf area (Bald 1969; Hornby and Fitt 1981) and biomass (Pullman and DeVay 1982; Ristaino et al. 1989; Neher 1990) are good estimates of growth and have been correlated with above-ground symptom severity. Leaf area can be measured using a variety of techniques including electronic leaf area meters, area diagrams, related area measurements, counting grids, and planimeters (Kranz 1988; Campbell and Madden 1990). Other estimates of host growth include plant height, canopy density and volume, percent ground cover, and plant growth stages (Berger 1988). Reduction in plant size is one of the first indications of water deficiency in the field (Hsiao 1982) and is evident as an early symptom of Phytophthora root rots (Duniway 1977; Ristaino et al. 1989), Fusarium wilts (Gordon et al. 1990), and Verticillium wilts (Pullman and DeVay 1982). Reduction in growth rate was detected about 2 weeks prior to the appearance of foliar symptoms of Verticillium wilt on cotton (Pullman and DeVay 1982). Reduction in plant size indirectly reduces the plant's ability to produce photosynthate. Within fields, plant water or nutritional stress may be measured using spectral radiometers (Nutter 1989) and for larger areas, infrared photography (Powell et al. 1976; Toler et al. 1981; Jackson 1986).

5.4.2.1 Visual Estimates

General verbal or pictorial descriptive scales subjectively divide levels of disease severity into discrete categories (O'Brien and van Bruggen 1992a). These scales describe or illustrate plants with arbitrary amounts or types of disease symptoms (Lindow 1983). Diagrammatic scales are used to assess plant organs and field scales are used to assess whole plants or plots (Large 1966). Field scales must refer to the area covered by disease lesions and tissue chlorosis/necrosis caused by disease (Large 1966). The first standard area diagrammatic key used to assess disease severity was the Cobb Scale (1892) which differentiated five degrees of severity ranging from 1–50% leaf area covered with rust pustules on cereal leaves in Australia (Large 1966; James 1971; Horsfall and Cowling 1978). The McKinney Index (1923) was the first "numerical rating" assessment key. It was developed with categories ranging from 0–3, with 0 equal to no disease and 3 referring to abundant lesions. There are several problems associated with these types of scales including potential nonreproducibility of assessment by different evaluators (Lindow 1983; Sherwood et al. 1983), unequal application to symptoms of different cultivars, difficult interpolation of intermediate levels of disease severity (Lindow 1983), difficulty in evaluating how close estimated severity is to actual severity (Tomerlin and Howell 1988), and the potential requirement of nonparametric statistical analyses if intervals between categories are unequal. Accuracy may also be in question with use of categorical scales, as identified earlier, because observers tend to overestimate disease intensity (Berger 1980; Kranz 1988), especially at low levels of infection (Sherwood et al. 1983). However, accuracy and precision can sometimes be improved by experience (James 1971, 1974; Berger 1980; Lipps and Madden 1989).

The Weber-Fechner law states that visual acuity is proportional to the log of the intensity of the stimulus with poorest acuity occurring in moderate ranges (40–60%) of disease intensity (Large 1966; James 1974; Horsfall and Cowling 1978). Consequently, several scales have been developed using logarithmic scales, e.g., the James's keys (James 1971) and the Horsfall-Barratt (HB) scale. The HB scale comprises 12 classes: 0, 0–3, 3–6, 6–12,

12–25, 25–50, 50–75, 75–87, 87–94, 94–97, 97–100, and 100% diseased tissue (Horsfall and Barratt 1945). Berger (1980) proposed a finer division of the middle range of the scale into smaller intervals, but there is no evidence that observers can readily distinguish between these categories (Horsfall and Cowling 1978; Campbell and Madden 1990). Horsfall-Barratt scales were originally designed for foliar diseases, but the same concepts can be applied to root diseases given appropriate diagrams, e.g., corky root of lettuce (O'Brien and van Bruggen 1992a).

Some argue that logarithmic scales are appropriate for all diseases, even if the relationship between disease and loss is linear, because the eye reads in logarithms (James 1974; Horsfall and Cowling 1978), whereas others argue that a logarithmic scale is not appropriate for assessment of all diseases (Hebert 1982). Even under the guise of straight-percentage scales, observers have the tendency to cluster observations of disease intensity around "knots" or common values, for example, 0.0, 0.05, 0.1, 0.25, and 0.5 (Koch and Hau 1980). Percent scales can be used to describe a variety of assessment types, for example, the proportion of plants infected, area damaged, and proportion of roots or fruits affected. Categorical classes can be converted to straight-percentage scales by using the midpoint value of each category (Berger 1980; Campbell and Madden 1990). The advantage of this transformation is that straight-percentage scales can be analyzed using parametric statistical analyses.

There are a very few studies that compare disease assessment methods for accuracy, precision, or correlation with disease loss of root diseases (Forbes and Jeger 1987). A recent study compared assessment scales for corky root of lettuce by evaluation purpose and plant age for accuracy, precision, and correlation with disease loss (O'Brien and van Bruggen 1992a). A variety of experienced and inexperienced personnel evaluated the same plants for the percentage of the tap root area corked using three different scales. Two discrete interval scales (7-level scale for mature plants and 10-level for seedlings) were compared to a 12-level Horsfall-Barratt continuous interval scale. The scales tested were diverse with respect to simplicity (number of levels), realism (photographs vs. drawings), intended use (mature plant vs. seedling), choice of levels (objective vs. continuous interval), and statistical restrictions (discrete vs. continuous interval). No single assessment aid was identified as the best for all situations. The scales performed best for the phenological stage(s) for which they were originally developed and were most useful for predicting yield loss at phenological stages when disease severities fell within moderate ranges.

Forbes and Jeger's (1987) evaluation of a Horsfall-Barratt scale supported the agreement with the Weber-Fechner law in the sense that the greatest variances of disease assessment were within the 25–75% range. However, the variance was greater at 25% than at 50 or 75%, which would not be expected with the Weber-Fechner law. Symptoms on true roots were overestimated more than any other plant structure, including tubers.

This error was attributed to the nonuniformity of root shape and the scarcity of good scales for root diseases (Forbes and Jeger 1987). It is also more difficult to accurately assess three-dimensional than two-dimensional structures (Kranz 1988); therefore, there may be reason to rate individual roots or orders of roots (Fig. 5.1) or lesion placement, rather than a percent of the whole root system that has lesions or is affected by symptoms.

There are circumstances that complicate assessment of both root and shoot symptoms associated with root diseases. First, there are diseases where defoliation, root loss, and/or subsequent compensation occur as a result of severe disease. For example, the majority of isolates of *Verticillium dahliae* from cotton are defoliating pathotypes (Pullman and DeVay 1982). In contrast, a flush of new foliage may develop on processing tomato plants with severe symptoms of Phytophthora root rot late in the season (Ristaino et al. 1989). This new foliage, however, will never yield mature fruit. In other examples, adventitious roots may grow near the crown if other roots are severely pruned by rots and lesions, for example, by Phytophthora root rots of alfalfa (Stuteville and Erwin 1990), tomato (Blaker and Hewitt 1987a,b), and avocado (Kelman and Coffey 1985).

Second, under natural field conditions, complexes of pathogens may occur, making it difficult to differentiate symptoms of individual diseases. Examples of disease complexes include root rot of horseradish (Percich 1990), sudden death syndrome of soybean (Roy et al. 1989), bean root rot complex (Huber and Watson 1974), and numerous examples of fungal and nematode disease complexes including: (1) root knot nematode (*Meloidogyne* spp.) with Fusarium wilts on cotton (Powell 1971; Shepherd and Huck 1989), tomato, peas (Powell 1971); Phytophthora diseases on tomato (Neher 1990) and tobacco (Powell 1971); and *Pythium aphanidermatum* on chrysanthemum (Powell 1971); (2) cyst nematodes (*Heterodera*) with Fusarium wilts of soybeans and sugar beets; (3) sting nematodes (*Belonolaimus*) with Fusarium wilt of cotton and *P. aphanidermatum* infections of chrysanthemum; and (4) lesion nematode (*Pratylenchus*) with Verticillium wilts on cotton, tomato, eggplant, peppers, and potatoes (Powell 1971). Depending on the objective, the contribution of individual pathogens to the symptoms may be assessed or, from a holistic epidemiological perspective, the total amount of tissue with symptoms may suffice.

Third, there are situations where pathogen colonization has occurred, but no disease symptoms were visible, i.e., the completion of the incubation period has not occurred. This phenomenon has been reported for pathogenic *Fusarium* spp. (Gordon et al. 1989), *Verticillium dahliae* (Gerik and Huisman 1988), and *Verticillium albo-atrum* in moderately resistant alfalfa cultivars (Newcombe and Robb 1988). Using specific immunoenzymatic staining techniques, Gerik and Huisman (1988) found that *V. dahliae* was able to colonize the entire depth of the cortex of cotton roots without breaching the endodermis or infecting the stele. *Fusarium oxysporum* is primarily an epiphytic colonizer, mostly confined to the root surface and outer cortex (Gerik and Huisman 1988).

Fourth, some fungi and nematodes may not cause much plant damage themselves, but cause damage indirectly by transmitting viruses. For example, infections by *Olpidium brassicae* on lettuce roots may be associated with symptoms of lettuce big vein (Grogan and Campbell 1966) and soilborne wheat mosaic virus is transmitted by the fungus, *Polymyxa graminus* (Wiese 1987). *Xiphenema index* is the vector of grapevine fanleaf virus (Pearson and Goheen 1988). Generally, *Xiphenema* and *Longidorus* spp. transmit nepoviruses and *Trichodorus* and *Paratrichodorus* spp. transmit tobraviruses, i.e., tobacco rattle and pea early browning (Taylor and Brown 1981).

5.4.2.2 Electronic Techniques

Leaf area meters, video image analysis systems, multispectral radiometers, and remote sensing devices with infrared photography or satellite technology are among electronic devices that can be used for assessing plant disease. Leaf area meters are useful in assessing the area of detached leaves rapidly, which is an important variable in calculation of a leaf area index or other applications of plant growth measurement. Video image analysis systems are capable of assessing the area of diseased tissue based on color contrasts between healthy and diseased plant tissues. Images can be digitized and changes in color analyzed (Nilsson 1980; Lindow and Webb 1983; Berger 1988). Lens filters can be utilized to increase contrast between healthy and diseased tissue, for example, a red filter in sunlight under field conditions or a yellow filter with a light gray background for distinguishing diseased tissue with darkly pigmented lesions (Lindow 1983). Multispectral radiometers have been successfully used to quantify green leaf area gradients by measuring percent reflectance for a wavelength band of interest e.g., 800 nm (Nutter 1989). Chlorophyll pigments are the unit reflecting a wavelength band of 800 nm (Nutter 1989). In principle, healthy plants have greater reflectance than plants under stress or with disease symptoms of wilting, chlorosis, necrosis, or defoliation.

Black-and-white, color, color-infrared, and false-color-infrared photography have been used in remote sensing to detect plant disease at scales ranging from fields (Powell et al. 1976; May et al. 1985; Blakeman 1990) to large geographical regions (Berger 1988). The sharpest contrast in reflectance between soil and green vegetation is in the near-infrared band (700–950 nm) (Toler et al. 1981; Kranz 1988; Campbell and Madden 1990). Root diseases and their associated stress symptoms can be captured and viewed as color differences on film, sometimes before they are visible to an observer on the ground (Rundquist et al. 1984/1985; Kranz 1988). Plant canopies with reduced leaf water content and thus higher temperatures, caused by wilting, drought, salinity, or nutritional stresses, reflect less near-infrared energy than healthy plants (Toler et al. 1981; Rundquist et al. 1984/1985; Jackson 1986; Kranz 1988). Canopy temperatures were 2.6–3.6°C and 3.3–3.5°C higher for sugar beets infected with *Pythium aphanidermatum* and cotton with *Phymatotrichum omnivorum*, respectively, than for healthy plants, even under conditions of water stress (Jackson 1986). Differences in reflectance, however, do not guarantee that a specific disease caused the stress. It is essential to accompany remote sensing with "ground truthing" (Horsfall and Cowling 1978; Kranz 1988). Once "ground truthing" is keyed to aerial photographs, surveys of expansive acreages can be conducted (Horsfall and Cowling 1978). Infrared photography may be useful for assessing systemic diseases but not for diseases with low severity that begin in the lower canopy hidden by upper healthy foliage (Berger 1980), or for diseases that develop before the leaf canopy is full, because exposed soil will further reduce reflectance in excess of diseased tissue (James 1974). Reviews by Jackson (1986) and Toler et al. (1981) are recommended for additional information.

Use of aerial photography has been reported more often for foliar than root diseases (Toler et al. 1981). Phymatotrichum root rot (*P. omnivorum*) on cotton and take-all (*Gaeumannomyces graminis*) on winter wheat are root diseases that have been assessed successfully with aerial photography. Sharp contrasts between diseased and healthy cotton plants were apparent because plants died quickly after the onset of symptoms (Toler et al. 1981). With take-all in the UK, aerial photography has been used since the 1980s to provide a visual record of case histories which demonstrate how soil and cultural interactions affect the development of the disease (Blakeman 1990). Aerial photography has also been used to detect Fusarium wilt on pear (Haglund and Jarmin 1978) and aerial photography may be useful for orchards and forests where disease development can be monitored and mapped on an annual basis (Toler et al. 1981). For example, Wallis and Lee (1984) used aerial photography to detect centers of root disease in Douglas-fir stands. However, successful assessments were dependent upon stand age, i.e., older stands with large openings caused by diseased or dead trees in the stand were easier to discern than younger stands with less distinct stand contrasts (Wallis and Lee 1984). Also, it was impossible to distinguish the two etiological agents, *Phellinus weirii* or *Armillaria ostoyae*, from aerial photographs using color or color-infrared film. However, distinct red or yellow foliage was discernible on both color and color-infrared film (Wallis and Lee 1984). MacNish and Lewis (1985) compared the relative ease and accuracy of aerial mapping, ground area mapping, and ground-based scoring methods for determining the extent of Rhizoctonia patch in cereals relative to yield. The three methods were highly correlated, but ground-based scoring was quicker, more convenient, and related better to yield losses than either aerial or ground-based mapping methods (MacNish and Lewis 1985) and was most likely more cost-effective (Toler et al. 1981).

5.5 Sampling Considerations

The challenge in sampling for disease assessments is to obtain a representative and statistically valid sample of the population on plants for an affordable cost with the greatest likelihood of preserving the integrity of the host population and, thus, the epidemic of interest. The questions that must be answered are "how many samples should be taken?" and "how should the samples be taken?" To answer these challenges, sampling must balance statistical, biological, and economic considerations.

5.5.1 Sampling Pattern

Most root diseases have an aggregated spatial pattern in the field (Campbell and Noe 1985; Gilligan 1988; Campbell and Madden 1990; Chap. 7, this Vol.). In simulation studies with aggregated disease patterns, systematic sampling provided better estimates of disease incidence than did simple random sampling (Lin et al. 1979). In a systematic sample, the number of so-called sampling "arms" or, basically, the number of changes in direction during a systematic sampling path through a field, is important to sampling success. As a result, a zigzag pattern, a W-shaped or diamond-shaped path is recommended for moving through a field to assess a root disease. Thus, once a random starting point is selected in a field or a stratum (e.g., x paces from one edge of the field and y paces from the adjacent edge of the field or a neighboring stratum), samples can be taken or sites of assessment selected every k paces or units along the selected sample pattern so that a large portion of the field is covered by the time n sample units (the number of sample units desired) have been assessed.

In another simulation study, stratified random sampling had a lower error rate or bias (difference between true and estimated mean) than other designs when incidence of disease was low (<10%) (Delp et al. 1986a). To determine inoculum density of soilborne pathogens, which also occur in aggregated patterns, a systematic pattern of sampling is also usually recommended (Goodell and Ferris 1981; Hau et al. 1982; Mihail and Alcorn 1987) and stratification of fields may improve estimates of disease at low incidence levels. A stratified random sampling design is implemented by dividing the field into equally sized sectors and sampling at randomly or systematically selected locations within each sector. This method has been viewed as cumbersome in the past, but is made simpler with a computer program called "Field Runner" designed for use on a portable personal computer, which directs the evaluator to specific locations in the field (Delp et al. 1986b).

5.5.2 Sample Numbers and Costs

The number of samples (n) taken for a disease assessment determines the quality or reliability and the cost of the assessment. Three factors that must be considered to determine sample number are the statistical estimates of the mean and variance obtained in a preliminary sample, the biological characteristics of the pathosystem, and the resources available for sampling. Most sampling texts have excellent presentations concerning the ways in which to estimate sample number based either on statistical criteria such mean, variance, and desired reliability or on statistical criteria and resource allocation (= costs). Several sources of such information provided by biologists should be consulted for readable presentations of such considerations (Karandinos 1976; Ives and Moon 1987; Kranz 1988; Campbell and Duthie 1989; Campbell and Madden 1990). However, few if any, strictly statistical approaches to sampling design make allowances for or provide recommendations how to account for the very real biological differences of sampling such as those that are encountered in assessment of root diseases, particularly when destructive sampling is necessary.

5.6 Recommendations

Several preliminary steps must be taken before successful disease assessment can be completed. First, be familiar with the symptoms of the disease(s) of interest. Second, define the objectives and goals of the disease assessment and determine the desired detection level. An assessment scale should be chosen to fit the detection level and maximum level of disease observed. Selection of an appropriate scale may require knowledge of the relationship of disease development to such factors as plant phenology and climate. Third, decide whether incidence and/or severity of disease is important in relationship to the study goals. Fourth, choose the plant organs that will be evaluated. This choice depends upon the disease and the study objectives, i.e., shoot symptoms may be more important for assessment of effects on yield for some diseases, whereas assessment of root and shoot symptoms may be necessary to determine physiological relationships that occur within a diseased plant. Once the choice of plant organs is made, the experiment can be designed to permit the type of sampling required to assess symptoms on the chosen plant organs at the desired frequency.

Estimation of disease intensity should begin early in disease development, generally when symptoms or signs first appear so that several estimates in time can be made. Berger (1980) suggested that observations be made every one-half incubation period with more assessments during rapid stages of epidemic development. The application of this guideline may be complicated if root and shoot symptoms are rated because of the potential difference in time of occurrence or detection of the incubation period for visible symptoms on shoots and roots. Assessment on a series of dates is more beneficial than a single rating at the end of a growing season (Watson et al. 1990). Also, with many statistical procedures, there is a gain in accuracy when the number of observations in large ($n > 15$) (Berger 1980).

It is advisable to use a scale with equal divisions, whether on a linear or logarithmic scale (James 1975; Kranz 1988). Such a scale makes the decision process during the disease assessments easier. Also, data derived from a scale with equally spaced divisions can be analyzed easily by use of parametric statistical analyses such as analyses of variance and regression.

Due to the potential influence of diurnal weather cycles, assessment of shoot symptoms associated with root diseases should be made at a similar time of day, preferably early in the day, because evapotranspiration demand can influence the expression of wilt symptoms. Sun angle and brightness may also influence the perception of color differences associated with disease symptoms.

Selection of an appropriate sampling design for assessment is critical. Systematic designs are preferred for aggregated diseases such as root diseases. A stratified random design can be used for diseases with a variety of spatial patterns (Delp et al. 1986a). Sample number should be based on information from preliminary samples and will be determined by balancing error and cost considerations.

The record of intensity of disease at each assessment should be accompanied with data on crop growth stage, the plant organ assessed (James 1974; Watson et al. 1990), the spatial pattern of disease (Watson et al. 1990), and the relevant sampling information (Campbell and Madden 1990). Also, environmental variables that may correspond to epidemic development, such as soil temperature and moisture, should be monitored. This information will allow meaningful evaluation of assessment scales and comparisons between epidemics.

Disease assessment should be followed by isolation of the pathogen of interest to confirm the cause of symptom development. Pathogen isolation and careful evaluation of all symptoms will permit (1) accurate diagnosis; (2) identification of multiple pathogens and diseases which may accentuate or diminish the symptom expression of the disease of interest; and (3) detection of interfering mineral deficiencies or toxicities. When a complex of diseases is present, initially assess the total integration of diseases on the plant(s), followed by assessment of effects of individual diseases as much as possible (Berger 1988). If multiple plant organs are effected by disease(s), assess one disease on one organ at a time; the error attached to any particular assessment will be less than when organs are evaluated collectively (James 1974). For survey or regional studies where remote sensing is used, "ground truth" activities should be adequate to provide evidence of the causal agents.

References

Amanat P (1977) Modellversuche zur Ermittlung individueller und objektabhängiger Schätzfehler bei Pflanzenkrankheiten. Dissertation, Univ Gießen

Analytis S (1973) Zur Methodik der Analyse von Epidemien dargestellt am Apfelschorf (*Venturia inaequalis* (Cooke) Aderh.). Acta Phytomed 1:1–75

Analytis S (1979) Die Transformation von Befallswerten in der quantitativen Phytopathologie. Phytopathol Z 96:156–171

Bald JG (1969) Estimation of leaf area and lesion sizes for studies on soil-borne pathogens. Phytopathology 59:1606–1612

Bateman DF, Lumsden RD (1965) Relation of calcium content and nature of pectic substances in bean hypocotyls of different ages to susceptibility to an isolate of *Rhizoctonia solani*. Phytopathology 55:734–738

Berger RD (1980) Measuring disease intensity. In: Proc E. C. Stakman Commemorative Symp Misc Publ No 7, Univ Minnesota, St Paul, pp 28–31

Berger RD (1988) Special report: measuring disease intensity. Biological and cultural tests, vol 3. APS Press, St Paul, pp 1–4

Blakeman RH (1990) The identification of crop disease and stress by aerial photography. In: Steven MD, Clark JA (eds) Applications of remote sensing in agriculture. Butterworths, London, pp 299–354

Blaker NS, Hewitt JD (1987a) A comparison of resistance to *Phytophthora parasitica* in tomato. Phytopathology 77:1113–1116

Blaker NS, Hewitt JD (1987b) Comparison of seedling and mature plant resistance to *Phytophthora parasitica* in tomato. HortScience 22:103–105

Böhm W (1970) Methods of studying root systems. Ecological studies, vol 33. Springer, Berlin Heidelberg New York, 188 pp

Bowers JH, Smerage GN, Mitchell DJ (1991) The development of a conceptual model of root growth. In: Keister DL, Cregan PB (eds) The rhizosphere and plant growth. Kluwer Academic, Dordrecht, pp 73–80

Boyd AEW (1967) The effects of length of the growth period and of nutrition upon potato tuber susceptibility to dry rot (*Fusarium caeruleum*). Ann Appl Biol 60:231–240

Campbell CL, Duthie JA (1989) Special report: sampling for disease assessment. Biological and cultural tests, vol 4. APS Press, St Paul, pp v–ix

Campbell CL, Madden LV (1990) Introduction to plant disease epidemiology. John Wiley & Sons, New York, 532 pp

Campbell CL, Noe JP (1985) The spatial analysis of soilborne pathogens and root diseases. Annu Rev Phytopathol 23:127–148

Campbell CL, Madden LV, Pennypacker SP (1980a) Structural characterization of bean root rot epidemics. Phytopathology 70:152–155

Campbell CL, Pennypacker SP, Madden LV (1980b) Progression dynamics of hypocotyl rot of snapbean. Phytopathology 70:487–494

Carley HE (1969) Factors affecting the epidemiology of pea (*Pisum sativum* L.) root rot caused by *Aphanomyces euteiches*. m.s. or Ph.D. Thesis, Univ Minnesota, St Paul

Clark CA, Moyer JW (1988) Compendium of sweet potato diseases. Am Phytopathol Soc Press, St Paul, 74 pp

Cobb NA (1892) Contribution to an economic knowledge of the Australian rusts (Uredineae). Agric Gaz NSW 3:60

Cook RJ, Papendick RI (1972) Influence of water potential of soils and plants on root disease. Annu Rev Phytopathol 10:349–374

Delp BR, Stowell LJ, Marois JJ (1986a) Evaluation of field sampling techniques for estimation of disease incidence. Phytopathology 76:1299–1305

Delp BR, Stowell LJ, Marois JJ (1986b) Field-runner: a disease incidence, severity, and spatial pattern assessment system. Plant Dis 70:954–957

Diomande M, Beute MK (1981) Relation of *Meloidogyne hapla* and *Macroposthonia ornata* populations to Cylindrocladium black rot in peanuts. Plant Dis 65:339–342

Duniway JM (1977) Changes in resistance to water transport in safflower during the development of Phytophthora root rot. Phytopathology 67:331–337

Duniway JM, Gordon JR (1986) Water relations and pathogen activity in soil. In: Ayres PG, Boddy L (eds) Water, fungi and plants. BMS Symposium 11. Cambridge Univ Press, Cambridge, pp 119–137

English JT, Mitchell DJ (1989) Use of morphometric analysis for characterization of tobacco root growth in relation to infection by *Phytophthora parasitica* var. *nicotianae*. Plant Soil 113:243–249

Fitter AH (1982) Morphometric analysis of root systems: application of the technique and influence of soil fertility on root system development in two herbaceous species. Plant Cell Environ 5:313–322

Fitter AH (1987) An architectural approach to the comparative ecology of plant root systems. New Phytol 106:61–77

Forbes GA, Jeger MJ (1987) Factors affecting the estimation of disease intensity in simulated plant structures. Z Pflanzenkr Pflanzenschutz 94:113–120

Gerik JS, Huisman OC (1988) Study of field-grown cotton roots infected with *Verticillium dahliae* using an immunoenzymatic staining technique. Phytopathology 78:1174–1178

Gilligan CA (1988) Analysis of the spatial pattern of soilborne pathogens. In: Kranz J, Rotem J (eds) Experimental techniques in plant disease epidemiology. Springer, Berlin Heidelberg New York, pp 85–113

Goodell PB, Ferris H (1981) Sample optimization for five plant-parasitic nematodes in an alfalfa field. J Nematol 13:304–313

Gordon TR, Okamoto D, Jacobson DJ (1989) Colonization of muskmelon and nonsusceptible crops by *Fusarium oxysporum* f. sp. *melonis* and other species of *Fusarium*. Phytopathology 79:1095–1100

Gordon TR, Jacobson DJ, May DM, Tyler KB, Zink FW (1990) Fruit yield, disease incidence, and root colonization of hybrid muskmelons resistant to Fusarium wilt. Plant Dis 74:778–781

Grogan RG, Campbell RN (1966) Fungi as vectors and hosts of viruses. Annu Rev Phytopathol 4:29–52

Haglund WA, Jarmin ML (1978) Aerial photography for the detection and identification of fusarium wilt of pear. Plant Dis Rep 62:570–572

Harris NE, Beute MK (1981) Histological responses of peanut germplasm resistant and susceptible to *Cylindrocladium crotalariae* in relationship to inoculum density. Phytopathology 72:1250–1256

Harssema H (1977) Root temperature and growth of young tomato plants. Meded Landbouwhogesch Wageningen 77:1–85

Hau FC, Campbell CL, Beute MK (1982) Inoculum distribution and sampling methods for *Cylindrocladium crotalariae* in a peanut field. Plant Dis 66:568–571

Hebert TT (1982) The rationale for the Horsfall-Barratt plant disease assessment scale. Phytopathology 72:1269

Hine RB, Aragaki M (1963) Influence of soil temperature on a crown rot disease of parsley caused by *Phytophthora parasitica*. Phytopathology 54:1287–1289

Hornby D, Fitt BDL (1981) Effects of root-infecting fungi on structure and function of cereal roots. In: Ayres PG (ed) Effects of disease on the physiology of the growing plant. Cambridge Univ Press, Cambridge, pp 101–130

Horsfall JG, Barratt RW (1945) An improved grading system for measuring plant diseases. Phytopathology 35:655

Horsfall JG, Cowling EB (1978) Pathometry: the measurement of plant disease. In: Horsfall HG, Cowling EB (eds) Plant disease, vol II. How disease develops in populations. Academic Press, New York, pp 119–136
Hsiao TC (1982) The soil-plant-atmosphere continuum in relation to drought and crop production. In: Drought resistance in crops, with emphasis on rice. International Rice Research Institute, Los Banos, Phillipines, pp 39–52
Huber DM, Watson RD (1974) Nitrogen form and plant disease. Annu Rev Phytopathol 12:139–165
Huisman OC (1988) Seasonal colonization of roots of field-grown cotton by *Verticillium dahliae* and *V. tricorpus*. Phytopathology 78:708–716
Inagaki H, Powell NT (1969) Influence of the root-lesion nematode on black shank symptom development in flue-cured tobacco. Phytopathology 59:1350–1355
Ives PM, Moon RD (1987) Sampling theory and protocol for insects. In: Teng PS (ed) Crop loss assessment and pest management. APS Press, St Paul, pp 49–75
Jackson RD (1986) Remote sensing of biotic and abiotic plant stress. Annu Rev Phytopathol 24:265–287
Jacobi WR, Main CE, Powell NT (1983) Influence of temperature and rainfall on the development of black shank. Phytopathology 73:139–143
James WC (1971) An illustrated series of assessment keys for plant diseases, their preparation and usage. Can Plant Dis Surv 51:39–65
James WC (1974) Assessment of plant diseases and losses. Annu Rev Phytopathol 12:27–48
Jeger MJ, Lyda SD (1986) Epidemics of Phymatotrichum root rot (*Phymatotrichum omnivorum*) in cotton: environmental correlates of final incidence and forecasting criteria. Ann Appl Biol 109:523–534
Jones KJ (1990) Components of resistance in *Nicotiana tabacum* to *Phytophthora parasitica* var. *nicotianae*. PhD Thesis, North Carolina State University, Raleigh
Karandinos MG (1976) Optimum sample size and comments on some published formulae. Bull Entomol Soc Am 22:417–421
Kelman MK, Coffey MD (1985) Quantitative comparison of the resistance to Phytophthora root rot in three avocado rootstocks. Phytopathology 75:230–234
Koch H, Hau B (1980) Ein psychologischer Aspekt beim Schätzen von Pflanzenkrankheiten. Z Pflanzenkr Pflanzenschutz 87:587–593
Kramer PJ (1983) Water relations of plants. Academic Press, New York, pp 384–387
Kranz J (1977) A study on maximum severity of plant diseases. Travaux dedies a G. Viennot-Bourgin, pp 169–173
Kranz J (1988) Measuring plant disease. In: Kranz J, Rotem J (eds) Experimental techniques in plant disease epidemiology. Springer Berlin Heidelberg New York, pp 35–50
Kuan TL, Erwin DC (1980) Predisposition effect of water saturation of soil on Phytophthora root rot of alfalfa. Phytopathology 70:981–986
Lapwood DH, Adams MJ (1973) The effect of a few days of rain on the distribution of common scab (*Streptomyces scabies*) on young potato tubers. Ann Appl Biol 73:277–283
Large EC (1966) Measuring plant disease. Annu Rev Phytopathol 4:9–28
Liddell CM, Davis RM, Nunez JJ, Guerard JP (1989) Association of *Pythium* spp. with carrot root dieback in the San Joaquin Valley of California. Plant Dis 73:246–249
Lin CS, Poushinsky G, Mauer M (1979) An examination of five sampling methods under random and clustered disease distribution using simulation. Can J Plant Sci 59:121–130
Lindow SE (1983) Estimating disease severity of single plants. Phytopathology 73:1576–1581

Lindow SE, Webb RR (1983) Quantification of foliar plant disease symptoms by microcomputer-digitized video image analysis. Phytopathology 73:520–524

Lipps PE, Madden LV (1989) Assessment of methods of determining powdery mildew severity in relation to grain yield of winter wheat cultivars in Ohio. Phytopathology 79:462–470

Lucas GB, Campbell CL, Lucas LT (1992) Introduction to plant diseases: identification and management. Van Nostrand Reinhold, New York, 364 pp

MacGuidwin AE, Rouse DI (1990) Role of *Pratylenchus penetrans* in the patato early dying disease of Russet Burbank potato. Phytopathology 80:1077–1082

MacNish GC, Lewis S (1985) Methods of measuring rhizoctonia patch of cereals in Western Australia. Plant Pathol 34:159–164

Marsh Ba'b (1971) Measurement of length in random arrangements of lines. J Appl Ecol 8:265–267

May MJ, Harvey RN, Wright KJ (1985) An infra-red photographic technique for the rapid assessment of weed and crop seedlings in field experiments. Ann Appl Biol 107:511–514

McKinney HH (1923) Influence of soil temperature and moisture on infection of wheat seedlings by *Helminthosporium sativum*. J Agric Res 5:195–217

Mihail JD, Alcorn SM (1987) *Macrophomina phaseolina*: spatial pattern in a cultivated soil and sampling strategies. Phytopathology 77:1126–1131

Neher DA (1990) Inoculum density, furrow irrigation and soil temperature effects on the epidemiology of Phytophthora root rot of processing tomatoes. PhD Thesis, University of California, Davis

Neher DA, Campbell CL (1992) Underestimation of disease progress rates with the logistic, monomolecular, and Gompertz models when maximum disease intensity is less than 100 percent. Phytopathology 82:811–814

Neher D, Duniway JM (1991) Relationship between amount of *Phytophthora parasitica* added to field soil and the development of root rot in processing tomatoes. Phytopathology 81:1124–1129

Neher DA, McKeen CD, Duniway JM (1993) Relationships among Phytophthora root rot, *Phytophthora parasitica* populations, and yield of tomatoes under commercial field conditons. Plant Dis 77:1106–1111

Newcombe G, Robb J (1988) The function and relative importance of the vascular coating response in highly resistant, moderately resistant and susceptible alfalfa infected by *Verticillium albo-atrum*. Physiol Mol Plant Pathol 33:47–58

Newman EI (1966) A method of estimating the total length of root in a sample. J Appl Ecol 3:139–145

Nilsson HE (1980) Application of remote sensing methods and image analysis at macroscopic and microscopic levels in plant pathology. In: Teng PS, Krupa SV (eds) Crop loss assessment. Misc Publ Agric Exp Stn Univ Minnesota, St Paul, pp 76–84

Nutter FW Jr (1989) Detection and measurement of plant disease gradients in peanut with a multispectral radiometer. Phytopathology 79:958–963

O'Brien RD, van Bruggen AHC (1992a) Accuracy, precision, and correlation to yield loss of disease severity scales for corky root of lettuce. Phytopathology 82:91–96

O'Brien RD, van Bruggen AHC (1992b) Yield losses to iceberg lettuce due to corky root caused by *Rhizomonas suberifaciens*. Phytopathology 82:154–159

Papendick RI, Cook RJ (1974) Plant water status and the development of Fusarium foot rot in wheat subjected to different cultural practices. Phytopathology 64: 358–363

Park EW, Lim SM (1985) Empirical estimation of the asymptote of disease progress curves and use of the Richards generalized rate parameters for describing disease progress. Phytopathology 75:786–791

Pataky JK, Beute MK, Wynne JC, Carlson GA (1983) A critical-point yield loss model for Cylindrocladium black rot of peanut. Phytopathology 73:1559–1563
Pataky JK, Black MC, Hollowell JE, Beute MK (1984) Effects of nitrogen fertilization on Cylindrocladium black rot of peanuts and peanut yield. Plant Dis 68:674–677
Pearson RC, Goheen AC (1988) Compendium of grape diseases. Am Phytopathol Soc Press, St Paul
Pegg GF (1981) The involvement of growth regulators in the diseased plant. In: Ayers PG (ed) Effects of disease on the physiology of the growing plant. Cambridge Univ Press, New York, pp 149–177
Percich JA (1990) A root rot complex of horseradish. Plant Dis 74:391–393
Populer C (1978) Changes in host susceptibility with time. In: Horsfall JG, Cowling EB (eds) Plant disease, vol II. How disease develops in populations. Academic Press, New York, pp 239–262
Powell NL, Griffin GJ, Garren KH, Pettry DE (1976) Use of aerial photography to detect diseases in peanut fields. II. Cylindrocladium black rot. Peanut Sci 3:25–29
Powell NT (1971) Interactions between nematodes and fungi disease complexes. Annu Rev Phytopathol 9:253–274
Pullman GS, DeVay JE (1982) Epidemiology of Verticillium wilt of cotton: effects of disease development on plant phenology and lint yield. Phytopathology 72: 554–559
Ristaino JB, Averre CW (1992) Effects of irrigation, sulfur, and fumigation on Streptomyces soil rot and yield components in sweet potato. Phytopathology 82:670–676
Ristaino JB, Duniway JM, Marois JJ (1989) The influence of irrigation schedule and Phytophthora root rot on growth, phenology, and yield of processing tomatoes. J Am Soc Hort Sci 114:556–561
Rowse HR, Phillips DA (1974) An instrument for estimating the total length of root in a sample. J Appl Ecol 11:309–314
Roy KW, Lawrence GW, Hodges HH, McLean KS, Killebrew JF (1989) Sudden death syndrome of soybean: *Fusarium solani* as incitant and relation of *Heterodera glycines* to disease severity. Phytopathology 79:191–197
Rundquist DC, Elmore RW, Roeth F, Eisenhauer DE, Doupnik B Jr (1984/1985) Color-infrared monitors crop changes. Farm Ranch Home 31:16–20
Rush CM, Lyda SD, Gerik TJ (1984) The relationship between time of cortical senescence and foliar symptom development of Phymatotrichum root rot of cotton. Phytopathology 74:1464–1466
Schroth MN, Hildebrand DC (1964) Influence of plant exudates on root-infecting fungi. Annu Rev Phytopathol 2:101–132
Seem RC (1984) Disease incidence and severity relationships. Annu Rev Phytopathol 22:133–150
Shepherd RL, Huck MG (1989) Progression of root-knot nematode symptoms and infection on resistant and susceptible cottons. J Nematol 21:235–241
Sherwood RT, Berg CC, Hoover MR, Zeiders KE (1983) Illusions in visual assessment of Stagonospora leaf spot of orchardgrass. Phytopathology 73:173–177
Shew HD, Lucas GB (1991) Compendium of tobacco diseases. Am Phytopathol Soc Press, St Paul
Shokes FM, Berger RD, Smith DH, Rasp JM (1987) Reliability of disease assessment procedures: a case study with late leafspot of peanut. Oleagineaux 42:245–251
Stuteville DL, Erwin DC (1990) Compendium of alfalfa diseases, 2nd edn. APS Press, St Paul
Taylor CE, Brown DJF (1981) Nematode-virus interactions. In: Zuckermann BM, Rohde RA (eds) Plant parasitic nematodes, vol III. Academic Press, New York, pp 281–301

Tennant D (1975) A test of a modified line intersect method of estimating root length. J Ecol 63:995–1001
Toler RW, Smith BD, Harlan JC (1981) Use of aerial color infrared photography to evaluate crop disease. Plant Dis 65:24–31
Tomerlin JR, Howell TA (1988) DISTRAIN: a computer program for training people to estimate disease severity on cereal leaves. Plant Dis 72:455–459
van Bruggen AHC, Jochimsen KN, Brown PR (1990) *Rhizomonas suberifaciens* gen. nov., sp. nov., the causal agent of corky root of lettuce. Int J Syst Bacteriol 40:175–188
Vivoda E, Davis RM, Nunez JJ, Guerard JP (1991) Factors affecting the development of cavity spot of carrot. Plant Dis 75:519–522
Voorhees WB, Carlson VA, Hallaver EA (1980) Root length measurement with a computer-controlled digital scanning densitometer. Agron J 72:847–851
Wallis GW, Lee YJ (1984) Detection of root disease in coastal Douglas-fir stands using large scale 70-mm aerial photography. Can J For Res 14:523–527
Watson G, Morton V, Williams R (1990) Standardization of disease assessment and product performance reporting: an industry perspective. Plant Dis 74:401–402
Wiese MV (1987) Compendium of wheat diseases, 2nd edn. APS Press, St Paul
Zadoks JC, Schein RD (1979) Epidemiology and plant disease management. Oxford Univ Press, New York
Zentmyer GA (1981) The effect of temperature on growth and pathogenesis of *Phytophthora cinnamomi* and on growth of its avocado host. Phytophathology 71:925–928

6 Temporal Aspects of the Development of Root Disease Epidemics

C.A. Gilligan

6.1 The Shapes of Disease Progress Curves

Rather like an iceberg, much of the temporal dynamics of root disease epidemics remains hidden. Beneath the surface, there is a rich mass of inter- and intraspecific interactions that determine the dynamics of infection, disease, loss of yield and the generation and carryover of inoculum. The temporal dynamics of some of these interactions are discussed in this chapter. The roles of primary and secondary infection, of root growth and of survival of inoculum as well as antagonism by other microorganisms are examined. Some of the practical constraints imposed on epidemiological analyses by the shortage of data are also considered.

The progress of epidemics over time is summarized in "disease progress" and "inoculum dynamic" curves in which the cumulative amount of disease or inoculum per unit area or volume (Y) is plotted against time. The shapes of a representative set of cumulative curves are summarized in Fig. 6.1, together with the rate curves (dY/dt). Monotonic curves for disease progress, in which there is a progressive increase in disease towards some upper, asymptotic level, have dominated work in the epidemiology of soilborne disease. Two examples are shown (Fig. 6.1A,B). In the first, infection of onions by *Sclerotium cepivorum* occurs from a large reservoir of inoculum in soil and the experimental conditions are such that there is no spread of infection between plants (Coley-Smith and Entwistle 1988). There is a lag period as roots explore soil and plants first show symptoms but thereafter the rate of increase of newly infected plants per unit time decreases progressively (Fig. 1A). Thus, as soon as the first plant in a population is infected, the epidemic begins to slow down as the density of remaining susceptible plants is decreased. Sigmoidal curves frequently result when secondary infection between infected plants or infected roots can occur as in the spread of *Phytophthora cryptogea* on *Rubus ideaus* (Washington 1988; Fig. 1B). The rate of increase of disease accelerates to a maximum and then decelerates as the amount of susceptible tissue becomes limiting.

Two analogous profiles for the survival of inoculum are shown in Fig. 1C,D, each for *Gaeumannomyces graminis*. In the first, the survival of *G. graminis* in colonized fragments of nodal wheat straw decreased exponentially in fallow soil under laboratory conditions (Scott 1969; Fig. 1C).

C.L. Campbell/D.M. Benson (Eds.)
Epidemiology and Management of Root Diseases

Although this exponential shape may be thought to characterize the survival of inoculum in soil, inspection and analysis of curves in the literature (Gilligan, unpubl.) suggest that a lag period frequently occurs before any appreciable loss in infectivity of inoculum is detected (Fig. 1D). The data for fragments of inoculum of *G. graminis* obtained by Butler (1959) under different, drier, environmental conditions illustrate this. Little thought has so far been given to the shapes of the rate curves for the survival of inoculum. Why for example should the rate of loss of inoculum accelerate and then decelerate as implied by the data in Fig. 1D? More experimental work on the dynamics of survival is needed to answer this.

The cumulative curves in Fig. 1A–D are monotonic: there are no maxima nor minima, only asymptotes, and there is no explicit allowance for cyclical epidemic behaviour. Yet epidemics of soilborne disease do exhibit periodicity, not only in perennial crops but also in monocultures of annual crops. Inoculum density, moreover, increases as new propagules are produced by parasitic or saprophytic activity and subsequently declines. Two examples of non-monotonic behaviour of inoculum are given in Fig. 1E,F. The density of propagules of *Pythium* spp. (Martin et al. 1983) under successive replantings of a susceptible crop of beans shows a single cycle as inoculum increases to a maximum and then declines to a background level in natural soil (Fig. 1E). Several cycles with periods of decreasing amplitude are exhibited in a smoothed curve from unpublished data of R.S. Smith (cited in Toussoun 1975) for the inoculum density of *Fusarium oxysporum* in a forest nursery soil.

6.2 The Dynamics of Infection and Disease

The rate of change in the average density of infected tissue in a field depends upon the interactions between host, pathogen and the biotic and abiotic environment. The biological components that affect the rates of change can be separated into parasitism of the crop plant and interspecific antagonism towards the pathogen from other soil microorganisms and mesofauna. Thus, taking $Y(t)$ as a generic variable to describe the density of infected or diseased roots, at time t:

$$\frac{\mathrm{d}Y(t)}{\mathrm{d}t} = \text{Parasitism} - \text{Interspecific antagonism}. \tag{1a}$$

The epidemic will continue to increase only as long as the net effects of parasitism exceed those of antagonism (see Chap. 7). Parasitism includes growth and production of inoculum on the host as well as host-parasite interaction, whereby the parasite affects the rate of growth of the host and hence the supply of susceptible tissue. The rate of change of inoculum density $P(t)$ depends also on saprophytism,

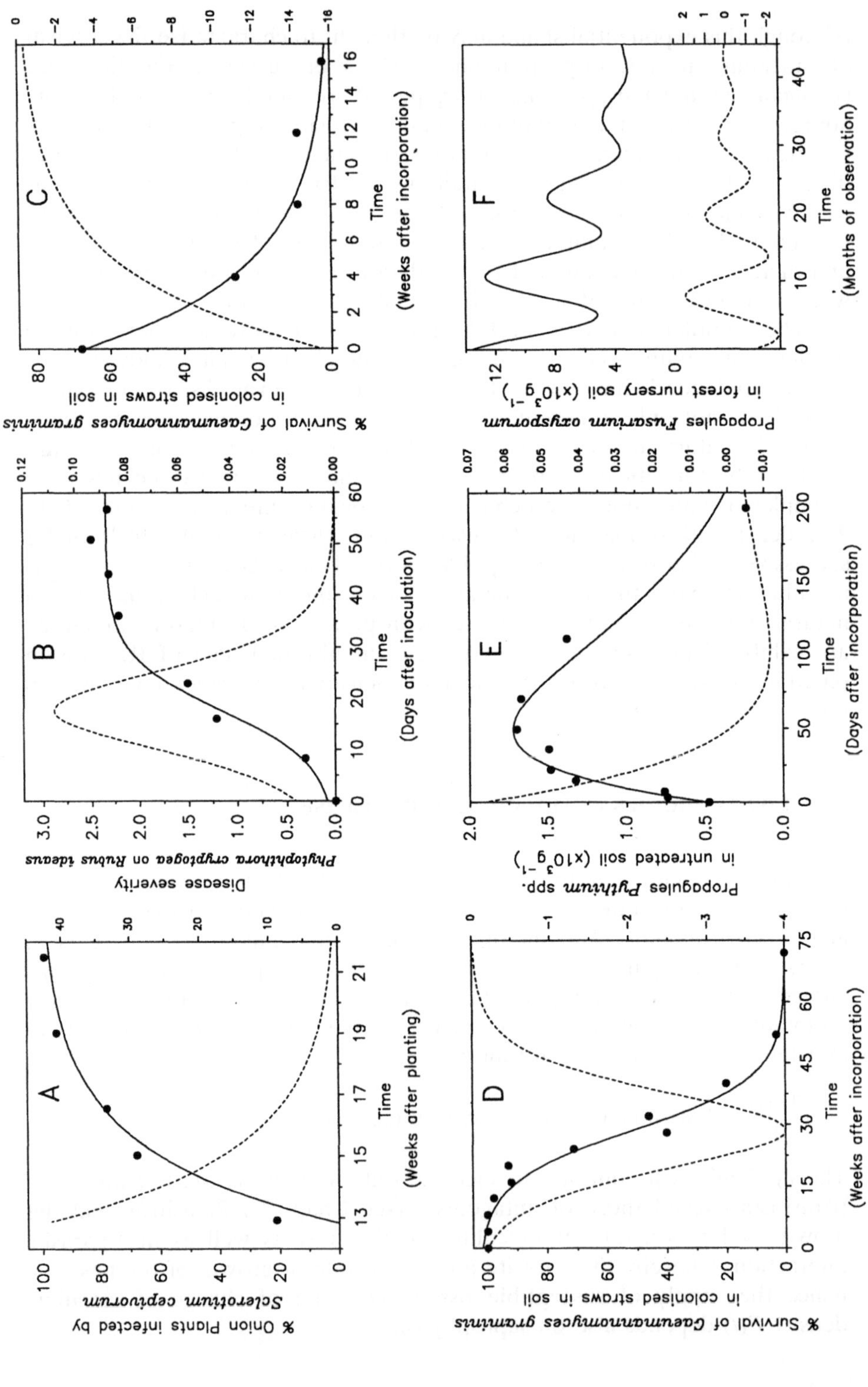

A
B
C
D
E
F
% Onion Plants infected by Sclerotium cepivorum
Time
(Weeks after planting)
Disease severity
Phytophthora cryptogea on Rubus ideaus
(Days after inoculation)
% Survival of Gaeumannomyces graminis in colonised straws in soil
(Weeks after incorporation)
Propagules Pythium spp. in untreated soil (x10^3 g^{-1})
(Days after incorporation)
Propagules Fusarium oxysporum in forest nursery soil (x10^3 g^{-1})
(Months of observation)

$$\frac{dP(t)}{dt} = \text{Parasitism} + \text{Saprophytism} - \text{Interspecific antagonism.} \qquad (1b)$$

Inoculum and hence infection and disease are seldom uniformly spread throughout a field. The rate of change in density of infected tissue depends not only on time but also on location within a field. The variable (Y) therefore depends not only on time (t) but also on location, usually measured along rows (x) and across rows (y):

$$\frac{\partial Y(t,x,y)}{\partial t} = \underbrace{\text{Parasitism} - \underset{\text{effects}}{\text{Interspecific}}}_{\text{[Temporal effects]}} + \underbrace{\underset{\text{spread}}{\text{Diffusive}} + \underset{\text{spread}}{\text{Convective}}}_{\text{[Spatial effects]}} \qquad (2)$$

Diffusion allows for spread of infection and disease from a focus by random growth of mycelium or dispersal of propagules. Convection includes transport of propagules, for example in mass flow of the soil water. All of the components in Eqs. (1) and (2) are influenced by abiotic variables, especially temperature and soil water potential. The control of plant infection and disease can be investigated by interpreting the effects of treatments on the components of Eqs. (1) and (2). Chemical control may reduce the parasitic activity of the fungus by diminishing the rate of production of secondary inoculum. Biological control is effected by changes in interspecific antagonism, for example, by competition, hyperparasitism and predation. Crop loss is encapsulated in the components for host-parasite interaction within the term for parasitism.

Most attention has so far been given to intraspecific components in the temporal dynamics of infection and disease. In the simplest form these can

Fig. 6.1A–F. Shapes of disease progress and inoculum dynamics curves, showing cumulative (—) and rate (--) curves. **A** Infection occurs from a large reservoir of inoculum in soil and the rate of increase in disease decreases progressively as the supply of susceptible hosts is exhausted, e.g. *Sclerotium cepivorum* (after Coley-Smith and Entwistle 1988, in Gilligan 1990b); **B** secondary infection can occur between hosts and the rate of increase in disease increases to a maximum and then decreases as the supply of susceptible host tissue becomes limiting, e.g. *Phytophthora cryptogea* on *Rubus ideaus* (after Washington 1988, in Gilligan 1990b); **C** inoculum decays exponentially, e.g. *Gaeumannomyces graminis* in fallow soil (after Scott 1969); **D** there is a short lag period before an accelerating followed by a decelerating decay or loss of inoculum, e.g. *G. graminis* (after Butler 1959); **E** increase followed by decrease in inoculum density, e.g. *Pythium* spp. in untreated soil (after Martin et al. 1983, in Gilligan 1990c); **F** periodicity in inoculum density, e.g. *Fusarium oxysporum* in forest nursery soil (data of R.S. Smith presented in Toussoun 1975)

be represented as the balance between the "birth" (b) and death (d) rates of infected roots. The rates define, respectively, the numbers of new infected roots per infected root per unit of time and the numbers of "removed" or dead roots per infected root per unit of time. In the absence of other effects,

$$\frac{dY}{dt} = bY - dY, \tag{3a}$$

and, if b and d are constant, then the density of infected roots grows ($b > d$) or declines ($b < d$) exponentially from an initial density (Y_0), giving,

$$Y = Y_0 \exp[(b - d)t]. \tag{3b}$$

When Y refers to the density of infected roots, than d^{-1} is an estimate of the infectious period of an infected root. While exponential decay may apply to inoculum (see Fig. 1C), exponential growth of infected tissue cannot occur for long and some form of density-dependent regulation occurs. Little is known about the behaviour of the death rate of infected roots and it has tended to be subsumed into the net reproductive rate within which the birth rate of new infections predominates. The simplest form of density-dependent regulation involves a linear decrease in the average number of healthy roots infected by each infected root as the epidemic spreads through a population of susceptible roots. Substitution for $b = b_0 - aY$, where b_0 and a are constants and letting $d = d_0$, in Eq. (3a), gives the familiar logistic equation,

$$\frac{dY}{dt} = \beta Y(1 - Y/\kappa), \tag{4}$$

in which $\kappa = (b_0 - d_0)/a$ is the carrying capacity, the upper limit to the density of infected tissue and $\beta = b_0 - d_0$ is the net reproductive rate. The rate of increase can now be seen (Turner et al. 1976) as the product of the net reproductive rate, the density of infected roots, which tends to speed epidemics up, and the proportion of susceptible roots, which is an intra-specific, density-dependent regulator that tends to slow epidemics:

$$\frac{dY}{dt} = \underset{[\text{Rate}]}{\beta} \cdot \underset{[\text{Infected roots}]}{Y} \cdot \underset{[\text{Susceptible roots}]}{(1 - Y/\kappa)}.$$

In the next sections, the effects of inoculum, root growth, latency and infectious periods on the development of epidemics are examined. Hitherto, Y appears, for notational convenience, as a generic term for infected tissue or diseased tissue. In succeeding sections, Y is restricted to refer to the *proportion* of infected or diseased tissue (as specified). The variable, N_i, is used to denote the density of infected host units and N is the density of host units. Host units may be entire plants or roots. The principal variables and parameters used in the text are summarized in the Appendix.

6.2.1 Primary Infection

6.2.1.1 Infection of Plants from a Large Reservoir of Inoculum in Soil

Some pathogens, notably the vascular wilts, exhibit little transmission between plants. New infections arise on previously healthy plants by primary infection from a reservoir of inoculum in soil. The rate of increase in infected plants per unit time depends upon the rate of contact of inoculum and roots,

$$\frac{dN_i}{dt} \propto (\text{Inoculum}) \cdot (\text{Susceptible tissue}) \qquad (5a)$$

If all uninfected plants are susceptible, the density of susceptible plants is simply given by $N - N_i$, and

$$\frac{dN_i}{dt} = r_p P(N - N_i), \qquad (5b)$$

in which r_p is the net rate of new primary infections per unit time per unit inoculum. If the reservoir of inoculum is large relative to the rate of decay of inoculum and the density of plant tissue is constant so that $N = \kappa$, the carrying capacity, Eq. (5b) can be solved analytically to yield the monomolecular model for the cumulative increase in the density of infected plants,

$$N_i = \kappa[1 - \exp(-\alpha(t - \delta))], \qquad (5c)$$

in which κ is the upper limit of the density of susceptible plants per unit area, α ($=r_pP$) is the rate of primary infection per unit time. The third parameter, δ, is a measure of the time to the occurrence of the first infection and depends on the density of infected plants at $t = 0$ (Gilligan 1990a). Epidemics that satisfy the monomolecular model have the characteristic shapes shown in Fig. 1A. The decelerating component ($N - N_i$) in Eq. (5b) dominates the dynamics of the epidemic as the progressive reduction in the density of susceptible plants slows the epidemic.

6.2.1.2 Root Growth and Decay of Inoculum

The notion of a "monomolecular" epidemic is highly simplified, even for wilt diseases in which there is no significant amount of alloinfection between plants. The monomolecular model fails to take account of the growth of roots or the decay of inoculum. Plants encounter more inoculum as the root system expands. This leads to an acceleration of the epidemic that may be counteracted by a decelerating effect of the loss of inoculum by decay or by exhaustion in infecting the host. The rate of change in the density of *infected plants* depends therefore not only on inoculum and susceptible tissue but on the density of root length per susceptible plant (Jeger 1987),

$$\frac{dN_i}{dt} \propto \text{(Inoculum)} \cdot \text{(Root length)} \cdot \text{(Susceptible tissue)}, \tag{6a}$$

hence, if the density of plants is fixed at $N = \kappa$, L is the mean root length per plant and P, as before, is the mean density of inoculum,

$$\frac{dN_i}{dt} = r_p'PL(\kappa - N_i). \tag{6b}$$

in which r_p' is the rate of primary infection, allowing for variation in L and P. Comparatively little is known about the relative dynamics of root growth and inoculum decay and Jeger (1987) suggested, with characteristic elegance, that we should concentrate on the net effect of the balance between the two. The problem then is one of selecting a simple function to summarize the product $P.L$. Suppose, for example, that an exponentially increasing length

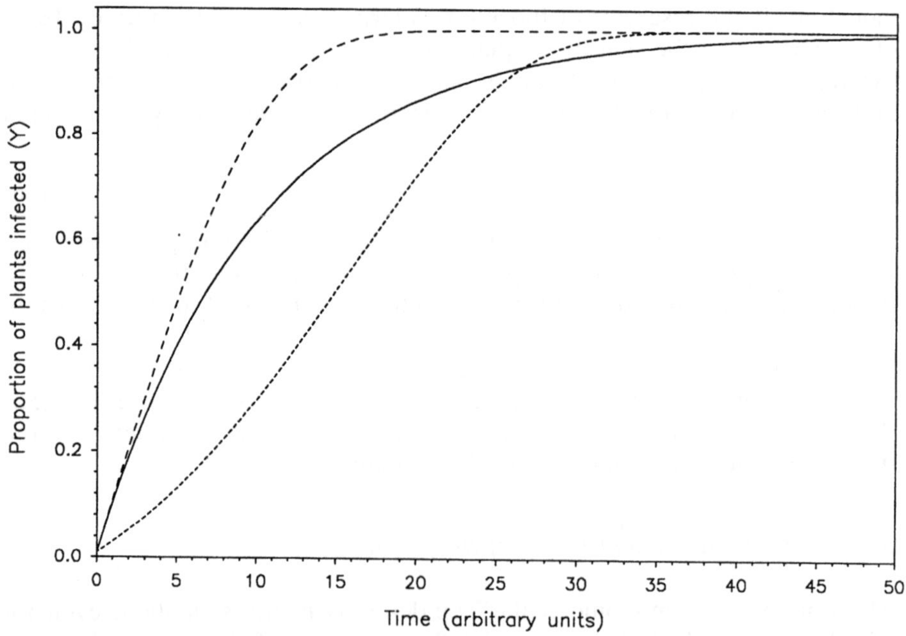

Fig. 6.2. The effects of exponentially expanding root tissue on the increase in the proportion of plants infected (Y) when infection is restricted to soil inoculum with no secondary infection. The data were generated from the model $dY/dt = r'_p(a \exp bt)(\kappa - Y)$, with solution $Y = \kappa\{1 - [(\kappa - Y_0)/\kappa]\exp[(r'_p a/b)(1 - \exp(bt))]\}$ after Jeger (1987). Three combinations of parameter values are shown $a = 1.0$, $b = 0$ (—) (equivalent to the monomolecular model), $a = 1.0$, $b = 0.1$ (--) and $a = 0.2$, $b = 0.1$ (--) with default parameters $\kappa = 1$, $r'_p = 0.1$ and $Y_0 = 0.01$

of root tissue is of over-riding importance, and that the density of inoculum is more or less constant, then,

$$\frac{dN_i}{dt} = r_p'(a\exp bt)(\kappa - N_i). \tag{7a}$$

This can be solved to give the cumulative density of infected plants,

$$N_i = \kappa\left\{1 - \left[\frac{(\kappa - N_i(0))}{\kappa}\right]\exp[(r_p'a/b)(1 - \exp(bt))]\right\}, \tag{7b}$$

in which $N_i(0)$ is the density at time $t = 0$. Three combinations of parameter values for a and b are shown in Fig. 6.2, including the default value of $b = 0$ (in Eq. 7a), equivalent to a net balancing effect of root growth and inoculum decay. A positive point of inflexion is now possible (see especially $a = 0.2$, $b = 0.1$) so that even for a system in which there is no secondary infection, the rate of an epidemic may first accelerate and then decelerate. We conclude therefore that the apparent simplicity of the dynamics of an epidemic in which all new infections are derived from soil inoculum is beguiling. Even when an epidemic appears to fit the monomolecular model, it may conceal a complicated balancing of the relative dynamics of root growth and inoculum decay. For further discussion of models of primary infection, including the role of the rhizosphere, see Jeger (1987) and Gilligan (1985a, 1990c).

6.2.2 Secondary Infection

6.2.2.1 Root-to-Root Spread

Transmission of infection by root-to-root spread between plants is arguably the principal mechanism by which healthy plants become infected by soil-borne pathogens. Thus, for example, zoospores of *Pythium* and *Phytophthora* spread between plants; mycelia of *Rhizoctonia* and *Gaeumannomyces* grow between plants; conidia of *Fusarium* are released from decaying roots and contact other susceptible roots. The rate of increase in infected roots depends upon the accelerating effect of the increasing amount of infected tissue and the decelerating component of the declining amount of susceptible tissue,

$$\frac{dN_i}{dt} \propto (\text{Infected tissue}) \cdot (\text{Susceptible tissue}). \tag{8a}$$

The logistic model and its close variants (Waggoner 1986), such as the Gompertz, Weibull and Richards' functions (Madden 1980; Madden and Campbell 1990), dominate the description of this form of epidemic, in which secondary infection predominates, thus,

$$\frac{dN_i}{dt} = r_s N_i \left(\frac{\kappa - N_i}{\kappa}\right), \quad \text{and} \quad N_i = \frac{\kappa}{\{1 + \exp[-r_s(t - \delta)]\}}, \tag{8b,c}$$

providing that the upper limit of root density is assumed to be constant ($N = \kappa$). The rate of increase in disease is negligibly slow at first. It speeds up towards a maximum, as more tissue becomes infectious and then subsides as the lack of susceptible tissue upon which new infections can arise slows the epidemic. The process is illustrated for *P. cryptogea* on *Rubus ideaus* in Fig. 1B.

The logistic model endures because of its robustness. It can be fitted to a large variety of data sets (Waggoner 1986; Gilligan 1990b). The model has only three parameters, each with biological meaning: r_s is the rate of secondary infection, κ is a measure of the upper limit of infected (or diseased tissue if disease rather than infection is being analysed) and δ is a measure of the delay or time of onset of the epidemic. Practical use of the model to interpret the epidemiological effects of treatments for the control of disease in designed experiments is described in Section 6.3.

Epidemics are clearly more complex than the logistic model or its analogues suggest. No explicit account is taken of inoculum. Primary infection is relegated to a subservient role of the carryover of inoculum from the previous crop to initiate the first infections. Thereafter, infection of roots from soil inoculum is assumed to be of negligible importance relative to the dominating influence of secondary, root-to-root spread. Infected roots are assumed to become immediately infectious and to remain so. The production and growth of roots and the effects of disease on the root population are all ignored.

6.2.3 Linking Primary and Secondary Infection with Allowance for Root and Inoculum Dynamics

6.2.3.1 Growth Equations

For many soilborne pathogens, the reservoir of inoculum will not simply serve as the "pump-primer" for infection, after which secondary infection takes over. Primary infection may continue alongside secondary infection, thus,

$$\frac{dN_i}{dt} \propto \underset{\text{[Primary infection]}}{(\text{Inoculum}) \cdot (\text{Susceptible tissue})} + \underset{\text{[Secondary infection]}}{(\text{Infected tissue}) \cdot (\text{Susceptible tissue})} \tag{9a}$$

One possible formalization of this (Brassett and Gilligan 1988) is

$$\frac{dN_i}{dt} = r_p P(N - N_i) + r_s N(N - N_i), \tag{9b}$$

in which r_p is the rate of primary infection per unit inoculum, as before and r_s is the rate of secondary infection. Equation (9b) can be solved analytically if inoculum and total root density remain constant (Brassett and Gilligan 1988). Letting $\alpha = r_p P$ and $\beta = r_s N$ and reparameterizing for the proportion of infected roots, $Y = N_i/N$ (see Gilligan 1990a for details), with allowance for a flexible carrying capacity, ($\kappa \leqslant 1$), yields

$$\frac{dY}{dt} = (\alpha + \beta Y)(\kappa - Y) \qquad \text{and if } Y(0) = 0,$$

$$Y = \frac{\alpha\kappa\{1 - \exp[-(\alpha + \beta\kappa)t]\}}{\alpha + \beta\kappa \exp[-(\alpha + \beta\kappa)t]}. \tag{9c,d}$$

There are just three parameters, all of which have some intuitive biological meaning: κ is the familiar carrying capacity; α is a coefficient of primary infection that incorporates inoculum (P) and the efficiency of inoculum in causing disease (r_p); β is a coefficient of secondary infection that incorporates host density (N) and the efficiency of transmission of infection between roots (r_s). The model can accommodate a range of shapes from a rising exponential to an asymmetrical sigmoid shape (cf. Fig. 1A,B). Brassett and Gilligan (1988) used the model to analyse epidemics of *Pythium ultimum* in populations of *Lepidium sativum*.

The model in Eqs. (9c) and (9d) approaches reality by allowing for dual sources of contemporary infection by root-to-root spread and from the reservoir of inoculum in the soil. Yet it falls short of reality by ignoring the increase in density of roots and loss of infectivity of inoculum. Fortunately, allowance for dynamically varying host and inoculum densities is conceptually easy (Gilligan 1990a) by combining the equation for the change in density of infected roots with equations for the change in total root density and inoculum density:

$$\text{Infected roots} \quad \frac{dN_i}{dt} = (r_p P + r_s N_i)(N - N_i); \quad N_i(t = 0) = N_{i0}, \tag{10a}$$

$$\text{Total roots} \quad \frac{dN}{dt} = r_N N\left(1 - \frac{N}{\kappa_N}\right); \quad N(t = 0) = N_0, \tag{10b}$$

$$\text{Inoculum} \quad \frac{dP}{dt} = r_d P; \quad P(t = 0) = P_0, \tag{10c}$$

in which r_N is the rate of production of new roots and r_d is the rate of decay of inoculum; κ_N is the upper limit to root density of a plant; N_0, N_{i0} and P_0 are the initial densities of roots, infected roots and inoculum. The equations for root density and inoculum density are solved analytically to yield,

$$N = \frac{\kappa_N}{\left[1 + \frac{\kappa_N - N_0}{N_0}\exp(-r_N t)\right]}, \tag{10d}$$

$$P = P_0 \exp(-r_n t). \tag{10e}$$

Thus the density of inoculum is assumed to decline exponentially (cf. Fig. 1C) and the density of roots to change logistically. The density of infected roots [Eq. (10a)] is obtained by numerical integration. The parameters of the model [Eqs. (10a–c)] suggest alternative epidemiological methods for the control of disease (Gilligan 1990a). These are:

1. Reduction of inoculum (P) by,
   a) "removal" of initial inoculum (P_0),
   b) increase in the rate of decay of inoculum (r_d);
2. Reduction of the rate of primary infection (r_p);
3. Reduction of the rate of secondary infection (r_s);
4. Alteration of the host density (N) by,
   a) change in initial host density (N_0),
   b) change in asymptotic root density (κ_N),
   c) change in the rate of production of roots (r_N).

For a fuller discussion and graphic illustration of the use of the model to compare alternative methods of control, see Gilligan (1990a). The role of host growth is discussed by Huisman (1982) and elsewhere in this volume.

A little reflection on Eq. (10b) again reveals a digression from reality. The density of roots certainly affects the dynamics of infection and disease: N occurs in Eq. (10c) for the change in infected root density. But no allowance is made for the effects of disease on the rate of appearance of new healthy roots. While many pathologists would assent to an interspecific effect, in which the presence of infection, and more importantly disease, affects root growth, little is known about such relationships. The relationship can be simply formulated as,

$$\frac{dN}{dt} = f(N) - g(N_i), \tag{11}$$

in which $f(N)$ is a measure of the intrinsic increase in root density. It represents the growth of the root population in the absence of infection and disease. It can be easily determined in the absence of the pathogen and will frequently be sigmoidal or exponential towards an asymptotic carrying capacity. The second term in Eq. (11), $g(N_i)$, is a measure of the effect of infected roots on the absolute rate of increase. But how will the effect change with infection? What shape will it have? While pathologists have made little progress in the quantification of these effects, the relationship has been widely studied in entomology, where it is known as the "functional

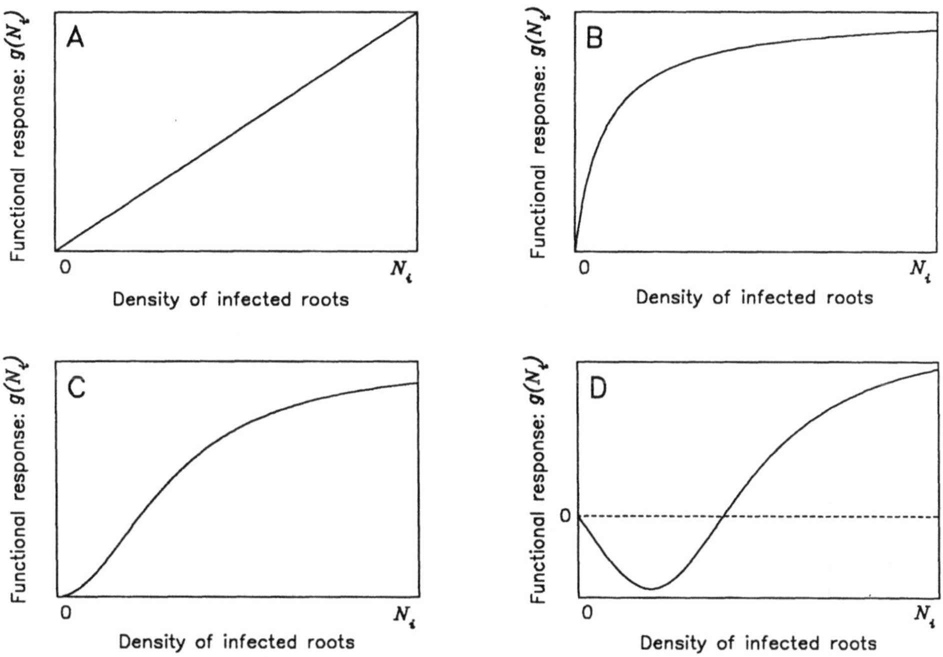

Fig. 6.3A–D. Functional responses ($g(N_i)$) for the effect of increasing density of infected roots (N_i) on the dynamics of host growth, $dN/dt = f(N) - g(N_i)$. **A** Linear inhibitory response, $g(N_i) = AN_i$; **B** exponentially diminishing response with an asymptote, $g(N_i) = (AN_i)/(B + N_i)$; **C** sigmoidally diminishing response towards an asymptote, $g(N_i) = (AN_i^2)/(B + N_i^2)$; **D** stimulatory response followed by inhibitory response towards an asymptote, $g(N_i) = A - B/[C + (N_i - D)^2]$

response" (Hassell 1981). The simplest assumption is that the relationship between $g(N_i)$ and N_i is rectilinear (Fig. 6.3A) but this implies that each additional infected root depresses the rate of production of new root tissue by an equivalent amount. It is perhaps more likely that the marginal effect of additional infected roots declines progressively after the first infection (Fig. 6.3B) or reaches a maximum and thereafter declines (Fig. 6.3C). A further possibility is suggested by work with *G. graminis*, namely that initial low levels of infection may stimulate the "over-production" of healthy root tissue so that marginally more roots are produced than are infected (Skou 1975). This may be reflected by the shape shown in (Fig. 6.3D), in which $g(N_i)$ is negative at low N_i, leading to a stimulation of growth but reverts to an inhibitory effect at higher levels of infection. Stimulation of yield at low levels of disease or infection followed by inhibition at higher levels is commonly regarded to occur in the field, although it has been little studied and still less quantified.

The equations for the functional responses are given in the legend to Fig. 6.3. I have used a rational polynomial function $g(N_i) = A - B/(C +$

$(N_i - D)^2)$, merely for consistency with the other forms. Further work in this area may lead to rather different functions. It should also be noted that it could be argued that Eq. (11) should be rendered as $dN/dt = f(N) - Ng(N_i)$ in which the inhibitory effect is represented by the product involving total root density and infected root density.

6.2.3.2 Computer Simulation: An Alternative Approach

The complexity of temporal models increases as more biological processes are introduced. More parameters are needed to describe the growth of host as well as pathogen populations [cf. Eq. (10)]. The parameters themselves are likely to vary with environmental conditions. Thus the net intrinsic rate of secondary infection, r_s, is unlikely to be constant throughout the season but to vary with temperature and soil water potential. Expressing the biological parameters as functions of environmental variables (e.g. $r_s = \gamma_0 + \gamma_1 T + \gamma_2 T^2$, where T is soil temperature) multiplies still further the numbers of parameters in the model. These conditions can be added to the growth equations discussed above. An alternative starting point that encourages greater biological complexity in the model is that of computer simulation. The distinction between computer simulation and growth equations is blurred (Hau et al. 1985; Campbell et al. 1988). It is possible to postulate a set of growth equations for a set of biological variables (X_i), such as diseased roots, healthy roots and inoculum, that incorporate environmental factors:

$$\frac{dX_i}{dt} = \left[f_i(X_i) - \sum_i \sum_j g_{ij}(X_i X_j) \mid \text{Environmental conditions} \right].$$

Simulators, however, tend to be focused on modelling of subprocesses such as the production of first, second or third order roots and lesion growth or sporulation. The subprocesses are then combined in a larger program. Amongst the notable simulators of soilborne disease are those of Bloomberg (1979) who simulated the infection of Douglas fir seedlings by *Fusarium oxysporum* and Reynolds et al. (1986) on the infection of Fraser fir by *P. cinnamomi*. Other simulation models by Guttierez et al. (1983) and Sanders and Sheikh (1983) are discussed elsewhere (Gilligan 1985b; Guttierez and DeVay 1986). The simulators by Bloomberg (1979) and Reynolds et al. (1986) model the dynamics of host, inoculum and disease in relation to environmental conditions and both incorporate stochastic components for host growth. Hence repeated runs of the simulator yield different results, even for the same initial and environmental conditions because of chance differences in the growth of the hosts and in the probability of successful infection. Bloomberg (1979) simulated the infection of up to 180 seedlings of Douglas fir by *F. oxysporum* for a period of 180 days. The model is driven by root growth which depends upon thermal time. It incorporates primary infection as invasion of growing root tips. Secondary infection is restricted to

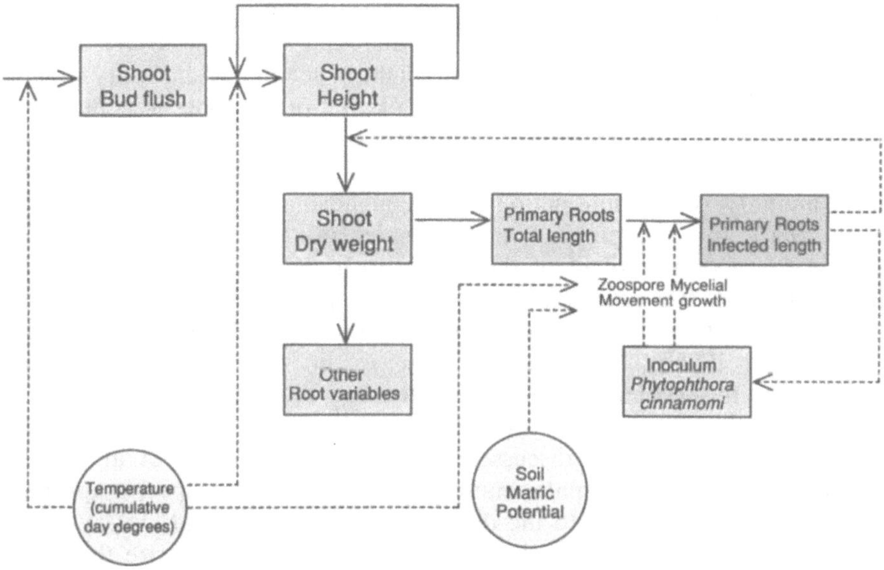

Fig. 6.4. Schematic representation of the simulation model of Reynolds et al. (1986), for the infection of seedlings of Fraser fir by *Phytophthora cinnamomi*. The model is centred on the partitioning of assimilates from shoot growth of the host seedlings and is driven by soil temperatures and water potential. Shoot growth is initiated at "bud flush", after satisfaction of a threshold for thermal time. Temperature also drives the rate of shoot elongation, together with the photosynthetic capacity of the shoot (estimated by shoot height). Shoot height determines the shoot dry weight which affects several root variables, including the total length of primary roots. Roots become infected by mycelial growth across soil between roots and by zoospore movement between roots under saturated soil conditions. The amount of infection on primary roots limits the rate of accumulation of shoot dry weight

lesion growth because cross-infection between seedlings is not thought to occur. The rates of primary and secondary infection also depend upon temperature (Bloomberg 1979).

Three-dimensional growth of roots in soil is also central to the model of Reynolds et al. (1986; Fig. 6.4). Root production of seedlings of Fraser fir depends upon the dry weight of the shoot. This is stimulated by elongation of the shoot height, which is a measure of photosynthetic activity, and is inhibited by infection of the roots by *P. cinnamomi*. Initiation of shoot growth occurs after "bud flush". Growth of the shoot and partitioning of reserves to the root are controlled by thermal time (cumulative and daily increments in day degrees). Infection also depends on soil temperature and its variation with depth as well as soil matric potential. Primary infection occurs from chlamydospores on previously infected root fragments in soil. Autoinfection occurs by lesion growth. Alloinfection between roots on the same or different plants occurs by mycelial spread or by zoospore movement, depending on the soil matric conditions. The model has approximately 20 parameters. It requires four input environmental variables. There are five principal output variables for the host; height of seedlings, shoot dry weight, length of tap root, length and density of first-order root branches. The principal output variable for the pathogen is the proportion of total root length colonized by *P. cinnamomi*. The model, therefore, comprises a complex summary of the interactions between host and pathogen (Reynolds et al. 1986). In common with most complex models, however, validation (testing the goodness of fit of the model to independently collected data) proved difficult. It is usually possible to adjust parameters in order to obtain a satisfactory fit to data but this does not truly validate the model (Gilligan 1985b; Jeger 1986b). In particular, there are no clearly defined statistical criteria for the estimation of parameters in these models, whether in pathology or in ecology more generally. Modelling of epidemics of soilborne diseases by simulators is discussed in detail by Campbell et al. (1988). Readers interested in the mathematical, rather than the computing approach to simulation are directed to the discussion by Guttierez and DeVay (1986) of the model of Guttierez et al. (1983) in which they summarize the use of the von Foerster equations to incorporate plant age in the dynamics of the interaction between host and pathogen. Jeger (1986b) also discusses the relative merits of analytical compared with simulation models for plant disease.

6.2.4 The Death of Roots, Latent and Infectious Periods

Little is known about the death of roots. Still less is known about the effects of infection and disease on the death rate, nor indeed on the rates of production of new roots on a diseased plant. Moreover, it is assumed in the models above that, once they are infected, roots are immediately infectious and remain so indefinitely. The dynamics of root growth is complicated by the branching patterns of the root system within plants. Nevertheless, it is technically possible to quantify the rates of production (birth) and death of root members on a plant in the absence and presence of disease. Quantification of the latent and infectious periods is more problematic, since these concepts are rather nebulous in relation to root disease. Strictly, the latent period is the time from the initiation of infection to the production of new propagules, such as the macroconidia of *Fusarium* or zoospores of *Phytophthora*. More conveniently, the latent period is tied to the time of initial contact between inoculum and roots rather than infection since this may be difficult to measure non-destructively. For pathogens with mycelial spread, such as *G. graminis* or *R. solani*, the criterion of the production of secondary propagules may be interpreted arguably as the mean time to grow between roots. The infectious period is the time during which the production of new propagules occurs from an infected, living root. Strictly, the pro-

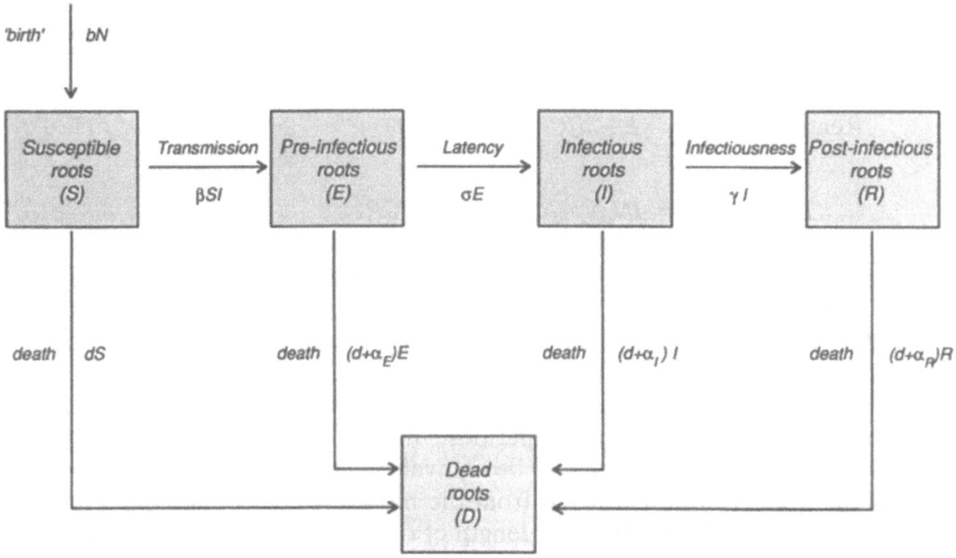

Fig. 6.5. Schematic representation of a compartmental model for the infection of a population of roots by a pathogen. As susceptible roots become infected, they pass through a number of compartments, each representing a different epidemiological status. Roots die through senescence and by the enhanced effects of infection

duction and release of propagules from a dead root are part of the soil inoculum but the distinction is probably of more theoretical than practical relevance. The incubation period is the time between initiation of infection and the appearance of the first symptoms of disease.

The progression of roots through a series of compartments for "susceptible" (S), "exposed" (infected but not infectious, E), "infectious" (I), "removed" (post-infectious, R) and "dead" (D) roots is shown in Fig. 6.5, together with the rates of gain and loss from each compartment. The symbols S, E, I, R are chosen for consistency with medical epidemiology (Hethcote 1989). Compartmental analysis has been widely used in medical epidemiology (Bailey 1975; Anderson 1982; Hethcote 1989) and most particularly in the work of Anderson and May (1979). Jeger (1984, 1986a) has also discussed the use of compartmental models for aerial botanical epidemics. The density of living roots (N) is given by $N = S + E + I + R$, that of infected roots is given by $N_{\mathrm{i}} = E + I + R$, and the equations for the dynamics of each compartment by,

$$\text{Susceptible} \quad \frac{\mathrm{d}S}{\mathrm{d}t} = bN - dS - \beta SI, \qquad (12a)$$

$$\text{Exposed} \quad \frac{\mathrm{d}E}{\mathrm{d}t} = \beta SI - dE - \alpha_{\mathrm{E}}E - \sigma E, \qquad (12b)$$

$$\text{Infected} \quad \frac{dI}{dt} = \sigma E - dI - \alpha_I I - \gamma I, \tag{12c}$$

$$\text{Removed} \quad \frac{dR}{dt} = \gamma I - dR - \alpha_R R, \tag{12d}$$

$$\text{Dead} \quad \frac{dD}{dt} = dN + \alpha_E E + \alpha_I I + \alpha_R R, \tag{12e}$$

The model given in Eq. (12) is known as an SEIR model with vital dynamics, because it incorporates the birth and death of roots. The parameters b and d are the per capita birth and death rates; α_E, α_I and α_R are the additional rates of death of each class of infected root induced by the parasite; β is the rate of transmission of disease, σ and γ are, respectively, the rates of passage through the latent and infectious periods. In practice, the additional rates of death of roots may be assumed to be equivalent, $\alpha = \alpha_E = \alpha_I = \alpha_R$. Several simple features can be extracted from the model: $1/d$ is the life expectancy of a healthy root; $1/\sigma$ is the mean length of the latent period; $1/\gamma$ is the mean length of the infectious period. Hence the expected lifespan of an infected root is $1/(d + \alpha + \sigma + \gamma)$ days. A dimensionless variable λ, sometimes known as the parasite's basic reproductive rate (Anderson and May 1981), may now be defined as the mean number of secondary infections per infectious individual. It is computed by,

$$\lambda = \underbrace{\{\beta S\}}_{\text{Rate of new infections}} \cdot \underbrace{\left\{\frac{1}{d + \alpha + \sigma + \gamma}\right\}}_{\text{Expected lifespan of an infectious individual}} \cdot \underbrace{\left\{\frac{\sigma}{d + \alpha + \sigma}\right\}}_{\text{Proportion that survive the latent period to become a second generation of infectious individuals}} \tag{13a}$$

S is used rather than N to denote universal susceptibility at the beginning of the epidemic, something that requires further thought and investigation. The epidemic will fail to establish if $\lambda < 1$, i.e. if,

$$\beta\sigma S < (d + \alpha + \sigma + \gamma)(d + \alpha + \sigma). \tag{13b}$$

Alternatively, by setting $\lambda = 1$, an expression can be derived for a threshold density of susceptible hosts (S_T) below which an epidemic cannot develop,

$$S_T = (d + \alpha + \sigma + \gamma)(d + \alpha + \sigma)/\beta\sigma. \tag{13c}$$

For many root diseases, the life expectancy of an infected organ is long relative to latent and infectious periods so $\sigma, \gamma \gg d$, whence d may be omitted from the expressions in the computation of relative orders of magnitudes for the components of λ. The expressions and conditions in Eqs. (13) are analogous to the empirical relationship proposed for a simpler system by Van der Plank (1963) in which he equated the "progeny-parent ratio" (cf. λ) to the product of the epidemic rate (r) and the infectious period (γ^{-1}) and stated the criterion ($r\gamma^{-1} > 1$) for an epidemic to proceed. For a fuller discussion of this early work, see Jeger (1982, 1986a) and Campbell and Madden (1990). Even with five parameters and three non-zero initial conditions, the models in Eq. (12) are still rather naïve. There is no density dependence for root growth so that in the absence of infection, root density would increase ($b > d$) or decrease ($b < d$) exponentially. More complicated behaviour is likely to apply to the effects of infection (and disease) on the birth and death rates. Parameters will not remain constant but will vary with environmental conditions, notably soil temperature and moisture status. The incorporation of environmental variation into epidemiological models is discussed by Hau et al. (1985) and Hau (1990). Much still needs to be done both experimentally and theoretically in these areas.

6.2.5 Antagonistic Interactions

6.2.5.1 Curves that Rise and Fall

Antagonistic interactions between pathogens and other microorganisms or the soil fauna include competition, parasitism, predation, antibiosis and lysis (Baker and Cook 1974). Considerable experimental work has been done on these forms of interaction, because of their importance in biological control of pathogens. Data sets for the temporal dynamics of pathogens and biocontrol agents, however, are unfortunately rather scarce. Nevertheless, there are sufficient data to show that the progress curves for disease or inoculum density are non-monotonic, in which there is an increase followed by a decrease in the dependent variable over time. See, for example, the dynamics of inoculum of *Pythium* spp. in natural field soil in which there is natural antagonism (Fig. 6.1E; Martin et al. 1983). Consideration of this shape, together with some elementary biological assumptions, suggests a simple starting point for modelling antagonism (Gilligan 1990c). The models are based upon a second-order, inhomogeneous differential equation with constant coefficients. Reverting to Y as a generic term for density of inoculum, of infected tissue or more generally fungal biomass, the equations are,

$$\frac{d^2Y}{dt^2} + 2\lambda\frac{dY}{dt} + \mu Y = \nu. \qquad (14)$$

There are three forms of solutions to this model that depend upon the relative magnitudes of λ and μ. These are for ($\lambda > 0$):

the over-damped model,

$$\lambda^2 > \mu \qquad Y = A + B\exp(\alpha_1 t) + C\exp(\alpha_2 t), \qquad (15a)$$

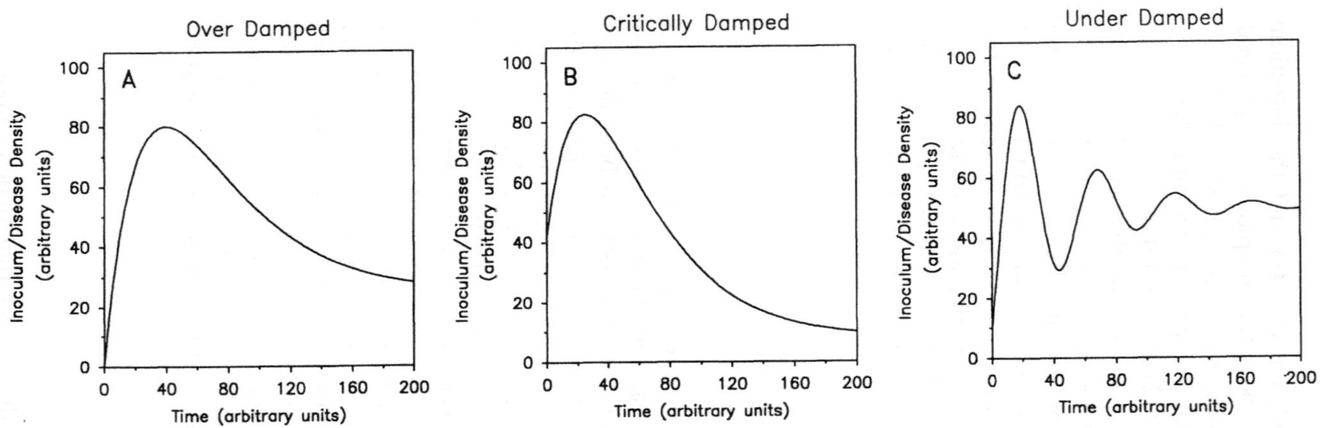

Fig. 6.6A–C. Illustration of the shapes of: **A** the over-damped, $Y = A + B\exp(\alpha_1 t) + C\exp(\alpha_2 t)$; **B** the critically damped, $Y = A + (B + Ct)\exp(-\lambda t)$; **C** the under-damped, $Y = A + (B\,\mathrm{Cos}\,\beta t + C\,\mathrm{Sin}\,\beta t)\exp(-\lambda t)$ curves for the relationships between inoculum disease or infection density with time

in which α_1 and α_2 are both negative and the signs of B and C differ, so the population density (Y) first rises and then falls towards an equilibrium density (A);

the critically damped model,

$$\lambda^2 = \mu \qquad Y = A + (B + Ct)\exp(-\lambda t), \tag{15b}$$

the under-damped model,

$$\lambda^2 < \mu \qquad Y = A + (B \operatorname{Cos} \beta t + C \operatorname{Sin} \beta t)\exp(-\lambda t), \tag{15c}$$

in which the population density rises and falls in a series of decaying oscillations about A.

The shapes of the models are illustrated in Fig. 6.6. The critically damped model (Fig. 6.6B) is best regarded as a special case of the over-damped model (Fig. 6.6A). Both have similar shapes, with a rise in inoculum density to a maximum followed by a "relentless" reduction towards some asymptotic, equilibrium density. The under-damped model (Fig. 6.6C), with its decaying oscillations, describes a more complex process in which the pathogen density may fluctuate about the asymptotic level. This summarizes the behaviour of the inoculum density of *F. oxysporum* in Fig. 6.2F, except that the mean level about which fluctuations occurred also declined with time.

The models contain up to five parameters. The parameters α_1, α_2 and λ are rate parameters; β is a measure of the periodicity of the change in inoculum density with time; A is an estimate of the equilibrium inoculum density (Y_∞). The remaining parameters B and C are "locational" parameters that depend upon the initial and boundary conditions. Thus for the over-damped model [Eq. (15a)]:

$$\text{Equilibrium density} \quad Y_\infty = A, \tag{16a}$$

$$\text{Initial density} \quad Y_0 = Y_\infty + B + C, \tag{16b}$$

$$\text{Initial absolute rate of increase} \quad Y'_0 = \alpha_1 B + \alpha_2 C, \tag{16c}$$

and

$$B = \frac{\alpha_2(Y_0 - Y_\infty) - Y'_0}{\alpha_2 - \alpha_1}, \tag{16d}$$

$$C = \frac{Y'_0 - \alpha_1(Y_0 - Y_\infty)}{\alpha_2 - \alpha_1}. \tag{16e}$$

Hence the initial density of inoculum, the equilibrium density and the initial absolute rate of increase in density all contribute to the behaviour of pathogen biomass. Consideration of the effects of changing these variables, for example by incorporation of antagonistic populations of microorganisms, allows some insight into the sensitivity and effectiveness of biological control.

To examine this, the effects of changing the parameters [Eqs. (16a–c)] on the temporal profiles of pathogen inoculum are shown in Fig. 6.7. The responses depend upon the default values selected for the non-varying parameters. Nevertheless, it may be seen that reduction in initial inoculum density (Fig. 6.7A) and of the initial absolute rate of increase in density (Fig. 6.7B) both led to marked reductions in the maximum density of inoculum as well as the area under the progress curve. Reduction of the equilibrium density had little effect on the maximum but did affect the area under the curve. Hence these curves provide a starting point for the analysis of biological control of inoculum and disease. For example, it may be important to assess the maximum density of inoculum or the time taken to achieve the maximum, t_{max},

$$t_{max} = \frac{1}{(\alpha_2 - \alpha_1)} \ln \left| \frac{\alpha_1 B}{\alpha_2 C} \right| . \tag{17}$$

Some examples of the effects of changing the rate parameters, the initial density (Y_0), initial rate of increase (Y_0) and ultimate density (Y_∞) of inoculum on t_{max} are illustrated in Fig. 6.8. The time to achieve maximum density increases as α_1 and α_2 approach zero (Fig. 6.8A); t_{max} of course does not exist when the rates are equal [see Eq. (17)]. Changes in the initial rate of increase have a marked effect on t_{max} below a threshold value of approximately 15 units of inoculum per unit time. The effect in delaying t_{max} is considerably enhanced when the initial amounts of inoculum are small (Fig. 6.8B). The effect of Y_∞ on t_{max} is comparatively small (Fig. 6.8C) when the initial density exceeds the asymptotic level. This effect is also evident in Fig. 6.7C. The effect increases as $Y_0 \ll Y_\infty$ but, overall, the sensitivity of the time at which maximum disease occurs is comparatively insensitive to the natural equilibrium level (Y_∞). Much more can still be done on these analyses but first it is essential that the mechanistic bases of these models is understood.

Although the under-damped model has not yet been fitted to data, the other two have (Gilligan 1990c). The over-damped model was used to describe the population dynamics of bacterial colonization of live- and heat-killed conidia of *Cochliobolus sativus* as well as to the population dynamics of *Laetisaria arvalis* in untreated and in pasteurized soil with and without subsequent reinfestation with *Pythium ultimum*. Jeger and Starr (1985) also fitted the model to the survival of second-stage juveniles of *Meloidogyne* spp. in soil. The critically damped model has been used to summarize pre-emergence damping off of *R. solani* in successive crops of radish, the population dynamics of *Pythium* spp. in untreated soil and pasteurized soil with added inoculum of *L. arvalis*, the population dynamics of *Trichoderma viride* in plant growth media, of *Macrophomina phaseolina* in soil and the population dynamics of *P. ultimum* with and without antagonism from *P. nunn* (Gilligan 1990c).

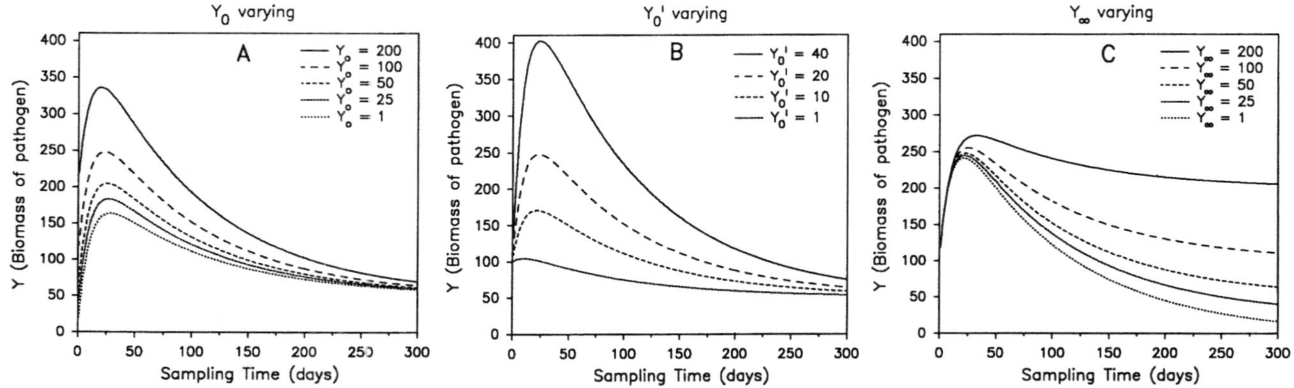

Fig. 6.7A–C. Effects of varying: **A** the initial density (Y_0); **B** the initial absolute rate of increase (Y'_0); **C** the asymptotic level (Y_∞) of pathogen biomass, on the over-damped model, of the form, $Y = A + B\,\mathrm{Exp}(\alpha_1 t) + C\,\mathrm{Exp}(\alpha_2 t)$ in which $A = Y_\infty$, $B = [\alpha_2(Y_0 - Y_\infty) - Y'_0]/(\alpha_2 - \alpha_1)$ and $C = [Y'_0 - \alpha_1(Y_0 - Y_\infty)]/(\alpha_2 - \alpha_1)$. Default values for parameters were $Y_0 = 100$, $Y'_0 = 20$, $Y_\infty = 50$, $\alpha_1 = -0.1$ and $\alpha_2 = -0.01$. (After Gilligan 1990c)

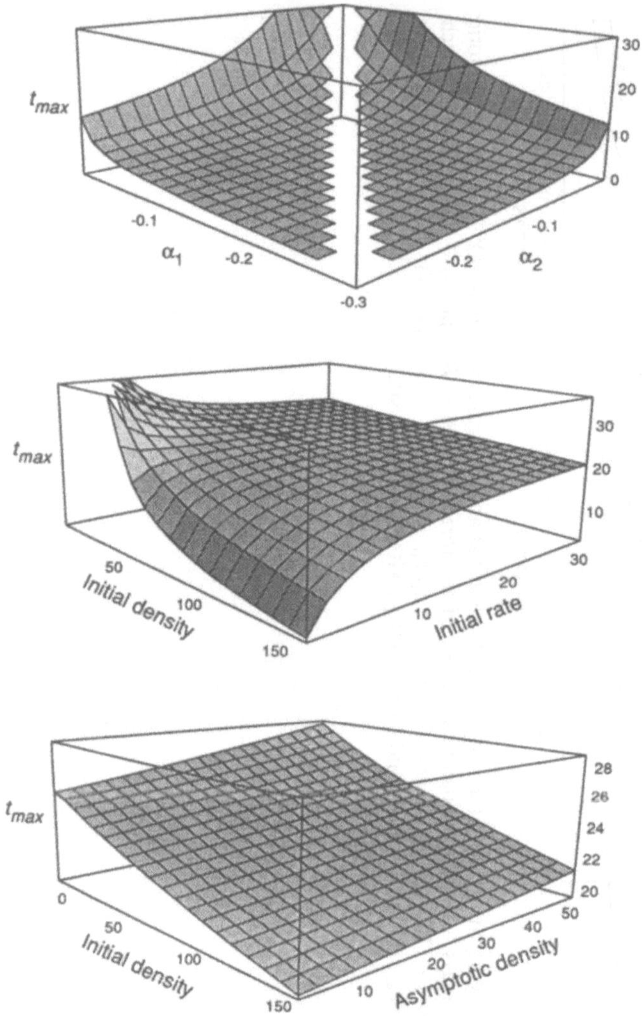

Fig. 6.8. Effects of varying combinations of the parameters of the over-damped model on the time to achieve the maximum amount of microbial biomass given by $t_{max} = \ln|(-\alpha_1 B)/(\alpha_2 C)|/(\alpha_2 - \alpha_1)$ in which $B = [\alpha_2(Y_0 - Y_\infty) - Y'_0]/(\alpha_2 - \alpha_1)$ and $C = [Y'_0 - \alpha_1(Y_0 - Y_\infty)]/(\alpha_2 - \alpha_1)$. *Above* Joint variation in α_1 and α_2; *centre* joint variation in initial inoculum (Y_0) and initial absolute rate of increase (Y'_0); *below* joint variation in initial inoculum and the asymptotic level (Y_∞). Default values for parameters were $Y_0 = 100$, $Y'_0 = 20$, $Y_\infty = 50$, $\alpha_1 = -0.1$ and $\alpha_2 = -0.01$. (After Gilligan 1990c)

6.2.5.2 *Mechanistic Analyses of Antagonism*

The models in Eqs. (15) are merely descriptive. They mimic the shapes of observed (Figs. 6.1E, 6.6A,B) or putative (Fig. 6.6C) profiles of the change in inoculum density or disease over time. Although the models do incorporate biologically meaningful components such as initial density, they do not, as they stand, represent epidemiological mechanisms for the interaction of pathogen and antagonist. This requires specification of the differential equations for the change in density of the pathogen ($\mathrm{d}Y/\mathrm{d}t$) and antagonist ($\mathrm{d}X/\mathrm{d}t$). This can be simply done. The inhomogeneous form of the second-order equation [Eq. (14)] can be shown to be derived from two first-order equations:

$$\frac{\mathrm{d}Y}{\mathrm{d}t} = aY + bX + e, \tag{18a}$$

$$\frac{\mathrm{d}X}{\mathrm{d}t} = cY + dX + f, \tag{18b}$$

in which Y is a measure of the pathogen and X a measure of the antagonist. The parameters a and d are the net per capita reproductive rates and b and c are measures of the per capita effects of one species on the increase in the other; e and f are constants. The solution of Eqs. (18a,b) is frequently given in the form;

$$\frac{\mathrm{d}^2Y}{\mathrm{d}t^2} - T\frac{\mathrm{d}Y}{\mathrm{d}t} + \Delta Y = \nu, \tag{18c}$$

where T and Δ are the trace and determinant of what is called the community matrix (May 1974). They are related to the parameters of Eqs. (18a,b), and the rate parameters of Eq. (14) by,

$$T = (a + d) = \alpha_1 + \alpha_2 = -2\lambda, \quad \text{and} \quad \Delta = (ad - bc) = \alpha_1\alpha_2 = \mu, \tag{19}$$

The magnitudes and signs of the trace and determinant allow inferences to be made about the outcome of competition between pathogen and antagonist. These are summarized in Fig. 6.9. Broadly, the outcome may be stable coexistence of the two species, or else elimination of one or the other. For stable coexistence, there may be an exponential approach to the equilibrium density or an oscillatory approach (Fig. 6.9). Similarly, there may be an exponential approach towards overwhelming dominance by one species or the approach may be oscillatory but now with cycles of population change of increasing amplitude (Fig. 6.9).

The models in Eqs. (18a,b) are biologically naïve: each species would grow exponentially in the absence of the other. Monomolecular growth can be readily incorporated for both species (Gilligan 1990c) but any attempt to incorporate density dependence in the reproductive rates of the species renders the equations analytically insoluble, though the existence of equili-

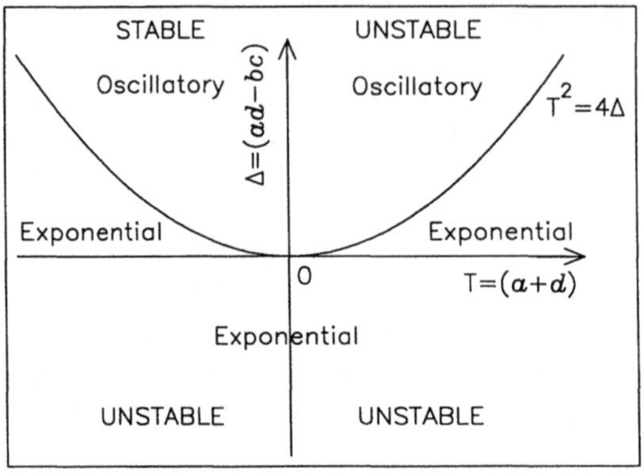

Fig. 6.9. Classification of the types of equilibrium inoculum densities of a pathogen in the presence of an antagonist according to magnitudes of the trace (T) and determinant (Δ) of the community matrix, which summarizes the interaction of the pathogen (Y) and antagonist (X) according to the differential equations, $dY/dt = aY + bX$ and $dX/dt = cY + dX$. These have a second-order form, $d^2Y/dt^2 - TdY/dt + \Delta Y = 0$, auxiliary equation $\alpha^2 - \alpha T + \Delta = 0$, with roots $\alpha_{1,2} = (T \pm \sqrt{T^2 - 4\Delta})/2$ and hence solutions which are exponential [$Y = B\exp(\alpha_1 t) + C\exp\alpha_2 t$), $Y = (B + Ct)\exp(Tt/2)$] or oscillatory [$Y = (B\,\mathrm{Cos}\,\beta t + C\,\mathrm{Sin}\,\beta t)\exp(Tt/2)$]. The trace ($T$) and determinant ($\Delta$) are related to the other parameters according to $T = (a + d) = \alpha_1 + \alpha_2$ and $\Delta = (ad - bc) = \alpha_1\alpha_2$

brium solutions can still be examined. Ecologists have long forsaken such naïve models in favour of elaborations of thc Lotka-Volettra models for predator-prey interaction and competition (see e.g. May 1981). The models can be represented in a rather general form as

$$\frac{dY}{dt} = aY + bXY + eY^2, \tag{20a}$$

$$\frac{dX}{dt} = cXY + dY + fX^2. \tag{20b}$$

The parameters a and d are per capita net reproductive rates, e and f are measures of density dependence while b and c represent the magnitude of interaction effects between the interacting species. The signs of b and c exemplify the type of relationship (Bazin et al. 1990). These range from competition ($b < 0$, $c < 0$) and predator (X) prey (Y) relationships ($b < 0$, $c > 0$), to amensalism ($b < 0$, $c = 0$). Co-operative associations such as mutualism ($b > 0$, $c > 0$) and commensalism ($b > 0$, $c = 0$) are also included. Remarkably, little progress has so far been made in the exploration and application of these models to the control of soilborne disease. This largely reflects the logistical difficulties of working with soil microorganisms but it also displays, as yet, a limited appreciation for this form of epidemiological insight into the mechanisms of biological control. It is essential, however, that exploratory modelling is accompanied by experimentation and rigorous attempts to fit models to data. This has certainly not been the case in ecology, where the two cultures of modellers and experimentalists have sometimes diverged alarmingly (Kareiva 1989).

It might be argued that the Lotka-Volterra types of model are also naïvely over-simplified. The dynamic behaviour of these models has been widely studied and numerous shortcomings revealed (Hallam 1986). They do not fit many data sets well; the models for certain predator-prey interactions ($e = f = 0$, $b < 0$, $d < 0$) give rise to neutral limit cycles that are seldom if ever observed in nature (May 1981); the interaction coefficients (b and d) cannot be accurately estimated (Hallam 1986); the models fail to take account of delayed responses of predators to prey, and of non-random encounters amongst antagonists. Why then should we consider the Lotka-Volterra models in botanical epidemiology when I have argued above for increasing biological reality in incorporating root growth, inoculum decay, latency, infectiousness and functional responses in models for disease dynamics? I suggest that there are two reasons. Firstly, the Lotka-Volterra models are a useful starting point for the construction of more realistic models. They exemplify the perceived importance of inter- and intraspecific density dependence in population growth. Moreover, familiarity with some of the elaborations of the Lotka-Volterra models in ecology may prevent pathologists, from "reinventing the wheel". Secondly, the "outrageous" simplicity of the models reminds us that we can only hope to fit comparatively simple models to data, given the logistical constraints on the collection of data. The mathematical exploration of more complicated dynamic behaviour, however, is not precluded since this serves to identify hypotheses about the components of the biological system that require experimental testing. One important area in which this is urgently needed is in the effects of spatial heterogeneity of soil physical conditions on the population dynamics of plant pathogens and biocontrol agents.

6.3 Epidemiological Comparison of Treatments for the Control of Disease

The emphasis in the previous section was to disentangle the complexities of subterranean epidemics into models with biologically meaningful parameters. But these models are only tools. The pathologist is more usually charged with devising methods of control for recalcitrant pathogens. Models can be used to optimize the timing of methods of control but here we consider how models can be used to compare the effects of treatments in statistically designed experiments on the progress of disease.

6.3.1 Comparison of Disease Progress Curves

6.3.1.1 Statistical Strategy

Suppose that two or more treatments are to be compared for the control of disease or inoculum. Given sufficient data, non-linear models can be fitted to each treatment. The effects of control measures can then be tested in relation to their effects on one or more of the parameters of the models. Then, if the parameters have biological meaning, it is possible to interpret the effect of the treatment in relation to dynamic biological processes. The statistical procedure involves fitting a series of hierarchical models to the data, then testing successive hypotheses (Table 6.1), first to determine whether or not one curve suffices for all treatments and, if not, to test whether certain parameters are common (Ross 1990). Acceptance of the null hypothesis for a single curve implies that the parameters are unaffected by the treatments and a single set of parameters common to both treatments suffices to describe the entire data set irrespective of treatment. Rejection implies that some or all of the parameters differ amongst the treatments. At the extreme, all parameters may differ. Consider the logistic model with an upper (κ_u) and lower (κ_l) asymptote, a rate (β) and displacement (δ) parameter,

$$Y = \kappa_l + \frac{\kappa_u}{1 + \exp(-\beta(t - \delta))}. \tag{21}$$

The lower asymptote is included to allow for background levels of infection. Then for an experiment with two treatments, the numbers of parameters required for a logistic description of the data set varies from four to eight. Some additional examples of differing numbers of common and varying parameters are summarized in Table 6.1.

Table 6.1. Numbers of parameters required to summarize data for a logistic model of disease progress in an experiment with two treatments that differentially affect none, one or more of the parameters

	Parameters needed to summarize the data	
Treatments affect:	No.	Parameters[a]
1. None of the parameters	4	$\beta, \kappa_l, \kappa_u, \delta$
2. One parameter, e.g. β	5	$\beta_1, \beta_2, \kappa_l, \kappa_u, \delta$
3. One parameter, e.g. δ	5	$\delta_1, \delta_2, \kappa_l, \kappa_u, \beta$
4. Two parameters, e.g. κ_l, κ_u	6	$\kappa_{l1}, \kappa_{u1}, \kappa_{l2}, \kappa_{u2}, \beta, \delta$
5. All parameters	8	$\kappa_{l1}, \kappa_{u1}, \beta_1, \delta_1, \kappa_{l2}, \kappa_{u2}, \beta_2, \delta_2$

[a] Suffixes: 1 and 2 relate to treatment 1 and treatment 2: β is the rate parameter, δ is a locational or delay parameter; κ_l is the lower and κ_u is the upper asymptotic level of disease.

Table 6.2. Sums of squares and associated hypothesis tests for selected models, with common parameters amongst treatments for the logistic model, $Y = \kappa_l + \kappa_u/\{1 + \exp[-\beta(t-\delta)]\}$, for n treatments each with r successive observations, where β is the rate parameter, δ is a locational or delay parameter; κ_l is the lower and κ_u is the upper asymptotic level of disease. (After Gilligan 1990b)

	Common parameters	Varying parameters	Degrees of freedom	Residual sum of squares
Hierarchical models				
Single curve	$\kappa_l, \kappa_u, \beta, \delta$	None	$nr - 4$	RSS_4
Common curve with vertical displacement	κ_u, β, δ	κ_l	$nr - 3 - n$	RSS_{3a}
Common curve with differing upper asymptote	κ_l, β, δ	κ_u	$nr - 3 - n$	RSS_{3b}
Common curve with horizontal displacement	$\kappa_l, \kappa_u, \beta$	δ	$nr - 3 - n$	RSS_{3c}
Common curve with differing rates	$\kappa_l, \kappa_u, \delta$	β	$nr - 3 - n$	RSS_{3d}
Common curve with variable linear parameters	β, δ	κ_l, κ_u	$nr - 2 - 2n$	RSS_2
Separate curves	None	$\kappa_l, \kappa_u, \beta, \delta$	$nr - 4n$	RSS_1
Hypothesis tests				
Overall comparison				Sum of squares
Single *vs*. separate curves:	Effect of varying $\kappa_l, \kappa_u, \beta, \delta$		$4(n - 1)$	$RSS_4 - RSS_1$
Residual (within curves)			$n(r - 4)$	RSS_1
Single parameter comparisons with implicit test for common curve				
Vertical displacement	Effect of varying κ_l		$n - 1$	$RSS_4 - RSS_{3a}$
Common parameters	Effect of varying $\kappa_u, \beta, \delta \mid \kappa_l$ varies		$3(n - 1)$	$RSS_{3a} - RSS_1$
Residual (within curves)			$n(r - 4)$	RSS_1
Upper asymptote	Effect of varying κ_u		$n - 1$	$RSS_4 - RSS_{3b}$
Common parameters	Effect of varying $\kappa_l, \beta, \delta \mid \kappa_u$ varies		$3(n - 1)$	$RSS_{3b} - RSS_1$
Residual (within curves)			$n(r - 4)$	RSS_1
Horizontal displacement	Effect of varying δ		$n - 1$	$RSS_4 - RSS_{3c}$
Common parameters	Effect of varying $\kappa_l, \kappa_u, \beta \mid \delta$ varies		$3(n - 1)$	$RSS_{3c} - RSS_1$
Residual (within curves)			$n(r - 4)$	RSS_1
Rate	Effect of varying β		$n - 1$	$RSS_4 - RSS_{3c}$
Common parameters	Effect of varying $\kappa_l, \kappa_u, \delta \mid \beta$ varies		$3(n - 1)$	$RSS_{3d} - RSS_1$
Residual (within curves)			$n(r - 4)$	RSS_1
Hierarchical model with progressively fewer common parameters				
Vertical displacement	Effect of varying κ_l		$n - 1$	$RSS_4 - RSS_{3a}$
Differing asymptote	Effect of varying $\kappa_u \mid \kappa_l$ varies		$n - 1$	$RSS_{3a} - RSS_2$
Common non-linear	Effect of varying $\beta, \delta \mid \kappa_l, \kappa_u$ vary		$2(n - 1)$	$RSS_2 - RSS_1$
Residual (within curves)			$n(r - 4)$	RSS_1

Intermediate cases are examined by allowing single parameters to be optimized with respect to each treatment while the rest are fitted as common to both treatments (Table 6.2). For example, it is possible to test whether or not curves differ by a vertical displacement (κ_l's differ; κ_u, β and δ are common) implying that levels of background infection differ amongst the treatments. Alternatively, the curves for the progress of disease may differ by a horizontal displacement (δ's differ; κ_l, κ_u, β are common). This implies that one treatment delays the epidemic relative to another. Several alternative models and the ways in which these can be compared to test different hypotheses about the effects of treatments on the parameters of the logistic model are summarized in Table 6.2. An approximate *F*-test based on the ratio of the mean square for the treatment effect to the within-curve variation is used to test the hypotheses. Of course treatments sometimes affect more than one parameter and it is possible to test for this by fitting successively fewer parameters as common. One example is shown in Table 6.2. A priori selection of hypotheses is essential, however, to avoid indiscriminate testing or tests suggested by the data, each of which violates the statistical assumptions underlying the inferences.

Initial work on the comparison of non-linear disease progress curves has revealed a marked conservatism in the rate parameter (Gilligan 1990b). Several experiments were analysed from the literature including the effects of soil solarization and artificial heating on the progress of tomato wilt caused by *Fusarium oxysporum* f.sp. *lycopersicae* (Greenberger et al. 1987) and the effects of chemical control and cultivar on *Phytophthora cryptogea* on raspberry (Washington 1988). Soil solarization affected the carrying capacity of disease with a reduction in the asymptote (κ_u). There was no evidence for any effect on the rate of development of the epidemic nor on the delay parameter (Fig. 6.10A,B). Application of phosphorous acid and fosetyl aluminium after transplanting seedlings also reduced the carrying capacity of raspberry for *P. cryptogea* relative to the control (Gilligan 1990b). Comparison of disease progress curves on different cultivars, in the absence of chemical control, revealed differences in carrying capacity and the time of onset of disease (Fig. 6.10C,D). More work now needs to be done in interpreting the physiological and epidemiological mechanisms associated with changes in the carrying capacity, the growth rate and the compound parameter involving delay. Further discussion of alternative methods for the comparison of disease progress curves are discussed by Madden (1986) and Madden and Campbell (1990).

6.3.1.2 The Contribution of Individual Observations

Data on disease progress are frequently collected at equally spaced time intervals. Although this reflects good experimental procedure for conventional experiments, it can lead to imprecision in the recording and com-

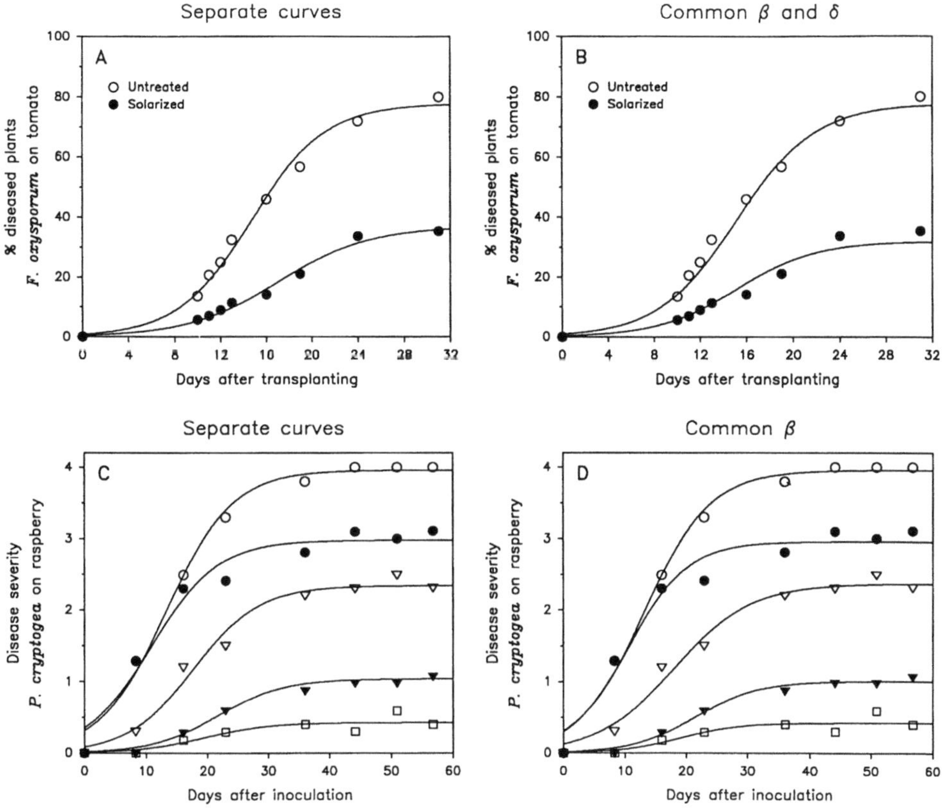

Fig. 6.10A–D. Comparison of disease progress curves, using the logistic model ($Y = \kappa_u/\{1 + \exp-[\beta(t - \delta)]\}$). **A** Separate curves and **B** curves with common rate (β) and delay (δ) parameters for the progress of disease due to *Fusarium oxysporum* f. sp. *lycopersicae* over time on tomato (data from Greenberger et al. 1987; analyses from Gilligan 1990b). **C** Separate curves and **D** curves with a common rate for the progress of disease due to *Phytophthora cryptogea* over time on five cultivars of raspberry (data from Washington 1988; analyses from Gilligan 1990b)

parison of progress curves. An example is given in Fig. 6.11, albeit for an aerial pathogen, *Phytophthora infestans*. The data, taken from an experiment by Hooker (1956), show how sampling at equal intervals provided little information on the important dynamic portion of the curve for the rapid epidemic in the untreated (control) plots. The amount of disease increased by more than 60% between the third and fourth observations, when the rate reached a maximum and then declined. By contrast, there were five observations on the ascending portion of the curve in plots treated with a fungicide, zineb.

The contribution of individual observations to the estimation of individual parameters can be estimated by the so-called parameter loadings (Ross

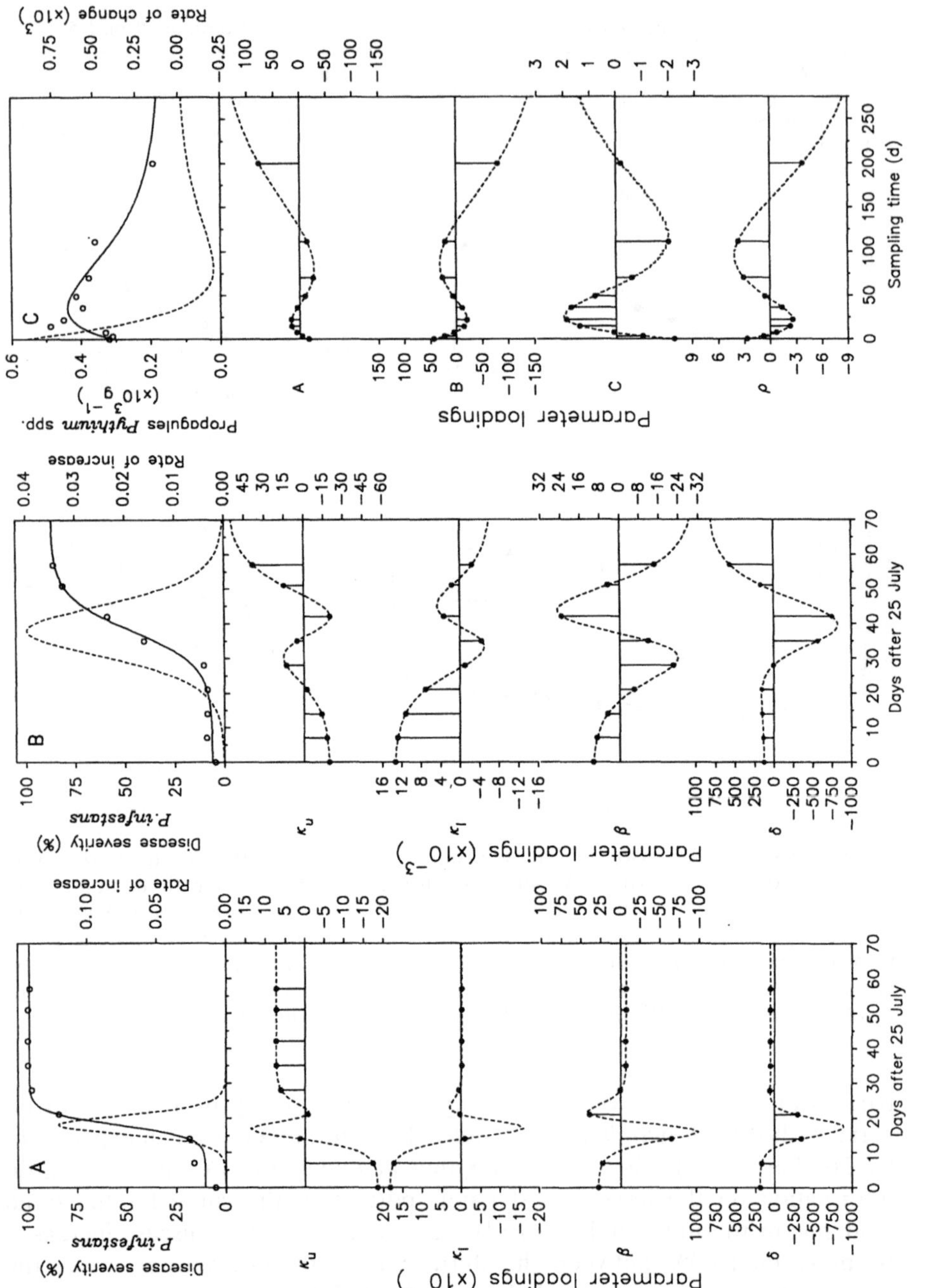

A
B
C
Disease severity (%)
P.infestans
Rate of increase
Propagules Pythium spp.
Rate of change
Parameter loadings
Days after 25 July
Sampling time (d)

1990). They are obtained by multiplying the dispersion matrix by the first derivatives of the cumulative curve for each parameter (i.e. $\partial Y/\partial\beta$, $\partial Y/\partial\delta$ etc.). The needle point curves show the contributions of each observation to the estimation of the parameters for the *P. infestans* data (Fig. 6.11A,B) and for data from Martin et al. (1983) for the population dynamics of *Pythium* in soil fitted to the critical exponential model (Fig. 6.11C). The contribution varies in direct proportion to the absolute magnitude of the parameter loadings. Interpolated and extrapolated values show the expected merit of taking intermediate and later observations. Late observations in the untreated plots for *P. infestans* made little contribution to the estimation of parameters other than κ_u. Even one more observation between the third and fourth observations would have greatly enhanced the estimation of the other parameters. The loading profiles for the fungicide-treated plots are more evenly dispersed but even so an additional later observation would have been advantageous. The value of additional later observations in fitting the critically damped model for the population dynamics of *Pythium* is apparent from the profiles in Fig. 6.11C. These properties are apparent, of course, only after the data are collected and analysed. How can we improve the collection of data? One way is to plot the data as soon as the plants or soil assays are assessed and to increase the frequency of observations when the rate of change of disease, infection or inoculum is rapid and to reduce the frequency when the rate is slower. The value of later observation may be judged empirically by the inspection of the asymptote. Two or three observations "on" the asymptote will usually be sufficient to estimate the carrying capacity for disease. Hence, where logistical constraints permit, different sampling intervals should be used for different treatments according to the dynamics of the epidemics.

6.3.2 Estimating Trends When There Are Few Data Points

Data for soilborne pathogens are expensive to collect. Often, therefore, few data points are available from which to estimate trends in disease progress. Fitting of non-linear models, such as those discussed above, is not possible

◄

Fig. 6.11A–C. Parameter loadings, cumulative and rate curves for disease progress over time of an aerial pathogen, *Phytophthora infestans*, on **A** untreated and **B** treated potatoes, fitted to the logistic model, $Y = \kappa_l + \kappa_u/\{1 + \exp -[\beta(t - \delta)]\}$. **C** Corresponding curves for the population dynamics of *Pythium* spp. fitted to the critical exponential model, $Y = A + (B + Ct)\rho^t$, where $\rho = \exp(-\lambda)$. Parameter loadings corresponding to observed data values are shown by the needle graph, interpolated and extrapolated values are shown by the dotted curve. The contributions of individual observations to the estimates of individual parameters vary in direct proportion to the absolute magnitude of the parameter loading. Data for *P. infestans* from Hooker (1956), for *Pythium* from Martin et al. (1983), analyses from Gilligan (1990b,c)

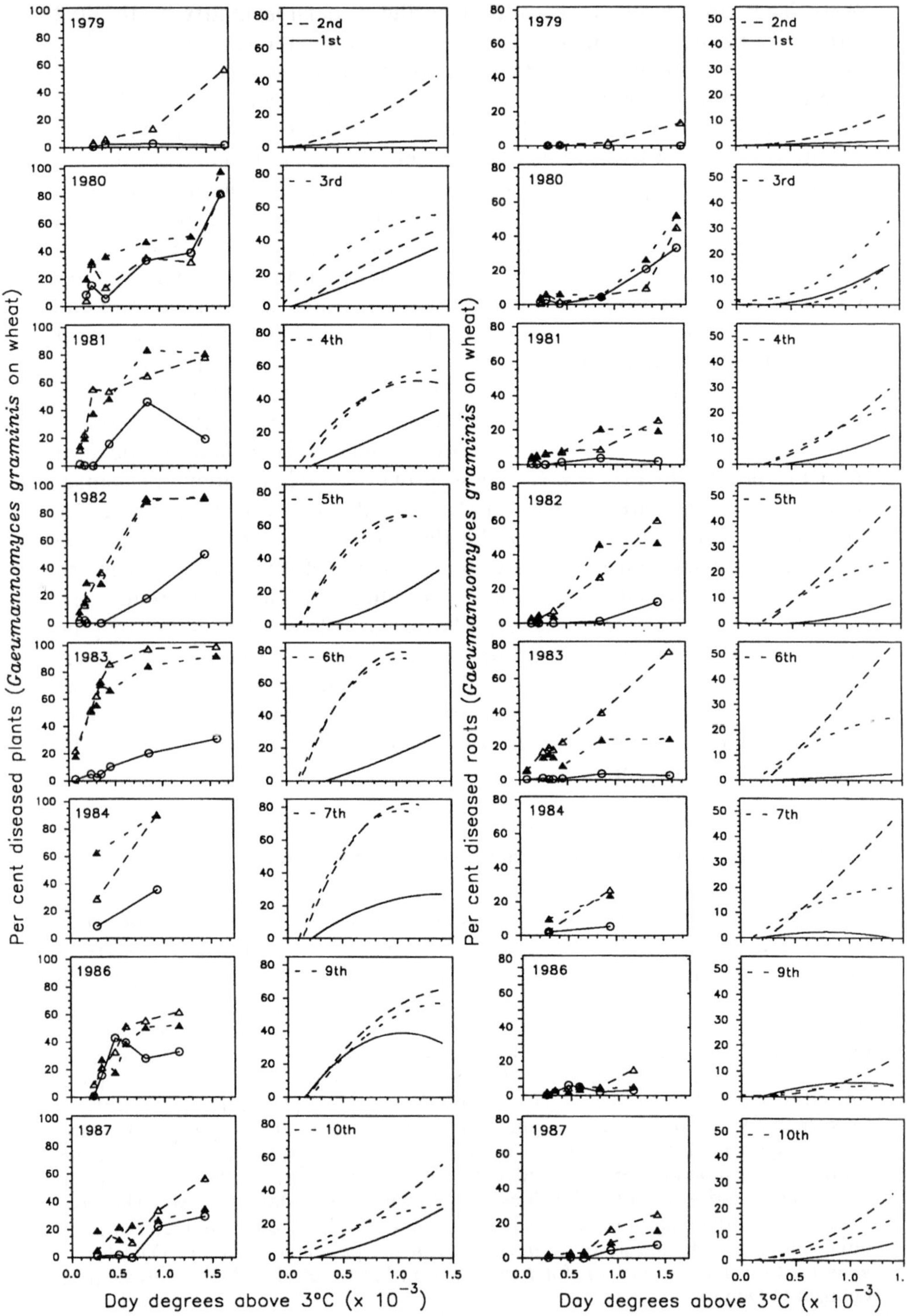
Per cent diseased plants (*Gaeumannomyces graminis* on wheat)
Per cent diseased roots (*Gaeumannomyces graminis* on wheat)
1979
1980
1981
1982
1983
1984
1986
1987
2nd
1st
3rd
4th
5th
6th
7th
9th
10th
Day degrees above 3°C (x 10^-3)
Day degrees above 3°C (x 10^-3)

because of scarcity of observations. Other problems arise when harvesting interrupts the progress towards an asymptote or when curves differ markedly in shape. An example of a parsimonious data set that exhibits all three of these properties is given in Fig. 6.12 for *G. graminis* on winter wheat (after Werker et al. 1991). Two variables are shown, percent diseased plants and percent diseased roots. Data were collected over nine seasons (1979–1987 with no observations in 1985) for the progress of disease in different sequences of wheat in large field plots (Hornby and Henden 1980). The principal objective was to compare the disease dynamics in first and second crops of wheat and a sequence of consecutive crops of wheat and so to infer epidemiological mechanisms for take-all decline (Slope and Cox 1964). The take-all disease is known to increase during early cereal sequences and then to decline with continued monoculture as the population of antagonistic microorganisms builds up. Paucity of data and variability in the shapes of the disease progress curves precluded the use of a simple non-linear model. Nevertheless, certain dynamic trends can be extracted from such data by the use of linear models. This can be done by fitting polynomial functions to the means for each curve. The expected curves for percent diseased plants and percent diseased roots in the *G. graminis* trial are summarized in Fig. 6.12. Quadratic curves sufficed to describe the trends. Note that account is taken in the analyses of the trend over seasons, thereby allowing predicted curves to be drawn for years in which there were few observations. These smoothed curves summarize the dynamic properties of the more ragged empirical curves but the parameters of the quadratic model do not have any obvious biological meaning. More information can be obtained by breaking up the curve into separate components for trend. The procedure involves the use of orthogonal polynomial contrasts (Neter and Wasserman 1974).

Initially, three components for trend of the take-all data were extracted by the use of polynomial contrasts for the linear, quadratic and cubic components. The mean amount of disease over the period of observation was also extracted. Unlike the parameters of a simple polynomial function, each of these variables has an intuitive biological meaning. The mean amount of disease is directly related to the area under the disease progress curve (Werker et al. 1991). The linear contrast is a measure of the average increase in disease over time. The quadratic contrast is a measure of the average degree of curvature; positive values imply an acceleration in the rate of disease increase, negative values imply a deceleration. The cubic

Fig. 6.12. Empirical and derived disease progress curves for the infection of first, second and continuous (≥3) crops of winter wheat by *Gaeumannomyces graminis* for 8 years annotated by year of harvest. Curves are shown for percent diseased plants and percent diseased roots. Derived curves were estimated from the mean amounts of disease per year and the linear and quadratic trends within years after allowance for trends across years (see Fig. 6.13). Data and analyses from Werker et al. (1991)

contrast is sensitive to inflexion points and hence the time at which an accelerating epidemic changes to a decelerating one. The mean amount of disease and contrasts also have the advantage of being statistically independent and so can be analysed separately by analysis of variance or generalized linear model, as appropriate. This is particularly useful in the epidemiological analysis of factorial experiments. For example, the effects of interactions amongst treatments, such as sowing time and fertilizer application, were analysed on the separate components of the trends in disease development (Werker and Gilligan 1990).

Computation of contrasts involves distillation of the data for each plot into a single value for that contrast. Thus suppose there were four successive values of disease (Y_i = 20, 35, 65, 70% diseased plants) at equally spaced intervals for a given plot in a given year. The coefficients (ζ_i) for a linear contrast are −3, −1, +1, +3 (Fisher and Yates 1974). The linear contrast is then obtained as,

$$\sum_{i=1}^{4} \zeta_i Y_i = (-3 \times 20) + (-1 \times 35) + (1 \times 65) + (3 \times 70) = 180, \qquad (22)$$

The expected curve can be reconstructed by the sum of the mean and the polynomial contrasts. The cubic components did not have any consistent effects in the *G. graminis* trial and so were omitted.

6.3.2.1 Analysing Trends Within and Between Seasons

Environmental and other conditions differ from one season to another and some seasons favour disease more than another. Averaging of within-season trends over seasons may mask an effect that is apparent only in seasons that favour disease. The use of thermal time (day-degrees above a threshold temperature) rather than chronological time as the independent variable removes some of the interseasonal variation. Similarly, allowance for variability in water potential could be incorporated into a rescaling of the sampling time. Ferrin and Mitchell (1986) used cumulative daily soil-water potential (-cbars) as an independent variable against which to express the cumulative mortality of tobacco plants by *P. parasitica* var. *nicotiana*. But interseasonal variation will almost certainly remain and treatment differences may be hidden by simple averaging over seasons. Werker et al. (1991) attempted to circumvent this problem by analysing the way that the components for trends *within* seasons (i.e. the mean, linear and quadratic contrasts) varied with the trend *across* seasons in the *G. graminis* rotational experiment. Curves for the change in mean, linear and quadratic contrasts for each treatment are shown in Fig. 6.13 for the data for percent diseased roots. Some important inferences about the dynamics of take-all decline emerge from these analyses.

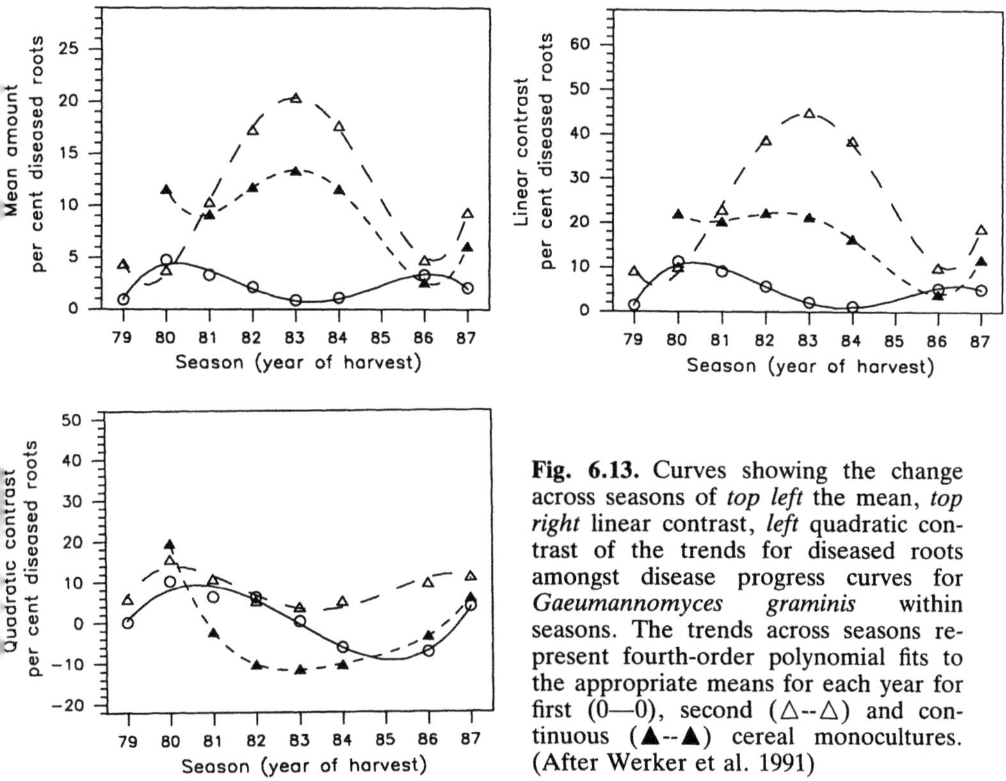

Fig. 6.13. Curves showing the change across seasons of *top left* the mean, *top right* linear contrast, *left* quadratic contrast of the trends for diseased roots amongst disease progress curves for *Gaeumannomyces graminis* within seasons. The trends across seasons represent fourth-order polynomial fits to the appropriate means for each year for first (0—0), second (△--△) and continuous (▲--▲) cereal monocultures. (After Werker et al. 1991)

The mean levels of disease were relatively low in 1979–1980 and 1986–1987 and relatively high in 1981–1984. The mean amount of disease and the average rate of increase (linear contrast) of diseased roots were closely correlated, having similar interseasonal patterns. First-year wheats showed little difference between years. The rate of progress of disease in continuous wheats was intermediate between the rate in second and first wheats. Differences in curvature (quadratic contrast) were also present. Overall, the progress of disease in second wheat crops was associated with a positive quadratic contrast, indicating an accelerating epidemic, whilst a negative quadratic contrast was more common in plots in continuous wheats. Unfortunately, because of limitations in the experimental design, the effect of additional years of monoculture in the continuous wheat treatment is still confounded with variability between seasons. Werker et al. (1991), however, noted that the average rate of increase in the diseased root population between second and continuous wheats differed little in early years of the experiment. In subsequent years, however, the rate of increase in disease in continuous wheat crops was approximately half that in second wheats. This occurred in seasons of high disease (for example, 1983, Fig. 6.12) and moderate disease (1987, Fig. 6.12).

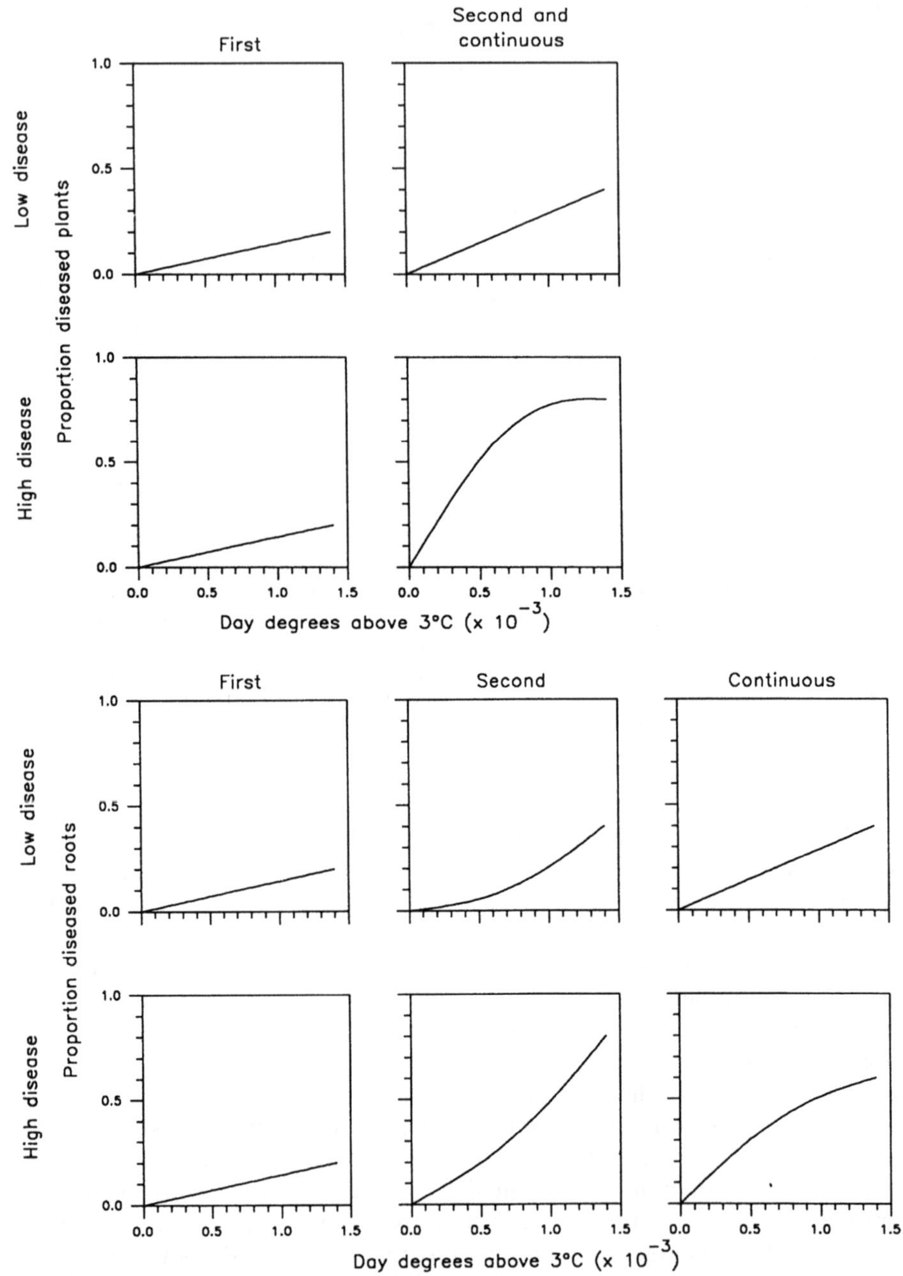

Fig. 6.14. Hypothesized shapes of disease progress curves for *Gaeumannomyces graminis* in successive crops of wheat in which suppression of disease occurs in later (continuous) crops. Curves are shown for the proportions of diseased plants and the proportions of diseased roots in years of high and low disease (after Werker et al. 1991). The suppression of disease under cereal monoculture is more evident in the root than plant population and in years of high rather than low levels of disease

Combination of the analyses for disease increase within and between seasons allows a summary of the rates of increase in disease in first, second and continuous wheats (Fig. 6.14). Progress amongst plants and roots in first wheats was slow and rectilinear, even in years of high disease. This was also the case for the progress of disease amongst plants in second and longer-term wheats in seasons in which epidemics developed comparatively slowly (Fig. 6.14). In years that were more conducive to disease, the epidemic started faster and then decelerated towards an asymptotic "carrying capacity" (>100% diseased plants), resulting in a monomolecularly shaped curve. The behaviour was different for the spread through a population of roots. Epidemics accelerated in second wheat crops (positive quadratic contrast), mimicking an exponential increase in disease. In continuous wheats, however, negative quadratic contrasts were more common, especially in years of high disease. So epidemics started fast and then slowed down (Fig. 6.14). Work is currently under way to interpret these differences in shapes of disease progress curves in relation to an epidemiological mechanism for take-all decline (see Sect. 6.3.2.3).

6.3.2.2 Allowance for Variable Sampling Intervals, Sampling Frequency and Period of Observation

Standardization of sampling interval and sampling frequency of plant disease epidemics in different seasons and treatments is not always possible, nor is it necessarily desirable (see Sect. 6.3.1.2). The period between be first and last observation may also vary amongst treatments, leading to further variation. The contrast coefficients (ζ_i) depend upon the sampling interval and frequency. Hence an additional observation will change the magnitude of the overall contrast in a way that is independent of the trend in disease progress. Scaling of the contrasts by dividing by the sum of the coefficients squared ($\Sigma\zeta_i^2$) removes this form of bias (Werker et al. 1991; Table 6.3). Account should be taken in analyses of the amounts of information contributing to each contrast. Accordingly, $\Sigma\zeta_i^2$ can also be used as a weighting variate that reflects the number of observations and sampling intervals that contributed to each contrast. More emphasis is therefore given to those plots that were more intensively sampled. The computation of linear, quadratic and cubic contrasts for three replicate plots is illustrated in Table 6.3.

6.3.2.3 Analysis of Trends to Interpret Suppression of Disease

Analysis and summary of trends in disease (Figs. 6.12–6.14) yield linear summaries of the dynamics of disease. In the take-all example, these suggested further hypotheses about the epidemiological mechanisms responsible for the suppression of disease (Werker et al. 1991). One simple mechanism

Table 6.3. Example showing the derivation of contrasts and appropriate weights for data for the progress of *Gaeumannomyces graminis* on a continuous (fifth) crop of winter wheat in three field plots (y_{1-3}) at Rothamsted Experimental Station in 1982 (after Werker et al. 1991). The contrasts are derived by summing the products of the contrast coefficients and observed disease values and dividing by the sum of squares of the contrast coefficients

		Sampling occasions							$\Sigma\xi_i^2$	$\Sigma(\xi_i \cdot y_r)/\Sigma\xi_i^2$
			1	2	3	4	5	6	(= weights)	(= contrasts)
Σ days from 15/10		0	43	131	152	197	243	288		
Σ day-degrees Above 3 °C ($\times 10^{-3}$)	t	0.00	0.13	0.19	0.21	0.36	0.85	1.46		
Observed values for % diseased roots in replicate plots	y_1		2.07	4.48	mv	4.08	mv	66.8		
	y_2		4.47	4.01	4.85	2.60	46.1	49.0		
	y_3		2.92	1.43	mv	5.98	mv	25.4		
ξ_i for y_1	Linear ξ_1	−0.428	−0.297	−0.242		−0.066		1.032	1.400	47.86
	Quadratic ξ_2	0.204	0.016	−0.052		−0.230		0.062	0.101	29.89
	Cubic ξ_3	−0.020	0.020	0.021	−0.248	−0.021		0.001	0.002	60.94
ξ_i for y_2	Linear ξ_1	−0.458	−0.326	−0.271	−0.032	−0.095	0.395	1.003	1.622	39.08
	Quadratic ξ_2	0.233	0.056	−0.007	0.049	−0.170	−0.296	0.215	0.221	−15.66
	Cubic ξ_3	−0.099	0.013	0.041		0.069	−0.101	0.028	0.030	−88.04
ξ_i for y_3	Linear ξ_1	−0.428	−0.297	−0.242		−0.066		1.032	1.400	17.58
	Quadratic ξ_2	0.204	0.016	−0.052		−0.230		0.062	0.101	1.78
	Cubic ξ_3	−0.020	0.020	0.021		−0.021		0.001	0.002	−11.18

mv = missing value; ξ_{1-3} = linear, quadratic and cubic functions of t

involves changes in the carrying capacities of populations of roots for *G. graminis* with continued monoculture. Roots may become resistant to infection by prior colonization by microorganisms antagonistic to *G. graminis* (Deacon 1976) as the population densities of these organisms build up. An alternative hypothesis, supported by independent work by Hornby et al. (1989) on the spatial pattern of take-all in successive crops, suggests that the balance of primary to secondary infection may be involved in the dynamics of take-all decline. Primary infections may assume a greater importance in continuous wheats due to high inoculum densities as the amount of *G. graminis* builds up in soil, particularly during early years of monoculture. The frequency of primary infections may be reduced in second wheats because of lower inoculum densities derived from the slow rates of disease progress in the first-year wheats in the previous crop. Hence, secondary infections may play a greater role in disease dynamics in second wheats. This can be considered in relation to the variables and parameters of the model for primary and secondary infection [Eq. (9b); Sect. 6.2.3]. Progress of disease on second wheats is initially slow [in the presence of relatively low inoculum densities (low P), although r_p may be high] and subsequently increases as secondary spread assumes a more important role (high r_s) as the inoculum for secondary infections builds up and as plant roots grow. Conversely, in continuous wheats, the higher levels of disease attained in the early part of the growing season may be caused by the presence of higher inoculum densities (high P). The rate of disease progress, however, is slow (low r_p and r_s). This is apparent later in the season when there is a decline in primary infections due to a shortage of susceptible tissue and loss in infectivity of soil inoculum (Hornby 1981).

6.3.3 Selecting the Disease Variable

Epidemics of root disease occur on a dynamically changing host population as roots grow, become infected and die. Quite different pictures of the temporal patterns of disease arise when alternative variables are used to summarize disease. This leads to an unsettling "twist in the tale" about modelling and analysis the temporal progress of disease. An apparently static epidemic, as assessed by the proportion of diseased plants, may still be undergoing active spread through the root system. This is shown in Fig. 6.12 for epidemics of *G. graminis*, especially in 1981–1983. Moreover, the rate of spread of infection and disease may be greater within diseased plants than through the entire (healthy and diseased) plant population. An example is shown in Fig. 6.15 for a different epidemic of *G. graminis* taken from Werker and Gilligan (1990).

Five variables were used to measure disease, together with the estimated density of roots per plant. The ratio of diseased to total plants is a measure of *disease incidence*. It is an estimate of the probability of occurrence of

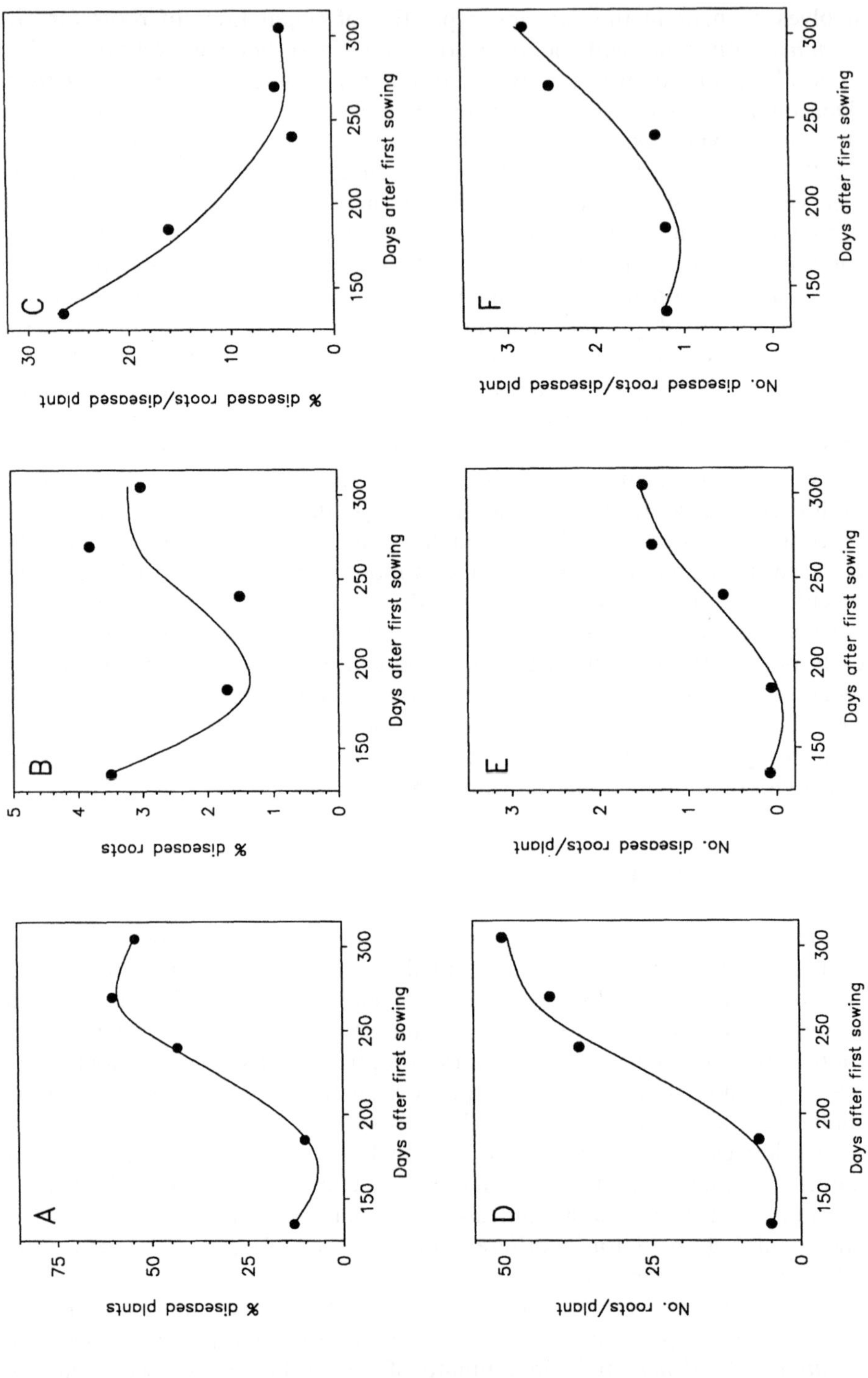
A
B
C
D
E
F
% diseased plants
% diseased roots
% diseased roots/diseased plant
No. roots/plant
No. diseased roots/plant
No. diseased roots/diseased plant
Days after first sowing

disease on a plant. Disease incidence amongst plants increased sigmoidally during the season (Fig. 6.15A). Since the plant population was static, the increase in incidence reflected the dynamics of root growth. Root growth affects the diseased plant population by increasing the probability of primary infection as more inoculum is encountered (cf. Sect. 6.2.1.2). Root growth also increases the probability of secondary (allo-) infection between plants as roots overlap. So more plants join the cohort of diseased plants. But further dynamic changes are masked as the disease approaches the carrying capacity (Fig. 6.12).

The ratio of diseased to total roots within samples is a measure of the *average severity of disease* per plant throughout the entire (diseased and healthy) plant population. Because it includes zeros (corresponding to healthy plants) and positive values for disease, it is actually a compound variable that combines presence or absence of disease with local aggregations. The ratio of diseased to total roots averaged over the diseased plants within a sample is a measure of the *local severity of disease* per diseased plant. It omits the zero values. It may, therefore, be used to assess the rate of spread of disease amongst roots on diseased plants.

Increase in the total root population density (Fig. 6.15D) can cause an apparent decline in the proportions of diseased roots (Fig. 6.15B,C). Analysis of the absolute numbers of diseased roots per plant (Fig. 6.15E,F) allows separation of the effects of treatments on the rate of production of roots and the rate of accumulation of diseased roots (Werker and Gilligan 1990). Much still, however, needs to be done to disentangle the interrelationships of disease dynamics with disease and host variables in the temporal analysis of disease.

Acknowledgements. I am grateful to a number of colleagues, particularly Drs. P.R. Brassett, D. Hornby and A.R. Werker for allowing me to quote extensively from joint work. I am also grateful to Mr B.V.D. Goddard for assistance in preparing the graphs.

Fig. 6.15A–F. Effect of using different disease and host variables to monitor a single epidemic of *Gaeumannomyces graminis* in winter wheat (Werker and Gilligan 1990): **A** percent diseased plants; **B** percent diseased roots averaged over the total plant population; **C** percent diseased roots averaged over the diseased plant population; **D** mean density of roots per plant; **E** mean density of roots per plant averaged over the total plant population (i.e. with zeros included); **F** mean density of roots per plant averaged over the diseased plant population (i.e. with zeros excluded). Trends were obtained by fitting simple polynomial models to the data

Appendix: Glossary of Symbols Used to Represent Principal Random Variables and Parameters in Text

Symbol	Explanation	Equation in which first used
A	Asymptotic parameter for equilibrium density of pathogen biomass	(15)
α	Coefficient of primary infection	(9)
$\alpha_{E,I,R}$	Infection-induced death rates of pre-infectious ("exposed", E), infectious (I) and post-infectious ("removed", R) roots	(12)
$\alpha_{1,2}$	Rate parameters for over-damped model	(15)
B	Locational parameter	(15)
b	Per capita "birth" rate of roots	(3)
β	1. Coefficient of secondary infection	(9)
	2. Rate of transmission of disease between infected and susceptible roots	(12)
	3. Rate parameter for logistic model ($=r_s$)	(21)
C	Locational parameter	(15)
D	Density of dead roots per plant or per unit area	(12)
d	Per capita "death" rate of roots	(3)
δ	Delay (time) in onset of epidemic	(5,8)
E	Density of pre-infectious ("exposed") roots per plant or per unit area	(12)
γ^{-1}	Infectious period	(12)
I	Density of infectious roots per plant or per unit area	(12)
κ	Carrying capacity	(4)
$\kappa_{(u,l)}$	Asymptotic (u = upper; l = lower) levels of disease, infestation or infection	(21)
L	Root length	(6)
λ	Basic reproductive rate of a parasite	(13)
N	Density of host units (i) roots per plant or per unit area, (ii) plants per unit area	(5)
N_i	Density of infected (i) roots per plant or per unit area, (ii) plants per unit area	(5)
P	Density of propagules of inoculum per unit area or volume soil	(5)
R	Density of post-infectious ("removed") roots per plant or per unit area	(12)
r_d	Rate parameter for decay of inoculum	(10)
r_N	Rate parameter for growth of new roots	(10)
$r_{(p,s)}$	Rate parameter for primary (p) and secondary (s) infection	(5,8)
S	Density of susceptible roots per plant or unit area	(12)
S_T	Threshold density of susceptible roots per plant or per unit area required for epidemic to occur	(13)
σ^{-1}	Latent period	(12)
t_{max}	Time to achieve maximum density of inoculum, infection or disease	(17)

X	Density of antagonist per unit area or volume	(18)
Y	1. Generic description of density of inoculum, infection or disease	(1)
	2. Ratio of (i) diseased plants to total plants per unit area, (ii) diseased roots to total roots per plant or per unit area	(9)

References

Anderson RM (ed) (1982) Population dynamics of infectious diseases: theory and applications. Chapman & Hall, London

Anderson RM, May RM (1979) Population biology of infectious diseases. Part I. Nature 280:361–367

Anderson RM, May RM (1981) The population dynamics of microparasites and invertebrate hosts. Philos Trans R Soc 291:451–524

Bailey NTJ (1975) The mathematical theory of infectious diseases and its applications. Academic Press, New York

Baker KF, Cook PJ (1974) Biological control of plant pathogens. Freeman, San Francisco

Bazin MJ, Markham P, Scott EM, Lynch M (1990) Population dynamics and rhizosphere interactions. In: Lynch J (ed) The rhizosphere. Wiley, Chichester, pp 99–127

Bloomberg WJ (1979) A model of damping-off and root rot of Douglas-fir seedlings caused by *Fusarium oxysporum*. Phytopathology 69:1072–1077

Brassett PR, Gilligan CA (1988) A model for primary and secondary infection in botanical epidemics. Z Pflanzenkr Pflanzenschutz 95:352–360

Butler FC (1959) Saprophytic behaviour of some cereal root rot fungi. III. Saprophytic survival in soils at high and low fertility. Ann Appl Biol 47:28–36

Campbell CL (1986) Interpretation and uses of disease progress curves for root diseases. In: Leonard KJ, Fry WE (eds) Plant disease epidemiology: population dynamics and management, vol I. Macmillan, New York, pp 38–54

Campbell CL, Reynolds KM, Madden LV (1988) Modelling epidemics of root diseases and development of simulators. In: Kranz J, Rotem J (eds) Techniques in plant disease epidemiology. Springer, Berlin Heidelberg New York, pp 253–265

Campbell CL, Madden LV (1990) Introduction to plant disease epidemiology. Wiley, New York

Coley-Smith JA, Entwistle AR (1988) Susceptibility of garlic to *Sclerotium cepivorum*. Plant Pathol 37:261–264

Deacon JW (1976) Biological control of the take-all fungus, *Gaeumannomyces graminis*, by *Phialophora radicicola* and similar fungi. Soil Biol Biochem 8: 275–283

Ferrin DM, Mitchell DJ (1986) Influence of soil water status on the epidemiology of tobacco black shank. Phytopathology 76:1213–1217

Fisher RA, Yates F (1974) Statistical tables for biological, agricultural and medical research. 6th edn. Oliver & Boyd, Edinburgh

Gilligan CA (1983) Modeling of soilborne pathogens. Annu Rev Phytopathol 21: 45–64

Gilligan CA (1985a) Probability models for host infection by soilborne fungi. Phytopathology 75:61–67

Gilligan CA (1985b) Construction of temporal models. III. Disease progress of soilborne pathogens. In: Gilligan CA (ed) Mathematical modelling of crop disease. Advances in plant pathology, vol 3. Academic Press, London, pp 67–105
Gilligan CA (1990a) Mathematical modelling and analysis of soilborne pathogens. In: Kranz J (ed) Epidemics of plant diseases: mathematical modelling and analysis. 2nd edn. Springer, Berlin Heidelberg New York, pp 96–142
Gilligan CA (1990b) Comparison of disease progress curves. New Phytol 115: 223–242
Gilligan CA (1990c) Antagonistic interactions involving plant pathogens: fitting and analysis of models to non-monotontic curves for population and disease dynamics. New Phytol 115:649–665
Gilligan CA (1990d) Mathematical models of infection. In: Lynch J (ed) The rhizosphere. Wiley, Chichester, pp 207–232
Greenberger A, Yogev A, Katan J (1987) Induced suppressiveness in solarized soils. Phytopathology 77:1663–1667
Guttierez AP, DeVay JE (1986) Studies of plant-pathogen-weather interactions: cotton and verticillium wilt. In: Leonard KJ, Fry WE (eds) Plant disease epidemiology: population dynamics and management, vol I. Macmillan, New York, pp 205–231
Guttierez AP, DeVay JE, Pullman GS, Frieberthauser GE (1983) A model of verticillium wilt in relation to cotton growth and development. Phytopathology 73:89–95
Hallam TH (1986) Population dynamics in a homogeneous environment. In: Hallam TH, Levin SA (eds) Biomathematics, vol 17. Mathematical ecology. Springer, Berlin Heidelberg New York, pp 62–94
Hassell (1981) Arthropod predator-prey systems. In: May RM (ed) Theoretical ecology: principles and applications. 2nd edn. Blackwell, Oxford, pp 105–131
Hau B (1990) Analytical models of plant disease in a changing environment. Annu Rev Phytopathol 28:221–245
Hau B, Eisensmith SP, Kranz J (1985) Construction of temporal models. II. Simulation of aerial epidemics. In: Gilligan CA (ed) Mathematical modelling of crop disease. Advances in plant pathology, vol 3. Academic Press, London, pp 31–65
Hethcote HW (1989) Three basic epidemiological models. In: Levin SA, Hallam TG, Gross LJ (eds) Applied mathematical ecology, Springer, Berlin Heidelberg New York, pp 119–144
Hooker WJ (1956) Foliage fungicides for potatoes in Iowa. Am Potato J 33:47–52
Hornby D (1981) Inoculum. In: Asher MJC, Shipton PJ (eds) Biology and control of take-all. Academic Press, London, pp 271–293
Hornby D, Bateman RW, Brown ME, Henden DR (1989) An experimental design and procedures for testing putative controls against naturally-occurring take-all in the field. Ann Appl Biol 115:195–208
Hornby D, Henden DR (1980) Take-all decline. Rothamsted Experimental Report for 1979, Part 1. Rothamsted Experiment Station, Rothamsted, pp 184–193
Huisman OJ (1982) Interrelations of root growth dynamics to epidemiology of root-invading fungi. Annu Rev Phytopathol 20:303–327
Jeger MJ (1982) The relation between total, infectious, and postinfectious diseased plant tissue. Phytopathology 72:1185–1189
Jeger MJ (1984) Relation between rate parameters and latent and infectious periods during a plant disease epidemic. Phytopathology 74:1148–1152
Jeger MJ (1986a) Asymptotic behaviour and threshold criteria in model plant disease epidemics. Plant Pathol 35:355–361
Jeger MJ (1986b) The potential of analytic compared with simulation approaches to modeling in plant disease epidemiology. In: Leonard KJ, Fry WE (eds) Plant

disease epidemiology: population dynamics and management, vol I. Macmillan, New York, pp 255–281
Jeger MJ (1987) The influence of root growth and inoculum density on the dynamics of root disease epidemics: theoretical analysis. New Phytol 107:459–478
Jeger MJ, Starr JL (1985) A theoretical model of the winter survival dynamics of *Meloidogyne* spp. eggs and juveniles. J Nematol 17:257–260
Kareiva P (1989) Renewing the dialogue between theory and experiments in population ecology. In: Roughgarden J, May RM, Levin SA (eds) Perspectives in ecological theory. Princeton Univ Press, Princeton, pp 68–88
Madden LV (1980) Quantification of disease progression. Prot Ecol 2:159–176
Madden LV (1986) Statistical analysis and comparison of disease progress curves. In: Leonard KJ, Fry WE (eds) Plant disease epidemiology: population dynamics and management, vol I. Macmillan, New York, pp 55–84
Madden LV, Campbell CL (1990) Nonlinear disease progress curves. In: Kranz J (ed) Epidemics of plant disease: mathematical modelling and analysis, 2nd edn. Springer, Berlin Heidelberg New York, pp 181–229
Martin SB, Hoch HC, Abawi GS (1983) Population dynamics of *Laetisaeria arvalis* and low-temperature *Pythium* spp. in untreated and pasteurized beet field soils. Phytopathology 73:1445–1449
May RM (1974) Stability and complexity in model ecosystems. Princeton Univ Press, Princeton
May RM (1981) Models for two interacting populations. In: May RM (ed) Theoretical ecology: principles and applications. 2nd edn. Blackwell, Oxford, pp 78–104
Neter J, Wasserman W (1974) Applied linear statistical models. Richard D Urwin, Homewood, IL
Reynolds KM, Gold HJ, Bruck RI, Benson DM, Campbell CL (1986) Simulation of the spread of *Phytophthora cinnamomi* causing a root rot of Fraser fir in nursery beds. Phytopathology 76:1190–1201
Ross GJS (1990) Nonlinear estimation. Springer, Berlin Heidelberg New York
Sanders FE, Sheikh NA (1983) The development of vesicular-arbuscular mycorrhizal infection in plant root systems. Plant Soil 71:223–246
Scott PR (1969) Control of survival of *Ophiobolus graminis* between consecutive crops of winter wheat. Ann Appl Biol 63:37–43
Skou JP (1975) Studies on the take-all fungus, *Gaeumannomyces graminis*. IV. Development and regeneration of roots in cereal species during the attack. K Vet Landbohojsk Arsskr 1975:142–160
Slope DB, Cox J (1964) Continuous wheat growing and the decline of take-all. Rothamsted Experimental Report for 1963, Rothamsted Experiment Station, Rothamsted, pp 108–115
Toussoun TA (1975) *Fusarium*-suppressive soils. In: Bruehl GW (ed) Biology and control of soilborne plant pathogens. Am Phytopathol Soc, St Paul, Minnesota, pp 145–151
Turner ME, Bradley EL, Kirk KA (1976) A theory of growth. Math Biosci 29: 367–373
Van der Plank JE (1963) Plant diseases: epidemics and control. Academic Press, New York
Waggoner PE (1986) Progress curves of foliar diseases: their interpretation and use. In: Leonard KJ, Fry WE (eds) Plant disease epidemiology: population dynamics and management, vol I. Macmillan, New York, pp 3–37
Washington WS (1988) *Phytophthora cryptogea* as a cause of root rot of raspberry in Australia; resistance of raspberry cultivars and control by fungicides. Plant Pathol 37:225–230

Werker AR, Gilligan CA (1990) Analysis of the effects of selected agronomic factors on the epidemiology of the take-all fungus. Plant Pathol 39:161–177
Werker AR, Gilligan CA, Hornby D (1991) Analysis of disease progress curves for take-all in consecutive crops of winter wheat. Plant Pathol 40:8–24

7 Spatial Aspects of the Development of Root Disease Epidemics

C.L. Campbell and D.M. Benson

7.1 Importance of the Spatial Dimension for Analyzing Root Disease Epidemics

Epidemics of root diseases, as well as epidemics of foliar diseases, have been studied and analyzed primarily from a temporal viewpoint. In part, this is due to the emphasis placed on temporal aspects of epidemics by Vanderplank (1963, 1968). The initial emphasis on the temporal dimension of plant disease epidemics worked well to illustrate the usefulness of epidemic analysis as a means of general disease control strategies. Also, Vanderplank (1963) wisely chose to introduce a simple framework for quantitative analysis of epidemics to plant pathologists, most of whom viewed mathematical equations as a foreign language that was at best suspect in its application.

Vanderplank was quite familiar, however, with the spatial dimension of plant disease epidemics and had been concerned with spatial patterns of disease occurrence even before his epic work was published on temporal analysis of epidemics. In his 1946 paper Vanderplank (1946) applied the methods of spatial analysis proposed by Cochran (1936) and Todd (1940) to a virus disease.

Freeman (1953) was also an early proponent of studying the spatial aspects of plant disease epidemics and applied spatial analysis to nettlehead, a virus disease of hops studied by Legg (1953). He acknowledged the limitations of the methods of Cochran (1936) and Todd (1940) as noted by Finney (1947) and, instead of doublets, used the methods of Moran (1948) and Krishna Iyer (1949, 1950) which address the pattern of diseased plants within a rectangular lattice. Freeman (1953) examined the pattern on a lattice in which vacancies occurred with nominal data, i.e., join counts for similar or dissimilar points on a lattice, in this case diseased or healthy plants. Interestingly, he also investigated a temporal dimension of the epidemic and concluded in the summary to his paper: "In this case the results achieved by the two methods [spatial and temporal] supplement each other, and, in general, both must be used before it can be concluded if the apparent spread [of the disease] corresponds to a spread of inoculum." Thus, 40 years ago the need for complementary spatial and temporal analysis of epidemics had been recognized.

C.L. Campbell/D.M. Benson (Eds.)
Epidemiology and Management of Root Diseases

Table 7.1. Classification of entities, sampling units, and random variables used in the quantification of inoculum, infection, and disease associated with soilborne pathogens. (Gilligan 1988)

	Entity	Sampling units		Random variables
		Natural	Artificial	
Inoculum	Propagule	Rhizosphere	Quadrat	Direct count
	Colony-forming unit (estimated by dilution plating)	Spermosphere	Cube	Estimated count
	Bioassay-response unit (estimated by number of lesions or infections in dilution series)			
Infection	Infection site	Organ (e.g., root, seed, hypocotyl)	Quadrat	Presence/absence
	Infected organ		Cube	Count
	Infected plant	Whole plant	Root length	Ratio
Disease	Lesion	Organ	Quadrat	Presence/absence
	Diseased organ	Whole plant	Cube	Count
	Diseased plant	Field	Root length	Ratio (e.g., proportion of total roots diseased; mean length of root diseased)
	Diseased focus or patch			
	Diseased field			

The spatial dimension is perhaps even more important in the analysis of root diseases than it is for foliar diseases. Many root diseases in temperate climatic zones are caused by pathogens that reside primarily in soil, have monocyclic or polycyclic life/disease cycles, and have only limited dispersal of propagules during the course of a growing season (Table 7.1). Thus, the initial spatial pattern of inoculum has a very large influence on the pattern of diseased plants which may change relatively slowly during a growing season. Changes in spatial pattern may occur within seasons if propagule dispersal occurs or between seasons with soil tillage. The between-season changes in spatial patterns of propagules are potentially important to the understanding of epidemic processes for root diseases and are potentially important to management strategies. In those cases where propagule dispersal occurs within a growing season, such dispersal is also of importance to the development of management strategies.

In order to address the issues associated with the spatial aspects of root disease epidemics with regard to understanding epidemic processes and to developing sound management strategies, we will examine four specific questions:

1. Which ecological and environmental aspects of root pathosystems are important to our understanding of spatial processes?
2. What are the spatial dispersal mechanisms of propagules of soilborne pathogens?
3. How can the relevant methods of techniques available for quantifying spatial patterns best be used to understand spatial processes associated with root disease epidemics?
4. How can information derived from spatial analyses be applied more effectively and efficiently to manage root diseases?

7.2 Aspects of Spatial Processes

Biological and ecological characteristics of pathogen life cycles and root diseases are the primary determinants of the spatial aspects of root disease epidemics. Pathogen reproduction, dispersal, and survival provide the initial spatial determinants during epidemics. Host root growth and architecture are also determining factors (see Chap. 2). The physical, chemical, and biological processes of soil also significantly influence the dispersal of propagules and the resulting patterns of pathogen inoculum and diseased plants that occur in fields (see Chap. 4). Each of these biotic and abiotic processes occurs at various scales ranging from that of a single root and rhizosphere to large, regional, geographic areas.

7.2.1 Biological, Chemical, and Physical Attributes

7.2.1.1 Inoculum Sources, Dispersal, and Infection

Propagules of soilborne pathogens serve as the starting point for root disease epidemics. Inoculum, and especially initial inoculum, the site and extent of infection, and the entities that are diseased provide a logical basis for the identification of natural and artificial sampling units as well as the random variables available for quantifying the various aspects of root disease epidemics (Table 7.1). The role of inoculum of soilborne pathogens as an initial determinant of disease pattern is clear (see Chap. 1).

A proposed, operational categorization of soilborne pathogens based upon life cycle, mode of initial infection, and dispersal mechanism comprises three primary categories with three subcategories in category 2 (Table 7.2). The initial categorization is based upon whether one (monocyclic) or more than one (polycyclic) life/disease cycle usually occurs during a growing

Table 7.2. Types of soilborne pathogens classified operationally based on life cycle and dispersal mechanisms with an annual growing season

Life cycle	Initial infection[a]	Dispersal mechanisms	Examples
Monocyclic	Direct	Movement of survival propagules in soil or plant tissue	*Cylindrocladium crotalariae* *Fusarium* spp. *Macrophomina phaseolina* *Verticillium* spp.
Polycyclic	Direct	A. Growth of mycelium	*Phymatotrichum omnivorum* *Rhizoctonia solani* *Sclerotium rolfsii*
		B. Localized dispersal by pathogen movement (e.g., swimming in soil water)	*Meloidogyne* spp. *Phytophthora* spp. *Pythium* spp.
		C. Longer distance movement in medium (water, air) outside of soil by physical forces	*Erwinia* spp. *Phytophthora* spp. *Pythium* spp.
Polycyclic	Indirect	Movement of spores (e.g., basidiospores, ascospores, zoospores) prior to initial infection with subsequent dispersal by mechanisms A, B, or C above	*Gaeumannomyces graminis* *Heterobasidion annosum* *Sclerotinia minor* *Sclerotinia sclerotiorum*

[a] Direct infection occurs primarily by mycelial growth from a resting/survival propagule; indirect infection occurs when an intermediate spore type is produced between the resting/survival propagule and infection (e.g., ascospores from sclerotia in *Sclerotinia* spp.).

season. Although many soilborne pathogens are monocyclic within a season, many also are or have the potential to be polycyclic within a single growing season. The second factor in the categorization is the mode of initial infection which can be *direct*, where infection occurs primarily by mycelial growth from a resting or survival propagule, or *indirect*, where an intermediate propagule type is produced between the original resting or survival propagules and infection. The third factor in classification is the type of dispersal mechanisms which operate during the epidemic. In the first category, movement of propagules occurs largely by physical movement in soil or plant tissue. For the second and third categories, dispersal occurs through pathogen growth (e.g., fungal mycelium) through soil, by localized dispersal which is brought about primarily by pathogen movement (e.g., zoospores swimming in soil water), or by longer distance movement in a medium external to the actual field soil by physical forces (e.g., bacterial aerosols).

7.2.1.2 Process and Pattern

After an initial pattern of propagules has been characterized, the dispersal processes, infection, and subsequent production of additional inoculum are the primary determinants of the pattern of propagules and disease that develops. After an analysis of these determinants, it should be possible to hypothesize the final expected patterns of propagules or root disease based upon the pathogen type or category and the initial pattern of primary propagules (Table 7.3). This critical analysis of process and pattern is essential to our understanding of how root disease epidemics develop and what types of epidemics may occur. It also allows insight into the relative importance of biological and physical forces in epidemic occurrence and into the spatial scales over which epidemic processes are expected to occur.

When studying root diseases as a spatial point process, it is generally not immediately clear whether significant clustering of propagules or disease is due to *true contagion* or to *apparent contagion*. *True contagion* arises from biological processes of growth, reproduction, and dispersal of propagules. With true contagion clusters are usually small with the size dependent on the localized dispersal mechanisms that are operating. The expansion of a diseased patch of turfgrass through the outward growth of mycelium of *Rhizoctonia solani* would be an example of true contagion. *Apparent contagion* occurs as a result of chemical and physical factors which provide a favorable environment in which propagules or disease can occur. In the case of apparent contagion, dispersal mechanisms may be operating; however, environmental factors are the primary determinants of pathogen or disease occurrence. For example, edaphic factors in soil affect the occurrence of many plant-parasitic nematodes and, thus, affect spatial patterns of nematodes (Goodell and Ferris 1980; Noe and Barker 1985). Thus, an understanding of the processes that result in specific patterns of propagules or diseases is critical to identify whether the observed patterns are the result of true or apparent contagion or both.

7.2.2 Temporal Stability of Pattern as an Epidemic Attribute

The temporal scale during which an epidemic occurs depends on the type of host and pathogen and the general environment in which the epidemic occurs. Root disease epidemics of most annual harvest, herbaceous crops, e.g., root rot on snap bean or wheat, can be characterized with a time scale of a single growing season; although, inoculum carryover among seasons may be an important factor. Root diseases of woody perennials, e.g., Phytophthora root rot on Fraser fir or rhododendron, or Annosus root rot on pines may, however, occur as epidemics over multiple-year periods.

The ability to predict epidemiological and ecological phenomena depends on the relationships between spatial and temporal scales of variation. With

Table 7.3. Hypothesized effect of type of soilborne and dispersal mechanism on initial and subsequent pattern of propagules and disease. Conclusions are based upon assumption of: (1) uniform physical and chemical environment in soil; (2) presence of uniformly susceptible host plants arranged in regular agronomic patterns; (3) random pattern of microbes antagonistic to pathogen in soil; (4) random death of pathogen propagules; (5) no influx of inoculum from outside the field during the period under consideration. The point of reference for comparisons is "monocyclic with direct infection." Direct infection occurs primarily by mycelial growth from a resting/survival propagule. Indirect infection occurs when an intermediate spore type is produced between the resting/survival propagule and infection (e.g., ascospores from sclerotia)

Pathogen type	Initial pattern of primary propagules (after soil tillage)	Initial pattern of disease foci (early season)	Final pattern and size of disease foci (late season)	Final pattern of propagules (before soil tillage)
Monocyclic with direct infection	Aggregated	Aggregated	Aggregated: integrity of initial foci maintained	Aggregated within area of initial foci
Polycyclic with direct infection				
A. Growth of mycelium	Aggregated	Aggregated	Aggregated; foci slightly larger than initial foci; focal number mostly constant over time	Aggregated within area slightly larger than initial foci
B. Localized dispersal by pathogen movement (e.g., swimming in soil water)	Aggregated	Aggregated	Aggregated; foci larger than initial foci; some new foci may be present over time	Aggregated, but larger area colonized
C. Longer distance dispersal in medium outside of soil by physical forces	Aggregated	Aggregated	Aggregated to random; initial foci may not be evident; new foci present at varying distances from initial focus	Aggregated to random over much larger area than area of initial inoculum
Polycyclic with initial indirect and subsequent direct infection	Aggregated	Random → some aggregation	Random to aggregated; foci may be larger and/or more numerous than initial foci in accordance with mechanism A, B, or C	Random to aggregated over area larger than area of initial inoculum

an increase in spatial scale, the time scale of important processes will also increase because these processes, such as dispersal of propagules, operate at slower effective rates, time lags between events increase, and indirect effects may become more important (Delcourt et al. 1983; Clark 1985; Wiens 1989). An epidemic of root rot may develop within an area of a few square meters within 1 or 2 weeks, whereas, depending on the mechanisms of dispersal that are operating, an epidemic within a field or among four to five fields may develop over the course of one or several growing seasons. A pandemic (i.e., an epidemic that expands over a continent; Gäumann 1946) of a newly introduced or discovered disease, such as Cylindrocladium root rot on peanuts when it was first found in Georgia in 1965, may occur in a geographic area over a period of decades. As with general ecology (Wiens 1989), there are no standard functions that define the appropriate units for such space-time comparisons in plant disease epidemiology.

7.2.3 Spatial Scale and Dimension

The study of root disease epidemics usually occurs in small plots or individual fields over a limited time period which may be a single growing season. The spatial dimension of concern is usually the horizontal plane of the fields; although the vertical distribution of propagules has been studied for several pathogens (Nakai and Ui 1977; Adams 1981; Mihail and Alcorn 1982; Bruton and Reuveni 1985; Boag et al. 1987; MacGuidwin and Stanger 1991). The study of root disease epidemics on a larger spatial scale, e.g., among fields or within an ecoregion, or a longer time frame is rare. The study of pandemics is extremely rare for root diseases.

The general framework or classification scheme for pandemics proposed by Heesterbeek and Zadoks (1987) has both spatial and temporal dimensions. Whereas it is clearly appropriate for most foliar diseases, it may need to be modified for root diseases. Three orders of epidemics were defined. A zero-order epidemic occurs when a disease spreads within the boundaries of a single crop field with the establishment of an initial focus and of daughter foci. A first-order epidemic involves the spread of disease over an area with many fields during a single growing season. A second-order epidemic entails the spread of a disease over a large area during consecutive growing seasons of the host crop. The mechanisms of pathogen spread, as discussed in Section 7.3, will determine the time scale over which a root disease epidemic moves from a zero- to a first-order epidemic. Also, for root diseases, the concept of a first-order epidemic may need to be modified to have a multiseasonal time scale.

Ecologists recognize matters of spatial scale, as defined by extent and grain of studies, as important in the description of the dispersion or distribution of species (Greig-Smith 1952; Addicott et al. 1987; Wiens 1989). Extent is the overall area encompassed by the study and grain is the size of individual units of observation such as quadrats. Extent and grain define the upper and lower limits of resolution of the study because inferences about pattern and process cannot be made legitimately beyond the extent or below the grain. In a survey of experiments in community ecology, Kareiva and Andersen (1988) found that half of nearly 100 field experiments were conducted in plots no larger than 1 m in diameter, despite the wide range in size and types of organisms studied. In an examination of 44 published (Table 7.4) studies on spatial patterns of soilborne pathogens (fungi and nematodes) and root diseases, we found that 14 were conducted with plots or sampling areas (=grain) less than 1 m^2 and 30 studies had plots less than 10 m^2 in area. The extent or total study area ranged from less than 1 m^2 to greater than 10 000 m^2, whereby most studies were conducted in a study area of 100–9999 m^2. Most studies on the dispersal of soilborne pathogens have also been conducted within single fields.

Table 7.4. Frequency of variable assessed, grain, extent, and type of analysis used in 44 papers[a] published between 1972–1993 on the analysis of spatial patterns of soilborne pathogens and root diseases

Variable assessed	Soilborne fungi and root diseases (n = 36) (number)[b]	Soilborne plant-parasitic nematodes (n = 8) (number)[b]
Propagule or population counts	22	8
Counts of infections or lesions	5	0
Disease incidence	11	0
Grain[c] (range 0.003–231) (m^2)		
<1	12	3
1–9.99	13	4
10–99.9	9	3
≥100	2	0
Extent[d] (range 0.4–43 840) (m^2)		
<1	2	1
1–9.99	2	0
10–99.9	6	3
100–999	12	3
1000–9999	12	0
≥10 000	2	3
Type of analysis		
Maps	21	2
Indices of aggregation		
k	5	2
v/m	17	2
Lloyd's index of patchiness	8	2
Ig (morisita)	5	1
Taylor's *b* (power law)	1	4
Morisita's index of clumping	2	1
Iwao's patchiness regression	2	1

Table 7.4. *Continued*

(cont'd)	Soilborne fungi and root diseases (n = 36) (number)[b]	Soilborne plant-parasitic nematodes (n = 8) (number)[b]
Quadrat variance	1	1
Spatial autocorrelation	7	1
Runs	4	0
Doublets	1	0
Geostatistics	1	0

[a] Adams (1986); Alby et al. (1983); Belmar et al. (1987); Benson and Campbell (1985); Boag and Topham (1984); Campbell and Pennypacker (1980); Campbell and van der Gaag (1993); Campbell et al. (1984); Chellemi et al. (1988); Culbreath et al. (1990); Dillard and Grogan (1985); Ferrin and Mitchell (1986); Ferris et al. (1990); Francl (1986); Frisina and Benson (1989); Goodell and Ferris (1980); Griffin and Tomimatsu (1983); Griffin et al. (1981); Hau et al. (1982); Jarvis and Hawthorne (1972); Jeger et al. (1987); Johnson et al. (1988); MacNish (1985); Marois and Adams (1985); Martin et al. (1983); McSorley and Dickson (1991); Mihail (1989); Mihail and Alcorn (1987); Noe and Campbell (1985); Olanya and Campbell (1988, 1989); Pennacchio et al. (1985); Phelps et al. (1991); Punja et al. (1985); Rittenhouse and Griffin (1985); Roth and Griffin (1981); Schuh et al. (1986); Shew et al. (1984); Smith and Rowe (1984); Stanghellini et al. (1982); Taylor et al. (1981); Tomasino and Conway (1987); Wheeler et al. (1987); Yang et al. (1991).
[b] The sum of the numbers within a category within a column may exceed 36 or 8 because of the assessment of several variables and use of more than one grain, extent, and type of analysis reported in a paper.
[c] Grain is the size of a sampling unit such as a quadrat.
[d] Extent is the total area studied, i.e., the sum of the area of all sampling units.

The scale of an investigation has profound effects on the spatial patterns detected. Our ability to discern and interpret the effects of biological and environmental (i.e., atmospheric and geological) factors on spatial processes also depends on the extent and the grain of the study (O'Neill et al. 1986; Wiens 1989). The perceived pattern of pathogen propagules found in a plant rhizosphere or on plant roots, where propagule reproduction actually occurs, will differ from that found at a broader scale such as a field or ecoregion. A random pattern of propagules at the scale of a root system may appear as an aggregated pattern among many root systems or over an entire field. For example, propagules of *Macrophomina phaseolina* had a random to slightly aggregated pattern within corn rhizospheres (Olanya and Campbell 1989) but generally have an aggregated spatial pattern within fields (Mihail and Alcorn 1987; Olanya and Campbell 1988; Mihail 1989; Campbell and van der Gaag 1993). Also, inoculum of *Rhizoctonia solani* and rice plants with sheath blight were found to have a random pattern with small sample spacings but were aggregated at sites where the quadrats sampled were greater than 1 m^2 in size (Belmar et al. 1987).

At the finer spatial scale, local biological interactions may introduce spatial or temporal lags into the system which make detection of direct physical determinants of pattern difficult. At a broader scale, however, physical processes may dominate or dissipate the biological effects (Levin 1989). For example, at the rhizosphere scale root exudates and microbial competition or antagonism are probably the factors that most influence growth and reproduction of many soilborne pathogens. At a large, regional scale, however, general soil characteristics and environmental factors such as precipitation and temperature may be the driving factors in pathogen occurrence and disease pattern. Thus, the physical, chemical, or biological factors that appear to dominate spatial processes may have apparently different impacts on pattern as the scale of interest changes.

When the scale of observation for a variable such as inoculum density is changed, the variance of that variable changes (Wiens 1989). In a heterogeneous area (i.e., a spatial autocorrelation exists among sample locations) if extent is held constant, an increase in the grain of measurements (i.e., larger quadrats) will generally decrease spatial variance because a greater proportion of the heterogeneity of the system is contained within the sample units. If grain is held constant and extent is increased, more spatial heterogeneity will be incorporated in the area of the study with a resulting increase in between-grain variance. A confounding factor in making inferences from spatial studies on pattern or dispersal is the usual logistic necessity of enlarging the grain or quadrat size when extent is expanded (Wiens 1989).

7.3 Dispersal of Pathogen Propagules

Propagules of soilborne plant pathogens are dispersed in time and space. Here, the primary emphasis is on the dispersal of propagules in space; temporal aspects of dispersal were discussed in Chapter 1 on inoculum (see Sects. 1.1.2 and 1.3). Dispersal of propagules during the time of soil fungistasis results in a mechanism for dispersal of pathogen propagules over time and is a rate-limited process governed by the soil environment through fungistatic and survival mechanisms. However, dispersal over time complements spatial dispersal, particularly for pathogens with limited dispersal ability.

As Hirst (1965) complained at the Berkeley Symposium over 25 years ago, the greatest challenge to describing dispersal of microorganisms in space is the lack of suitable methodology to quantify dispersal gradients. Although significant progress has been made, methodologies for quantification of dispersal gradients still limit our understanding of pathogen dispersal. The goal here is to examine the mechanisms of dispersal which result in pathogen propagules being released to the environment where they can cause further infection of roots. Dispersal may be active, e.g., motile zoospores, or inactive, e.g., aerosols for bacteria. For some pathogens, both active and passive mechanisms may play a role in dispersal, e.g., forcibly ejected spores that are subsequently dispersed by wind currents.

7.3.1 Dispersal of Foliar vs. Soilborne Pathogens

Pathogens, such as the rust and powdery mildew fungi that infect aboveground portions of host tissue and rely on wind dispersal, have adapted to the problems inherent in projecting propagules above the laminar boundary area of host surfaces into the turbulent air above for dispersal. The modeling of both short- (Fitt and McCartney 1986; Aylor 1990; Aylor and Ferrandino 1990) and long-distance (Pedgley 1986; Davis 1987) dispersal of foliar pathogens is quite complex and challenging.

Propagules of soilborne pathogens are often formed and released within the solid matrix of soil. Thus, these propagules may be dispersed only very short distances if dispersal occurs below ground. However, many organisms that cause root diseases have evolved mechanisms through which propagules can be moved out of the soil matrix and the resulting aboveground dispersal may be long ranged.

7.3.1.1 Wind Dispersal of Soilborne Pathogens – Crop Debris

The operations associated with crop harvest can facilitate the localized dispersal of soilborne fungal pathogens by wind. For example, combining peanut plants infected with *Cylindrocladium crotalariae* can eject pathogen-infested crop debris into the air. Rowe et al. (1974) detected peanut root fragments of a size capable of containing microsclerotia of *C. crotarlariae* at a distance of 234 m downwind from a combine on a day when wind speed did not exceed 4 m/s. However, the number of particles large enough to contain microsclerotia decreased rapidly beyond 60 m from the combine. Dispersal of *C. crotalariae* with peanut root residues may be unique because peanuts are first dug to expose pods, then combined to remove pods from plants. Alternatively, the ubiquitous pathogen *Rhizoctonia solani* has a strong saprophytic phase, such that colonization of aboveground, senescent plant tissues may result in wind dispersal over short distances during harvesting activities.

Wind storms in crop production areas may be important factors in the dispersal of soilborne plant pathogens. Easton et al. (1969) collected dust samples 6.1 m above ground in an irrigated potato-growing area of Washington. Over 100 propagules of *Verticillium albo-atrum* per gram dust were enumerated. The nature of the propagules and their ability to initiate outbreaks of Verticillium wilt were not determined. However, the dispersal of propagules of *Verticillium* spp. on alfalfa mowing equipment (Isaac 1957) suggests that conidia or microsclerotia dispersed during dust storms might infect aboveground host tissue that was wounded by particulates during a dust storm.

In production of greenhouse tomatoes, growers normally steam-treat ground beds between crops to control weed seeds, soil insects, nematodes, and plant pathogens. In the mid-1970s this practice failed to control a crown and root rot caused by *Fusarium oxysporum* (Rowe et al. 1977). Trapping of microconidia on petri plates of *Fusarium*-selective agar medium suspended up to 1.5 m above the soil surface demonstrated the occurrence of numerous airborne microconidia of *F. oxysporum* (Rowe et al. 1977). Crown and root rot developed when airborne microconidia recolonized the steam-treated beds in which tomatoes were planted. Even though only 4% of the *F. oxysporum* isolates trapped from the air in greenhouses were pathogenic to tomato, the biological vacuum created when ground beds were steam-treated favored rapid recolonization by the pathogen and subsequent disease development (Rowe et al. 1977).

7.3.1.2 Wind Dispersal of Soilborne Pathogens – Perfect States

The teleomorph or sexual stage of some soilborne fungal pathogens may be important in dispersal of these pathogens. For instance, perithecia of *Calonectria crotalariae*, the perfect state of *Cylindrocladium crotalariae*, form on peanut stems at the soil line during the growing season. However, the sensitivity of ascospores to desiccation would seem to limit dispersal to short-distance, in-field dispersal (Rowe and Beute 1975). Ascospores of *Calonectria theae*, conversely, may have an important role in the epidemiology of Cylindrocladium root rot and blight on azalea (*Rhododendron* sp.) and other susceptible ornamentals in propagation beds and production areas. Normally, propagation beds are misted to keep cuttings moist until rooted. Perithecia formed on infected tissue can forcibly discharge ascospores up to 2 cm if relative humidity decreases prior to ascospore discharge (Linderman 1974). Ascospores dispersed by wind over relatively short distances in the propagation bed would have a good chance of landing on susceptible, moist tissue for germination and infection.

Although *Gaeumannomyces graminis* usually spreads over the root system of wheat plants from infected plant debris in soil and from plant to plant through soil, the ascospores of this pathogen can also play a vital role in the dispersal of the pathogen, particularly into new areas. The function of ascospores as infective propagules was unclear (Garrett 1939, 1970) until Brooks (1965) demonstrated that ascospores could readily infect the proximal part of seminal roots of wheat seedlings that were produced by allowing seeds to germinate on the soil surface. Because the proximal part of the seminal roots, between the seed and the soil, was not in direct contact with the soil, the usual competition provided by root surface microflora for nutrients was not present. Thus, ascospores of *G. graminis* could infect "volunteer" wheat plants which grew from seed deposited on the soil surface at harvest (Garrett 1970). Gerlagh (1968) demonstrated the role of ascospores in the introduction and establishment of *G. graminis* into the virgin soils of new polders in The Netherlands.

A tobacco leaf spot disease on flue-cured tobacco caused by *Rhizoctonia solani* developed in unusually wet and cool weather when basidiospores of *Thanatephorus cucumeris* were dispersed by wind from hymenial layers that formed on infected tobacco leaves (Shew and Main 1985). A unique aspect of the epidemiology of the leaf spot disease was the observation that all isolates that anastomosed were AG 2 type 2. This anastomosis group consists of isolates that cause crown and root rots rather than aerial blights typical of isolates in AG 1. Thus, a pathogen that is normally associated with root diseases had adapted under favorable weather to an aerial environment mediated by wind dispersal.

White mold of bean caused by *Sclerotinia sclerotiorum* develops when windborne ascospores dispersed from apothecia are deposited on bean flowers (Abawi and Grogan 1975). Even though the pathogen survives as sclerotia in soil, the primary inoculum consists of ascospores ejected forcibly from apothecia. The apothecia develop through the carpogenic germination of sclerotia buried in the top 2–3 cm of infested soils when adequate and continuous soil moisture is present. Ascospores must be forcibly ejected to reach turbulent air above the apothecium for wind dispersal. Presumably, changes in turgor pressure in the apothecium mediated by diurnal changes in moisture result in ascus discharge (Abawi and Grogan 1979). Wind dispersal of ascospores varies from less than 30 m, to several hundred meters, to several kilometers (see refs. 10, 35, and 36 in Abawi and Grogan 1979) depending on experimental design. Ascospores deposited on senescing bean flowers adhere with a mucilaginous substance present on the ascospore (Abawi and Grogan 1979).

7.3.1.3 Wind Dispersal of Soilborne Pathogens – Aerosols

The soft rot erwinias, *Erwinia carotovora* ssp. *atroseptica* and *E. c.* ssp. *carotovora*, cause black leg and tuber rot in potato. *E. c. atroseptica* is more common on potatoes at cooler temperatures and *E. c. carotovora* is more prevalent at warmer temperatures near 15 °C (Perombelon and Kelman 1980). Both organisms are common in the rhizospheres of hosts and non-hosts and bacteria from aboveground infections may also wash down plant stems during rain storms to infect tubers below ground. However, aerosols also have a demonstrated role in dispersal of these pathogens.

Graham and Harrison (1975) demonstrated that rain impacting on potato stems infected with *E. c. atroseptica* could disperse the bacterium in aerosols created from rain impaction. Although experiments were done in wind tunnels, aerosols persisted up to 60 min in still air after simulated rain. As rain droplet size increased from 2 to 4 mm in diameter, the number of colonies of *E. c. atroseptica* recovered in aerosol particles increased. In subsequent work, the number of propagules released in aerosols of *E. c. atroseptica* was in direct proportion to the number recovered per gram of infected stem tissue, and based on the power law model for dispersal (Gregory 1973). Graham et al. (1977) estimated that the propagules could be dispersed in aerosols up to 1000 m from the source inoculum. Venette and Kennedy (1975) sampled the margin of soybean field plots and were able to detect the foliar pathogen, *Pseudomonas glycinea*, in aerosol particles generated by rain storms and irrigation events. Bacteria were most common in aerosols of 2.1 to 3.3 μm in diameter. The upper size limit of aerosols containing *P. glycinea* was 4.7 to 7.7 μm and the lower limit was 0.65 to 1.1 μm, as determined by colony development on a selective medium placed in an Anderson air sampler (Venette and Kennedy 1975).

Pulverization of potato vines by mechanical means prior to harvest can produce aerosols of *E. carotovora* (Perombelon et al. 1979). Even in healthy fields, it was estimated that each hectare of potato released 10^7 to 10^9 cells of *E. c. carotovora* in aerosols by September. An additional 10^8 cells were estimated to be released per hectare at 1% disease incidence. Aerosols of 3.3 to 4.7 μm in diameter (one class size larger in the Anderson sample than were most common for *P. glycinea*) occurred in the greatest frequency with *E. c. carotovora*. The probability of aerosols of *E. c. carotovora* remaining suspended in the atmosphere decreased exponentially over time, but 3% of the aerosol particles were still suspended in the atmosphere after 120 min (Perombelon et al. 1979). Downwind dispersal of aerosols was measured by exposing plates of selective medium at various distances from areas where potato vines were being actively pulverized. When the source tissue contained 10^8 cells, up to 10^3 colonies at 50 m, and up to 10^2 colonies at 100 m were recovered. Perombelon et al. (1979) used diffusion models to estimate long-range dispersal of *E. carotovora* from areas where potato haulm was pulverized. The most optimistic model (i.e., the power law model of Gregory 1973) predicted up to 1.4×10^2 cells 10000 m downwind, while the least optimistic model (Pasquill 1974) predicted only 0.02 cells at 10000 m. Differences in assumptions used to develop the models and more detailed information on weather parameters in the Pasquill model led to the large differences in the estimated dispersal of *E. carotovora* in aerosols with the two models. Perombelon et al. (1979) estimated that rain-generated dispersal of *E. carotovora* in aerosols might be effective up to 100 m from source tissue and up to several hundred meters from mechanically pulverzied haulm.

Long-distance transport of *E. carotovora* in aerosols may be a significant factor in the epidemiology of soft rot diseases. Franc et al. (1985) speculated that dispersal of soft rot erwinias from Pacific Ocean waters to snow packs in Colorado mountain passes occurs by means of bacterial aerosols transported in the atmosphere by prevailing winds. Transported bacterial aerosols may serve as ice nuclei or cloud condensation nuclei, thus enhancing the probability of this dispersal mechanism. Subsequent contamination of drainage systems and eventually irrigation water used for potato production in Colorado may occur based on the distribution of the same serotypes in both locations (Franc et al. 1985; Cappaert et al. 1988).

Dispersal of *E. carotovora* in aerosols is important epidemiologically only if dispersed bacterial cells can survive and infect potato tissue upon deposition. Perombelon (1978) found poor survival of *E. carotovora* on leaves at 10 or 20 °C and a relative humidity of 80%; however, the bacterium did survive on leaves for up to 10 days at 10 and 20 °C when relative humidity was 100%. Survival was best on senescent leaves at 20 °C when foliage was wet (100% RH) and poor on young leaves, conditions that occurred during September and October in Scotland.

7.3.2 Dispersal in Water

Soilborne pathogens can be introduced into a field or crop in irrigation water. Contaminated water sources provide some pathogens with a suitable environment for survival and provide the necessary moisture needed for epiphytic growth or infection of the host. *E. c. carotovora* was detected by an anaerobic liquid enrichment technique or by direct culture of surface water samples from potato-growing areas and nongrowing areas in Scotland and Colorado (McCarter-Zorner et al. 1984). In an extensive sampling project, Harrison et al. (1987) detected *E. c. carotovora* in 66 rivers, springs, creeks, streams, lakes, reservoirs, and ponds in 16 states, plus water samples from the Pacific Ocean and samples taken 29 km offshore and 8.5 m deep in the Gulf of Mexico. The majority of strains recovered were *E. c. carotovora* (98.8%), but *E. c. atroseptica* (1.2%) was also found. Fortunately, in Colorado, *E. c. atroseptica* is the predominate strain causing blackleg.

Cappaert et al. (1988) investigated the relationship between erwinia soft rot strains in irrigation water and epiphytic growth on potato leaves and disease incidence. A serotyping scheme was used to categorize individual strains into serogroups. Soft rot erwinias were recovered from surface and well-water samples in Oregon but only from surface water in Colorado. Populations as high as 575 cfu (colony-forming units)/ml were detected in surface waters in Oregon, but usually >1 to 20 and >1 to 100 cfu/ml were recovered depending on year. The predominate bacterium was *E. c. carotovora* (92–100% of samples). Well-water samples in Oregon but not Colorado also yielded *E. c. carotovora*, but the numbers were so low that an enrichment technique was required to detect the bacteria and quantiative population estimates could not be made. Only one potato stem developed aerial soft rot symptoms in Colorado in 1 year of the 2-year study, but potatoes in all locations in Oregon had aerial stem rot. Incidence of aerial stem rot was not correlated with well-water or surface-water sources.

Significantly, serogroups present as epiphytes on potato leaves in Oregon and Colorado and in diseased stems in Oregon were the same serogroup as those in irrigation water but different from seed tuber and soil serogroups (Cappaert et al. 1988).

Zoospores of *Phytophthora* spp. have been detected in surface water used for irrigation in apple regions of British Columbia (McIntosch 1966), in citrus regions of Arizona and California (Klotz et al. 1959; Thomson and Allen 1974), in ohia forest areas of Hawaii (Kliejunas and Ko 1976), in orchard fruit and nut tree regions of California (Mircetich et al. 1985), in flue-cured tobacco regions of Georgia (Dukes et al. 1977), and in the major river systems of South-Western Cape Province in South Africa (Von Broembsen 1984). Most irrigation water assays have been based on exposure of a suitable bait suspended in an irrigation pond or canal for several days followed by tissue isolations on semiselective media from symptomatic baits, although Shokes and McCarter (1979) and Von Broembsen (1984) filtered water samples prior to culturing.

Klotz et al. (1959) suspended lemon fruit in mesh bags in headwater sources and in irrigation runoff water from citrus areas in southern California. No *Phytophthora* spp. were recovered from headwater sources, but isolates were obtained from citrus orchard runoff water samples. The detection of *Phytophthora* spp. depended on water temperature, in cooler months when water temperature was 6–17 °C, *P. syringae* was isolated, whereas *P. parasitica* was found when water temperatures were between 17–31 °C. Normally, *P. syringae* is a cool season pathogen and *P. parasitica* is a warm temperature pathogen, thus the correlation with water temperature and frequency of recovery in irrigation sources. *P. citrophthora*, on the other hand, was recovered throughout the year in irrigation runoff (Klotz et al. 1959).

Thomson and Allen (1974) studied dispersal of *Phytophthora* spp. in irrigation sources in citrus regions of Arizona. Samples from 20 sites contained isolates of *Phytophthora* spp. in 93 of 599 assays. *P. parasitica* and *P. citrophthora* were recovered in about 4% of the water samples. *P. parasitica* was found when water temperatures were 20–30 °C as Klotz et al. (1959) had reported, whereas *P. citrophthora* occurred in samples when the water temperature was above 23 °C. *Phytophthora* spp. were never recovered in water samples from noncitrus regions in Arizona (Thomson and Allen 1974).

McIntosh (1966) monitored seasonal recovery of *Phytophthora* spp. with green pears in irrigation sources for apple-growing areas in British Columbia. From 23 of 31 irrigation sources he recovered *Phytophthora* spp. most frequently in July, August, and September. *P. cactorum*, the apple collar rot pathogen, occurred most frequently of all *Phytophthora* spp. isolated.

Mircetich et al. (1985) recovered 1633 isolates of *Phytopthora* spp. from irrigation sources in orchard fruit and nut tree regions around Sacramento, California. The isolates represented eight identifiable species and seven unidentified *Phytophthora* spp. Selected isolates were pathogenic to almond, apple, cherry, peach, plum, and walnut seedlings.

Pythium spp. have also been recovered from irrigation ponds with root systems of potted plants serving as the bait (Gill 1970). *Pythium myriotylum*, pathogenic to tomato, pansy, and sweet william in greenhouse pathogenicity tests, was isolated more frequently in the period May–June than in the period June–July, but water sample temperatures were not reported (Gill 1970). Shokes and McCarter (1979) recovered *Pythium irregulare* in 13.9% of the water samples from south Georgia irrigation ponds. *Pythium aphanidermatum*, as well as 21 other *Pythium* spp. in population densities ranging up to 700 propagules per liter of water sample, were also recovered with a high volume filtration system.

Dispersal of *Phytophthora* spp. by furrow irrigation is a rapid means to introduce inoculum across a field. When inoculum of *P. parasitica* was buried 5 cm below the surface of an irrigation furrow, leaf disks of tomato at distances of up to 32 m downstream became infected by the fourth irrigation (2-week irrigation intervals); and tomato seedlings as inoculum traps in furrows 50 m downstream became infected after each irrigation (Neher and Duniway 1992). Although movement of inoculum can be rapid during introduction, tomato plants in rows as close as 2 m from the point of original inoculum introduction, developed only slight symptoms of buckeye rot at harvest (Neher and Duniway 1992). Nonetheless, furrow irrigation water can be a significant source for introduction and dispersal of pathogen propagules within a field of an agricultural region.

Zoospores of *Phytophthora* spp. are the most likely dispersal unit in irrigation water. As reported above, isolations of *Phytophthora* spp. were successful with baits suspended in water sources, whereas other pathogen structures would not be expected to infect submerged or floating baits. Thomson and Allen (1974) used screens of different sizes but recovered *Phytophthora* spp. only on a 8-μm sieve, a size known to trap zoospores. Kliejunas and Ko (1976) collected water from streams and puddles in areas of ohia forest in Hawaii naturally infested with *P. cinnamomi*. Colonies developed only from lupine baits or in filtrates passed through a 20-μm-mesh sieve to exclude propagules larger than zoospores.

Dispersal of *Phytophthora* spp. in irrigation water may be a primary means of introducing pathogens into new orchards in established growing districts as well as reintroducing the pathogen to established orchards during irrigation. For water dispersal to be a significant part of the ecology and epidemiology of a pathogen, the microorganisms must be adapted to survival in irrigation water for periods sufficient to insure dispersal to new host tissues. Thomson and Allen (1976) found that zoospores of *P. parasitica* were liberated from infested soils within 10 min after water was added and that zoospores remained motile for up to 20 h at 20 °C. Once zoospores geminated a contracted protoplasm was formed in hyphae which survived up to 40 days in untreated waste water at 25 °C. Survival of zoospores of *P. aphanidermatum* in pond water was short-lived, however (Shokes and McCarter 1979). Recovery of zoospores placed in 200-ml containers with filters to allow exchange of pond water with water within the container was only 4.2% after 6 days and 0% after 12 days. However, when oospores or ryegrass seeds colonized by *P. aphanidermatum* were placed in the containers and added to an irrigation pond in September, 42% of the ryegrass seeds had viable propagules of *P. aphanidermatum* after 185 days in the pond, while 17% of the oospores germinated after 185 days.

Shokes and McCarter (1979) isolated *Fusarium oxysporum*, *F. solani*, 11 other fusaria, *R. solani*, *R. zeae*, *Macrophomina phaseolina*, and a *Sclerotium* sp. from irrigation ponds. Several of the isolates were pathogenic on selected crops. Survival of *R. solani* and a *Sclerotium* sp. was tested in the filtered containers in pond water during September to March. Population density of *R. solani* decreased gradually over time and failed to survive beyond 100 days regardless of whether inoculum was introduced as sclerotia or colonized cotton stem segments to the containers, whereas 92% of the sclerotia of the *Sclerotium* spp. produced colonies after 133 days. Water temperatures at 1.2–1.4 m were 26 °C in September, 7 °C in December, and up to 12 °C by March.

Easton et al. (1969) collected irrigation water as it came into a potato field in Washington infested with the microsclerotial form of *Verticillium albo-atrum*, then collected another water sample at the end of the irrigation furrow. Samples in June and July had 37 950 propagules/l and 301 290 propagules/l, respectively. An irrigation pond, about 0.9 km from an infested potato field had water samples ranging from 4388–15 410 propagules/l of water. Unfortunately, microsclerotia, the primary infective propagule for this pathogen, and conidia were not differentiated during isolation. Given the relatively low inoculum density of *V. albo-atrum* as microsclerotia in soil and the relatively high propagule count of propagules in irrigation water, most of the propagules recovered may have been short-lived conidia that would not function in infection of host tissue.

In the North Platte region of western Nebraska, muck samples collected in July and August from the bottom of irrigation canals contained sclerotia of the white mold pathogen of bean, *Sclerotinia sclerotiorum* (Steadman et al. 1975). Sclerotia occurred most frequently in muck samples from canals that collected bean field runoff, but also in muck samples from canals that collected corn field runoff. Over a 3-year period 23% of 252 samples contained sclerotia. Survival of sclerotia in irrigation water was 85–99% after only 10 days but declined precipitously between 21 and 51 days. Whether or not sclerotia contained in bottom muck samples in canals and irrigation ditches could ever become resuspended in moving water and dispersed into fields with irrigation water was not examined.

Steadman et al. (1975) also studied the occurrence of plant pathogenic nematodes in water samples at various depths in irrigation canals. *Ditylenchus* spp. was recovered most frequently (26%) followed by *Aphenlenchoides* spp. (14%). *Pratylenchus* spp. (9%), and *Heterodera* (5%) in all 110 locations sampled and at most water depths.

7.3.3 Dispersal in Soil

Movement of propagules in soil by aboveground means including plant shipments, cultivation equipment, and other human activities may result in both short- and long-range dispersal of soilborne plant pathogens. In an ohia decline area of Hawaii, Kliejunas and Ko (1976) detected *P. cinnamomi* in mud samples adhering to tires of vehicles, boots of workers, and hooves of feral pigs. The importance of people's activities in dispersal of propagules of *P. cinnamomi* in jarrah forests of western Australia was recognized and several general policies and preventative measures were employed to prevent further dispersal of this devastating pathogen (Weste and Marks 1987).

Dispersal of soilborne plant pathogens in the below-ground soil matrix is dependent on many physical factors including temperature, moisture, particle size distribution, aggregation of particles, channels produced by activities of soil arthropods and small burrowing animals, etc. most of which have not been studied extensively. Unfortunately, information on dispersal of pathogens in undisturbed soil is limited due to difficulties in experimental techniques. Much of the available data is based on experiments that used sieved and reconstituted soil materials. Even when these materials are packed to constant bulk densities, the experimental soil may not reflect conditions in undisturbed soil.

Dispersal of zoospores of oomycetous fungi, particularly *Phytophthora* spp., has been studied fairly extensively in both reconsituted soils and in ideal soils, e.g., glass beads. Zoospores liberated from sporangia near 0 millibars (mb) matric potential (saturation) (MacDonald and Duniway 1978) disperse in the soil matrix to find host roots through chemotactic mechanisms (Allen and Newhook 1973).

Zoospore dispersal is limited by matric potential and soil texture. Duniway (1976) compared zoospore dispersal in UC mix (peat : sand) and four loam soils under flooded and nonflooded conditions. Zoospores of *P. cryptogea* swam 25–35 mm in static surface water or through the coarse-textured UC mix at >-1 mb to infect safflower seedlings. As matric potential became more negative, dispersal was limited at -10 mb and not observed at -50 mb. In finer-textured soils, dispersal was limited compared to coarse-textured soils. In a silt loam and fine sandy loam, zoospore dispersal was only 5 mm at -1 mb, while in a clay loam soil zoospores swam this same distance at -10 mb (Duniway 1976). In a coarse sand, zoospore dispersal of *P. cinnamomi* from a point source under static water conditions followed a simple diffusion pattern over time with some zoospores detected up to 6 cm from the source after 10 h (Newhook et al. 1981).

Zoospore dispersal may also be limited by motility under various soil conditions as well as the inherent motility of a given species. For instance, MacDonald and Duniway (1978) found that zoospores of *P. megasperma* lost motility quickly after release into soil. Zoospores of *P, cryptogea*, on the other hand, remained motile for several hours depending on soil texture.

In fine-textured soils with a range of 38–60 μm particles, motility diminished rapidly with only 25% of the zoospores motile after 2 h. In a coarse-textured soil with particles >250 μm, 30% of the zoospores were still motile after 6 h. Texture effects were so dominant that a change in temperature from 27 to 12 °C did not affect motility (MacDonald and Duniway 1978).

Zoospore dispersal under static water conditions provides only a partial description of the influence of matric potential, texture, and the other physical factors associated with dispersal. In nature, zoospores may be dispersed by autonomous movement over short distances as well as passive transport as rainfall or irrigation water percolates into the soil profile. Under these conditions, motility and subsequent dispersal may be greater than the distances suggested by experiments under static water conditions. In an "ideal soil", e.g., glass beads of uniform size, zoospores did not pass a column of beads with pore neck diameters less than the length of the zoospore (15 μm), but zoospores did pass columns with pore neck diameters of 25–30 μm (Newhook et al. 1981). Percent motility of zoospores passing a 25–30 μm column was not different than that of zoospores passing a 150–200 μm column after 1 h, suggesting that pore neck diameter and hence texture do not affect motility (Newhook et al. 1981). Dispersal of zoospores of *P. cinnamomi* in soil water was 81–87% in a sand-soil mix with pore neck diameters greater than 180 μm, a diameter equivalent to the space needed for the helical swimming path of a zoospore to make one cycle (Newhook et al. 1981). As soils drain to −10 mb, pore neck diameters are about 300 μm in diameter, a distance adequate for zoospore swimming. At −50 mb, water-filled pore necks are only 60 μm, zoospore swimming is restricted, and zoospore dispersal is limited (Duniway 1976).

Once in the vicinity of a host organ, zoospores are aided in dispersal by a chemotactic response across nutrient concentration gradients. Allen and Newhook (1973) found that zoospores of *P. cinnamomi* were attracted across ethanol gradients of increasing concentration. In subsequent experiments, however, the chemotactic response was overcome when the flow rate of water through horizontal columns of glass beads exceeded the swimming speed of zoospores (Newhook et al. 1981).

Dispersal of motile zoospores through undisturbed soil columns 12 cm deep was diminished in each subsequent aliquot of surface water applied (Newhook et al. 1981). When a 200-ml suspension of zoospores was added to undisturbed soil columns, 4.8 colonies per 5-μl drop were recovered after passage through the column. This represented 75% of the original sample based on recovery of zoospores in additional aliquots. Motile zoospores are dispersed through soil columns by passive transport. Duniway and McKeen (1987) found that as many as 40% of motile zoospores added to the top of 20-cm soil columns were dispersed through a reconsituted soil. In contrast, encysted zoospores were dispersed only 4 cm under the same flow rate conditions. Without water flow through the columns, motile zoospores were only dispersed 4 cm deep. Apparently, motility is important for dispersal in soil under water flow conditions. Encysted zoospores must be trapped in pore necks too small to allow their passage, while motile zoospores are able to avoid these potential trap sites in pore necks and remain in larger pore channels that permit dispersal to greater depths.

Benson (1990) determined the dispersal of motile and encysted zoospores of *P. parasitica* in saturated 15-cm-high columns of three different media used in the ornamentals industry. Zoospores were added to the tops of columns, then the columns were allowed to drain, and the effluent containing zoospores was collected and cultured. Additional void volumes without zoospores were added and effluent collected and assayed for zo-

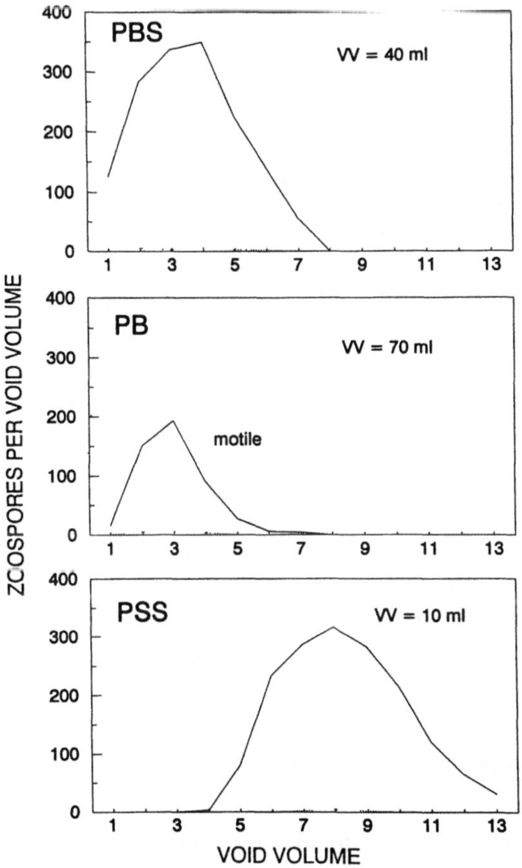

Fig. 7.1. Dispersal of zoospores of *Phytophthora parasitica* through 15-cm-high columns of *PBS*, pine bark:sand (3:1, by volume); *PB*, pine bark, or *PSS*, peat: sand:soil (1:1:1, by volume). Void volume (vv) refers to the amount of water needed to replace that held in the column of mix after saturated columns were allowed to drain freely. Zoospores moved more rapidly through coarse-textured mixes like *PBS* and *PB* than through the fine-textured *PSS*. *Solid lines* denote motile zoospores; *dashed lines* denote encysted zoospores

ospores. Encysted zoospores did not pass through the media columns or were detected in only very small numbers (Fig. 7.1). The greatest number of motile zoospores moved through a pine bark medium that had the largest pore size and hence the largest void volume in three void volumes, compared to four void volumes for a pine bark:sand medium and eight void volumes for a peat:sand:soil medium with the smallest pore size (Fig. 7.1)

Other factors such as soil salinity may affect zoospore dispersal through effects on zoospore discharge, motility, and encystment. Swiecki and MacDonald (1991) found that zoospores of *P. parasitica* were released from sporangia at one-fourth the rate in soils at an electrical conductivity (EC) of $8\,dS\,m^{-1}$ as in nonsaline controls, even though more sporangia formed in vitro at EC values up to $11.8\,dS\,m^{-1}$ compared to a nonsaline control. Motility of zoospores decreased with increasing salt concentration in vitro. In vivo, the number of zoospores recovered from baiting of leachate from pots with tomatoes inoculated with *P. parasitica* and grown under constant saline conditions was significantly less than the number recovered from nonsaline or saline-leached pots (Swiecki and MacDonald 1991). Under saline conditions zoospores would be motile for shorter periods of time and thus not able to disperse over as wide a soil volume as zoospores in nonsaline soils.

Although dispersal of motile zoospores has been studied fairly extensively, dispersal of conidia and other large-spored structures has not received as much attention. Dispersal of conidia of *Penicillium* spp. and spores of *Streptomycetes* spp. in soil was affected by spore size and spore surface (Ruddick and Williams 1972). The relatively large conidia of *Penicillium* (3 μm diameter) were limited to dispersal in soils with 24.4-μm pores. Two species of *Streptomyces* that differed only in short versus smooth spines on the surface of 1-μm-diameter spores were dispersed in soils having 2.3-μm pores, but 14% more smooth-spined spores were detected in the 2.3-μm pore size soil. As with zoospores, dispersal of conidia and other nonmotile spore structures in soil by passive water transport is probably limited by the size of the propagule and the average pore neck diameter of a given soil.

7.4 Spatial Pattern of Propagules and Diseases

Plants and soilborne pathogens are associated to varying degrees in a spatial framework. This framework is established by the occurrence of initial inoculum and modified extensively through pathogen dispersal. This spatial framework or pattern varies along a continuum from a high degree of aggregation to randomness for most plant diseases (Campbell and Noe 1985; Campbell and Madden 1990) and is a determining, ecological characteristic of species of pathogens and plants (Taylor 1984). The quantitative description of the spatial association between plants and pathogens, because it is an ecological characterization, is essential for understanding and managing plant disease epidemics.

7.4.1 Spatial Pattern of Soilborne vs. Foliar Pathogens and Associated Diseases

The environmental and biological factors that affect the spatial patterns of diseases caused by soilborne pathogens are often quite different than those affecting diseases caused by foliar or airborne pathogens. The infection courts for soilborne pathogens, primarily roots and lower stems, are different from those of airborne or foliar pathogens. The soil environment is much more stable than that of plant canopies. The physical and chemical characteristics of the media in which the propagules are dispersed may be different for the different types of pathogens. The biological environment may be variable and may be more hostile for soilborne pathogens than for airborne pathogens. Dispersal mechanisms (see Sect. 7.3) may also vary widely between soilborne and airborne pathogens.

Because of the differences in the physical, chemical, and biological environments in which root diseases and foliar diseases occur, their respective spatial patterns may have little similarity. Although the data in the literature are not yet sufficiently general to fully support or refute specific hypotheses about the difference between the spatial patterns of diseases caused by soilborne and foliar pathogens, the following hypotheses are proposed:

1. Initial spatial patterns of diseases caused by soilborne pathogens are generally more aggregated than patterns of diseases caused by foliar pathogens.
2. Spatial patterns of inoculum of soilborne pathogens and associated diseases will be better conserved and more recognizable over multiyear periods than will spatial patterns for airborne pathogens and the resultant diseases.
3. Geographic extent of studies on dispersal and patterns for soilborne pathogens and root diseases can be smaller than that for airborne pathogens and foliar diseases and still be ecologically meaningful and interpretable; grain (plot size) can also be smaller for studies on many soilborne pathogens because of more limited distances of dispersal and expected smaller cluster size for most soilborne pathogens and root diseases than for strictly airborne pathogens and foliar diseases.

7.4.2 Data, Sampling Units, and Approaches

Selection of the random variable and sampling unit for use in analyzing spatial patterns is critical to the success of the analysis. Gilligan (1988)

Table 7.5. Types of analyses available for quantifying the spatial pattern of soilborne propagules of pathogens and root diseases and the information gained from each type of analysis

Type of analysis	Random variable[a]	Type of sample unit	Information gained or application
Mapping	Count, ratio presence/absence, nominal	Any	Visual display of pattern
Fitting probability distribution	Count (unmapped)	Quadrat or other discrete area	Random/nonrandom indication; mechanism underlying distribution if sample size and number of data sets are sufficient
Indices of dispersion	Count (unmapped)	Quadrat	Degree of aggregation
Blocked quadrat variance	Count (mapped)	Quadrat	Mean size of clusters
Spatial autocorrelation	Count, ratio, nominal (mapped)	Quadrat or sample points at specific locations	Occurrence, size, and shape of clusters, suggestion of type of dispersal
Transect analysis (runs analysis or doublets)	Presence/absence (mapped)	Plant organ, plant, or quadrat	Random/nonrandom indication
Two-dimensional lattice or two-dimensional distance class	Presence/absence (mapped)	Plant organ, plant, or quadrat	Size and shape of clusters; suggestion of type of dispersal

[a] Count is synonymous with discrete and ratio is a type of continuous data.

identified the random variables and sampling units available for use in quantification of inoculum, infection, and disease associated with soilborne pathogens (Table 7.1). Direct or estimated counts of propagules, infections or diseased organs, and presence/absence or ratio data for infection and disease are the random variables available for spatial analysis. Quadrat or whole plants are often the sample units selected.

The decision as to what type(s) of data to collect and from which sample units is not trivial. Both factors will influence the cost of collecting data, the type of spatial analysis that can be performed, and the interpretations that are possible from the spatial analysis. Some analyses, e.g., fitting probability distributions and indices of dispersion, require count data collected from quadrats or other discrete areas (Table 7.5). Other analyses, such as transect analysis or two-dimensional lattice analysis, are appropriate for use only with presence/absence data, but the data can be derived from a range of sample unit types (Table 7.5).

Three general approaches have been taken for the analysis of spatial patterns of plant pathogens and diseases. In the first, the pattern is quantified as a *spatial point process*; in the second, *spatial autocorrelation analysis* is used; and in the third, *distance-based analysis* is employed. Each approach has merit and is appropriate for certain types of spatial data and for answering certain questions (Table 7.5).

If a field is divided into quadrats and the number of points (e.g., propagules or diseased plants) are counted in each unit, the spatial pattern of the points can be displayed on a map and analyzed as a frequency distribution. The fitting of statistical distributions such as the Poisson or negative binomial to spatial data and the use of indices of dispersion have formed the basis of many of the spatial analyses in ecology (Pielou 1977) and in plant pathology (Noe and Campbell 1985; Madden 1989; Campbell and Madden 1990). The specific question addressed is: "Does the collection of frequencies of occurrence for the variate correspond with a random expectation for a given mean value?" Although this type of analysis concentrates on the variance and not the two-dimensional spatial location of the points, it provides an indication as to whether the points are scattered randomly relative to an identified area, e.g. relative to the plane of a field (Sakai and Oden 1983).

With spatial autocorrelation analysis the spatial relationships among the points among quadrats is preserved and forms the basis of the analysis. Comparisons can be made in several directions and the degree of similarity between adjacent quadrats is examined. This approach to the analysis of spatial data addresses the question: "Are the variates (e.g., counts of propagules) scattered randomly on the points in space relative to each other?" The spatial autocorrelation approach examines the relation of the variate of interest and associated characteristics to each other (Cliff and Ord 1981) and provides more insight into the spatial processes that are operating (Jumars et al. 1977; Sokal and Oden 1978a,b; Sokal and Thomson 1987) than does the analysis of spatial point processes.

The third approach, distanced-based analysis, as it has been applied in plant pathology (Campbell and Noe 1985; Noe and Campbell 1985; Campbell and Madden 1990), is an extension of the analysis of spatial point processes but is based upon presence/absence data and preserves some of the spatial information in the data. With the two-dimensional distance class analysis (Gray et al. 1986a,b; Frisina and Benson 1989; Nelson et al. 1992) the number of pairs of diseased plants at each specified distance is compared to the expectation of that number with a random spatial point pattern. The question addressed is: "Is the occurrence of pairs of diseased plants over all distance classes similar to that expected for a random pattern of disease occurrence?

7.4.3 Descriptive and Analytical Methods

Seven types of analyses within the three general approaches have been used in characterizing spatial patterns of root diseases and soilborne pathogens (Table 7.5; Noe and Campbell 1985; Gilligan 1988; Campbell and Madden 1990; Jeger 1990). Mapping is the most elementary type of analysis and should be considered as the first element of any spatial analysis (Fig. 7.2). The other analyses can then be divided on the basis of whether unmapped or mapped random variables are utilized. For analyses, such as fitting probability distributions and calculating indices of dispersion, the data are

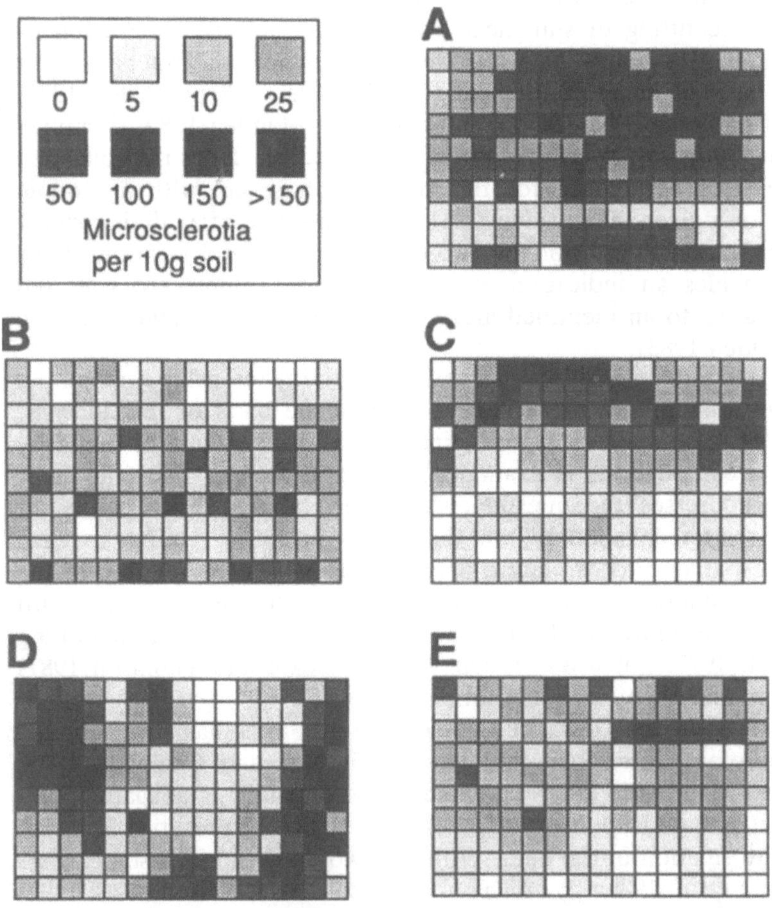

Fig. 7.2A–E. Maps of spatial pattern for microsclerotia of *Macrophomina phaseolina* in soil in five agronomic fields (**A–E**) in Wayne County, North Carolina. Relative density of microsclerotia is represented visually by intensity (*darkness*) of shading. Each quadrat was 6.1 × 6.1 m (C.L. Campbell and S.C. Nelson, unpubl. data)

used in unmapped form which means that the spatial framework of the data is not preserved. For the other analyses, e.g., blocked quadrat variance analysis, spatial autocorrelation analysis, transect analysis, and two-dimensional lattice or distance class analysis, the spatial framework of the data is largely preserved in the analysis. With transect analyses, such as runs or doublet analysis, the two-dimensional structure of data is usually lost but a one-dimensional relationship (i.e., along rows) is maintained.

7.4.3.1 Fitting Probability Distributions

The goodness-of-fit to probability distributions has been widely used as an indicator of spatial pattern for soilborne pathogens and root diseases (Campbell and Noe 1985; Gilligan 1988; Campbell and Madden 1990; Jeger 1990). The most often used probability functions have been the Poisson, negative binomial, and Neyman type A. The discrete distributions such as those mentioned are only valid on a theoretical basis for discrete variables. The procedure is usually to test the null hyphothesis (H_o) that the sample count of propagules, infections, or disease is distributed randomly among the sample units by examining the goodness-of-fit of the data to the Poisson distribution. The test is most often one-sided with the alternative hypothesis (H_a) of aggregation or clustering. If the null hypothesis is rejected, then goodness-of-fit to one of the contagious distributions, such as the negative binomial or Neyman type A, is tested. Microcomputer software is available to apply the goodness-of-fit tests to a variety of probability distributions (Gates and Ethridge 1972).

Although the fitting of probability distributions has been used frequently, several problems are associated with the type of spatial pattern analysis (Gilligan 1988; Madden 1989; Campbell and Madden 1990). One problem involves the use of the chi-square test to test goodness-of-fit between observed and expected frequency data. Because the sample size is often small, the chi-square test lacks power and it is not unusual for good fits to be obtained for more than one contagious distribution or even to the Poisson and one of the contagious distributions. This dilemma poses a problem in interpretation that is not easily resolved. Another problem is that each of the probability distributions used may be derived from a number of biologically realistic mechanisms and a single data set is usually insufficient to distinguish among the mechanisms. Because locational data are not included in the examination of the goodness-of-fit to probability distributions, even observational insight into how the spatial pattern was generated (i.e., by biological, environmental, or edaphic influences) is not possible (Pielou 1977; Cliff and Ord 1981; Nicot et al. 1984).

7.4.3.2 Indices of Aggregation

Indices of aggregation provide a measure of the degree of spatial dispersion or aggregation in a population. Of the many indices used in the ecological literature (Pielou 1977; Southwood 1978), Gilligan (1988) proposed five heuristically distinct indices: (1) the variance-to-mean ratio (v/m or s^2/m); (2) the aggregation parameter (k) of the negative binomial distribution; (3) Lloyd's index of mean crowding ($\overset{*}{m} = m + s^2/m - 1$); (4) Lloyd's index of patchiness (LIP $= \overset{*}{m}/m$); and (5) Taylor's index b which is derived from a power law $s^2 = am^b$. Morisita's index (I_d), which has also been used with root diseases and soilborne pathogens, is closely related to LIP. For v/m, LIP, and Taylor's b the expectations are <1, 1, >1 for regularity, randomness, and aggregation, respectively. Each of these indices is considered by Campbell and Madden (1990).

Indices of dispersion have been used frequently in analyses of the spatial pattern of soilborne pathogens and root diseases (Table 7.4). Three of the five types of indices, v/m, k, and LIP, can be calculated from a single data set, i.e., one set of counts or observations from a grid of quadrats. The v/m has historically been the most popular index of aggregation; however, LIP is probably a better choice because it is independent of density, widely applicable, and also easy to calculate. The index k has the disadvantage that it should be used only if there is satisfactory agreement with the negative binomial distribution (Taylor et al. 1978).

Morisita's index of clumping (Morisita 1959) relies on data of counts of propagules, lesions, etc. from a grid of contiguous quadrats. Whereas Morisita's index (I_d) alone provides an indication of degree of aggregation, the index of clumping, which is computed as IC = (I_d for quadrat of area x)/(I_d for quadrat of area 2x), provides an indication of cluster sizes. Several cluster sizes were detected for propagules of *Macrophomina phaseolina* (Mihail and Alcorn 1987), *Peronosclerospora sorghi* (Schuh et al. 1986), and for *Criconemella aphaerocephalus* (Wheeler et al. 1987) with Morisita's index of clumping.

Taylor's b and a regression model based upon Lloyd's index of mean crowding proposed by Iwao (1968, 1979) are indices of aggregation based upon multiple sets of quadrat-count data. Taylor's b, which was originally proposed (Taylor 1961, 1971, 1984) to examine aggregation as a species characteristic, has been used in more cases with nematode data than with data on soilborne fungi and associated diseases (Table 7.4). In addition to characterizing aggregation for specific nematode species (Ferris et al. 1990), it has also been used in sampling studies with nematodes (Wheeler et al. 1987; McSorley and Dickson 1991).

Iwao's model (1968, 1977, 1979) utilizes Lloyd's index of mean crowding ($\overset{*}{m}$) in the model $\overset{*}{m} = a + bm$, where $\overset{*}{m}$ is the mean, a is an index of basic contagion and indicates whether the basic units of a population are individuals or clusters of individuals, and b is the density contagiousness coefficient or describes the distribution of the basic units of the population within their habitat. Although the two methods may initially appear to be quite different, the results obtained from the use of Taylor's power law and Iwao's regression can be similar (Taylor 1984; Campbell and Madden 1990). Iwao's regression method has been used to describe spatial attributes of infections on peanut by *Cylindrocladium crotalariae* (Griffin and Tomimatsu 1983), *Pratylenchus dianthus* (Pennacchio et al. 1985) in greenhouse soil, and *Macrophomina phaseolina* (Olanya and Campbell 1989) in field soil. The *a* and *b* parameters from Iwao's model have great potential for assisting in the interpretation of the biological factors that contribute to spatial patterns of plant pathogens and diseases and this method merits greater attention.

Although a large number of indices of aggregation have been proposed and used, the actual analysis of spatial data with such indices can be relatively simple and straightforward. LIP can be used if a single descriptor of degree of aggregation is desired, with the caveat that the spatial relationship or dimension of quadrats and counts is lost. If a grid of quadrats of sufficient size is used (e.g., 128 or 256 quadrats), Morisita's index of clumping can be used to suggest cluster sizes. If data are available from several sets of quadrats for counts of a single organism or disease, Taylor's *b* or Iwao's regression can be used to investigate the biological basis for aggregation.

7.4.3.3 Blocked Quadrat Variance

The blocked quadrat variance technique and its variations provide information about the average size of clusters of propagules or diseased plants. The Greig-Smith procedure (1952) is the most commonly used blocked quadrat variance method for data derived from grids of quadrats. With this technique successively larger blocks of quadrats, i.e., 1, 2, 4, 8, 16, 32, . . . , are aggregated and the variance among the blocks is calculated. A peak or local maximum in the graph of variance vs. block size indicates an approximate cluster size. Data to be analyzed in this way must be collected from a transect or grid of quadrats (Cliff and Ord 1981; Upton and Fingleton 1985). An example of this analysis was provided by Campbell and Madden (1990) for propagules of *Macrophomina phaseolina*; the mean cluster size detected was 16 quadrats within a grid of 128 continguous, 6.1 × 6.1 m quadrats.

Two variations of the blocked quadrat variance method have been used with soilborne pathogens. Francl (1986) used the two-term local quadrat variance (TTLQV) method (Hill 1973; Ludwig 1979) in an analysis of the spatial pattern of soybean cyst nematodes (*Heterodera glycines*). With the TTLQV method blocks of quadrats containing 1, 2, 3, 4, . . . , n/2 original quadrats are formed and variances calculated for each block size. Clusters of *H. glycines* 1–3 m long were detected in plant beds prior to cultivation

(Francl 1986). Yang et al. (1991) used a paired-quadrat variance analysis to examine cluster sizes for foliar blight of soybeans caused by *Rhizoctonia solani* and found different cluster sizes in different fields.

7.4.3.4 Spatial Autocorrelation Analysis

Spatial autocorrelation is the dependence of the values of the variable, such as propagule counts or disease intensity, on values of the same variable at geographically adjoining locations (Sokal 1979; Sokal and Thomson 1987). It is a property that is exhibited in mapped data whenever an organized pattern is present (Upton and Fingleton 1985). Originally developed by statisticians and geographers, spatial autocorrelation analysis has been widely applied in biology and ecology following the initial summarization of concepts by Cliff and Ord (1973) and initial applications in marine biology (Jumars et al. 1977). Subsequently, numerous applications have been made in population biology (Sokal and Oden 1978a,b; Cliff and Ord 1981; Sakai and Oden 1983; Sokal and Thomson 1987). This type of analysis has also been employed in research on soilborne pathogens and root diseases (Shew et al. 1984; Noe and Campbell 1985; Mihail 1989; Campbell and van der Gaag 1993) and holds great promise for research designed to elucidate spatial processes during root disease epidemics.

Biological variables associated with root disease epidemics may be spatially autocorrelated for two reasons: inherent forces such as limited dispersal or vegetative growth of pathogens tend to make neighboring areas resemble each other and pathogens may be restricted by or may actively respond to certain environmental factors such as temperature, soil moisture, soil type, etc. which are spatially autocorrelated. The first reason relates to the concept of true contagion discussed earlier, whereas the second reason relates to apparent contagion of pathogens and diseases. Spatial autocorrelation analysis can, therefore, be used to investigate the underlying spatial relationships associated with an observed pattern and to seek inferences from an observed pattern to the process that has produced the pattern (Sokal 1979; Sokal and Thomson 1987).

The data used in spatial autocorrelation analysis may be counts of soilborne propagules or diseased plants (ordinal or discrete data), ratios such as disease severity or disease incidence (interval or continuous data), or nominal data in which a quadrat or sample point is classified as diseased or healthy. For nominal data the analysis is based upon the calculation of join counts between points that are like or unlike with respect to the categorical variable (i.e., diseased or healthy). Although the methodology for the analysis of nominal data is well established (Cliff and Ord 1973; Sokal and Oden 1978a), this approach has not yet been applied to root diseases. Autocorrelation analysis has, however, been applied to count and ratio data associated with root diseases and pathogen propagules.

Methodology. With count and ratio data Moran's I and Geary's coefficient c are often used as autocorrelation coefficients (Cliff and Ord 1973; Sokal and Oden 1978a). Both are analogous to correlation coefficients in that the numerator is an estimator of covariance and the denominator is an estimator of variance of the observations. Of the two, Moran's I has been used more frequently in plant disease epidemiology (Madden 1989). Moran's I is calculated as

$$I = n \sum_{ij} w_{ij} z_i z_j / W \sum_{i=1}^{n} z_i^2,$$

where n is the number of localities in the study; Σ indicates summation over all localities i from 1 to n, and j from 1 to n, i is not equal to j; w_{ij} is the weight given to the geographic connection between localities i and j (w_{ij} need not equal w_{ji}); $z_i = Y_i - Ybar$, where Y_i is the value of Y for locality i, and $Ybar$ is the mean of Y for all localities; and $W = \Sigma_{ij}\ w_{ij}$ is the sum of the matrix of weights, i = not j. Values of Moran's I vary from -1 to $+1$ with the expected value approaching zero for large sample size in the absence of autocorrelation.

To compute spatial autocorrelation coefficients a system of assigning weights w_{ij} must be specified to connect the locations sampled. The locations of observations in root disease studies correspond to quadrats or samples in a field. Coordinates are given by row and column designations i and j such that first-order (or lag one) correlations are calculated by relating the value in each quadrat (i, j) with the values in the adjacent [e.g., $(i - 1, j)$, $(i, j + 1)$] quadrats. Second- and higher-order correlations are based on quadrats separated by one or more quadrats.

For Moran's I an autocorrelation coefficient is calculated in accordance with a specified directional comparison such as across-row, within-row, rook's case, or bishop's case. Autocorrelation coefficients are then plotted in relation to the lag order to provide a graph called a spatial correlogram. The correlograms describe the underlying spatial relationships for a surface rather than the appearance of that surface. This method was used by Mihail (1989) in analyzing the spatial pattern of propagules of *Macrophomina phaseolina* and mortality of plants of the highly susceptible host, *Euphorbia lathyris*.

An alternative weighting system for the calculation of spatial autocorrelation coefficients was specified by Modjeska and Rawlings (1983). This method allows decomposition of the single autocorrelation coefficient and the resulting spatial correlogram into direction-associated autocorrelation coefficients and a directional spatial correlogram. The advantage of the correlogram is that autocorrelations across and within rows and along diagonals can be viewed simultaneously and directional asymmetry of autocorrelations within fields can also be detected. Microcomputer software is available for the Modjeska and Rawlings method (Gottwald et al. 1992).

Shew et al. (1984), Noe and Campbell (1985), and Campbell and van der Gaag (1993) utilized this method in investigations on the spatial pattern of Southern stem rot (caused by *Sclerotium rolfsii*), of plant parasitic nematodes, and of propagules of *Macrophomina phaseolina*, respectively. The method has also been utilized in studies on several foliar pathogens (Madden 1989; Gottwald et al. 1992).

Applications. Spatial correlograms provide a description of the spatial properties of a surface or grid of variables. Because correlograms describe the underlying spatial relationships of the area from which the samples were taken (e.g., a two-dimensional, "horizontal slice" of soil 20 cm deep over an x by x m area in a field) and not the spatial map of, for example, propagule counts as in Fig. 7.2, they may be guides to the processes that resulted in the specific, observed pattern. Different processes may, however, give rise to the same pattern and, conversely, two realizations of the same process may result in different patterns. The goals in using spatial autocorrelation analysis with root diseases and soilborne pathogens are, then, to identify patterns associated with specific pathogens or diseases and to hypothesize what physical, chemical, and biological processes resulted in that pattern. The order (high or low lags) and sign (positive or negative) of autocorrelation coefficients can be used to propose mechanisms or processes by which specific patterns develop (Table 7.6).

Table 7.6. Ecological interpretations of spatial autocorrelation coefficients appropriate for root diseases and soilborne pathogens. (After Sokal 1979)

Sign of autocorrelation	Order of autocorrelation (lag)	
	Low	High
Positive	Dispersal by movement into adjoining area or radiation (colonization) by growth	Symmetrical surfaces or symmetrical ordering of patches
	Large homogeneous patches with diameter > interlocality distances and separated by gaps > patch diameter; patches may be favored by environmental factors	
	Gradients of ecological factors that affect population density	
Negative	Heterogeneous, small patches	Gradient which may correspond to gradient of ecological (physical, chemical, or biological) resources
	Aggregation into small clumps with empty areas between clumps	

Table 7.7. Spatial correlations of *Meloidogyne incognita* population densities, data set A. (Noe and Campbell 1985)

Distance across rows (m)	Lag quadrat distance (m) within plant rows						
	0	1	2	3	4	5	6
0	1.00	0.56*	0.50*	0.47*	0.43*	0.53*	0.04
1	0.39*	0.23	0.18	0.17	0.15	0.29	0.03
2	0.18	0.18	0.18	0.20	0.13	0.16	0.15
3	−0.19	0.01	−0.06	−0.07	−0.24	−0.09	−0.05
4	−0.35*	−0.17	−0.20	−0.32	−0.19	−0.20	−0.03
5	−0.11	0.05	0.14	0.08	0.25	−0.10	0.16
6	−0.14	−0.16	−0.20	−0.28	0.11	−0.52	−0.66

* Correlation significantly different from zero at the 0.05 level. Degrees of freedom = $[(8 - L)(8 - K)] - 2$, where L is the distance across columns and K is the distance across rows.

In an analysis of the spatial pattern of several plant-parasitic nematodes associated with tobacco roots, Noe and Campbell (1985) utilized an 8 × 8 grid of 1 × 1 m quadrats in which the sampling grid was aligned with plant rows. For one nematode, *Meloidogyne incognita*, in Johnston County, North Carolina, in October they found a mean density of 1890 juveniles per 500 cm^3 soil (standard deviation = 1430). The value of Lloyd's index of patchiness was 1.6, which indicated a nonrandom pattern; the spatial correlation matrix for lag quadrat distance across and within plant rows is given in Table 7.7. The matrix element (0,1) with a value of 0.56 represents the correlation of all possible adjacent quadrat pairs within plant rows, whereas matrix element (1,0) with a value of 0.39 represents the correlation of all possible adjacent quadrat pairs across plant rows. The correlations within rows remained high, but correlations across rows decreased rapidly. Thus, clusters of *M. incognita* in the soils studied were ellipsoidal with long axes oriented in the direction of plant rows. Noe and Campbell (1985) hypothesized that the higher spatial correlation of nematode densities within plant rows was due to cultural practices which led to dispersal of nematodes into adjoining areas within rows because most management activities (e.g., planting, cultivation) were oriented in the direction of plant rows.

Comparison with Indices of Dispersion. Indices of dispersion which describe point pattern processes are based upon the location of points, but in the analysis only the counts per sample (quadrat) are actually analyzed. Thus, these measures of pattern focus on measures of variance and not on the spatial location of the variate values. Sokal and Oden (1978b) and Jumars et al. (1977) discussed the point that the heterogeneity or dispersion of a sample can be considered as independent of its spatial pattern. Thus, two

Table 7.8. Ecological interpretation of different combinations of results of spatial autocorrelation analysis and indices of dispersion based on population counts for root diseases and soilborne pathogens. (After Sokal 1979)

Spatial autocorrelation[a]	Indication of population counts based on indices of dispersion	
	Heterogeneous[b]	Homogeneous[b]
Significant	A synchronous population growth among local areas	Unlikely to occur
	Autocorrelation of physical, chemical, or biological resources	
Not significant	Random arrangement of resources which vary in favorability for population growth	Spatially random population growth occurs
	Random deposition of dissimilar numbers of propagules among sites with little subsequent dispersal	Random patterns of environmental resources

[a] A significant autocorrelation indicates the presence of a discernible spatial pattern.
[b] Heterogeneous counts are indicated by values of indices such as Lloyd's index of patchiness or Morisita's index greater than one; homogeneous counts are indicated by values of such indices not different from one.

samples of counts may have the same variance (and mean) but represent very different spatial patterns (Sokal 1979). This dilemma has been brought to the attention of plant pathologists (Nicot et al. 1984).

Information from point pattern analysis, such as indices of dispersion, and autocorrelation analyses provides answers to two different questions; however, the interpretation of a given spatial data set based upon autocorrelation analysis can be aided by a simultaneous examination of an index of dispersion of the sample of counts. An index such as Lloyd's index of patchiness addresses the question "Are the points scattered randomly in space?", whereas a spatial autocorrelation coefficient addresses the question "Are the variates scattered randomly on the points in space?" (Sakai and Oden 1983).

The ecological interpretations of different combinations of results for spatial autocorrelation analysis and indices of dispersion (Table 7.8) can provide insight into characteristics of pathogen population growth, operational mechanisms of propagule dispersal, and arrangements of resources. Thus, the combination of spatial analysis techniques can provide inferences on processes that resulted in specific, observed patterns. This more holistic view of spatial pattern analysis has not been applied to root diseases but has been applied to virus diseases of tobacco (Madden 1989).

7.4.3.5 Transect Analysis

This category of spatial pattern techniques uses discrete, binary, or presence/absence data and generally looks at the occurrence of healthy or diseased plants within rows. Examples of the type of data used include disease incidence or the presence/absence of propagules. Disease severity data can be transformed to binary data by coding all values at or above the median severity as 1 or + and all values below the median severity as 0 or −.

Doublet analysis is a type of trajectory analysis that has been used with soilborne pathogens. Doublet analysis was discussed by Vanderplank (1946) as a means of detecting whether diseased plants in a homogeneous field have a random or nonrandom occurrence within plant rows. Doublets occur when two infected plants occur adjacent to one another. In the sequence +−−+++−−++− there are three doublets; +++ is counted as two doublets and + is not a doublet. Doublets are often counted in alternate directions along adjacent rows with the first plant of a row forming a doublet with the first plant of the next row. Jarvis and Hawthorne (1972) used doublet analysis in a study on an epidemic in lettuce caused by *Sclerotinia minor*. The authors indicated that the initial source of inoculum was believed to be soilborne sclerotia which produced airborne ascospores. Interestingly, these authors also used the method of Roach (1968) to ascertain whether interplant spread alone accounted for disease increase or whether additional randomly occurring infections also contributed to the epidemic.

Madden et al. (1982) questioned the reliability of doublet analysis and corrected doublet analysis (Converse et al. 1979) and proposed ordinary runs analysis for detecting clustering of infected plants in rows. A run of infected plants is defined in the same way as a doublet; however, the test for nonrandomness differs (Campbell and Madden 1990). Runs analysis was used in two studies to examine plant-to-plant spread of mycelia of *Sclerotinia minor* on lettuce (Dillard and Grogan 1985; Marois and Adams 1985). Dillard and Grogan (1985) did not find a significant departure from randomness for infected plants within rows in five plots; however, Marois and Adams (1985) found that all six fields examined had clustered or nonrandom populations as determined by the ordinary runs analysis.

Two additional applications of runs analysis have been applied to epidemics of root diseases. Campbell et al. (1984) performed runs analysis for plants with tobacco black shank (caused by *Phytophthora parasitica* var. *nicotianae*) within each row and then calculated the percentage of rows with clustered patterns. Such an approach permits a description of changes in spatial pattern over the course of an epidemic and may allow an interpretation of whether or not pathogen spread mechanisms are operational during different phases of the epidemic. Although true two-dimensional runs analysis is intractable for most situations, Jeger et al. (1987) utilized runs analysis both within and across rows to analyze the pattern of cotton plants infected with *Phymatotrichum omnivorum*. In general, there was a greater degree of clustering within rows than across rows, which suggested that most pathogen spread was occurring from plant to plant within rows.

7.4.3.6 Two-Dimensional Lattice or Distance Class Analysis

Two-dimensional distance class analysis (DCA) was proposed by Gray et al. (1986a) for use with binomial or presence/absence data for variables such as disease incidence in which the actual position of the sample unit is known within a particular quadrat or lattice position within a rectangular field plot. The distance between infected plants is defined in terms of both horizontal (X) and vertical (Y) units. Pairs of infected plants are thus grouped into two-dimensional (X,Y) distances that refer to the absolute distance between plants in a pair. Because the total possible number of pairs varies among (X,Y) distance classes, the number of pairs of infected plants in each distance class is standardized by the total number of pairs of living plants in that distance class. This allows for the direct comparison of standardized count frequency values in any (X,Y) distance class. The standardized count of pairs of infected plants in each distance class is then compared directly to expected values obtained from a large number (e.g., 400) of computer-generated simulations of a lattice containing the same number of infected plants arranged under the assumption of a random pattern. This method allows the detection of a departure from randomness and preserves the two-dimensional spatial information needed to determine cluster size and arrangement.

This analysis has been used for several foliar pathosystems. Gray et al. (1986b) used DCA to study spatial pattern of cantaloupe plants infected with watermelon mosaic virus 2. Frisina and Benson (1989) utilized DCA in the study of the spatial pattern of web blight (caused by binucleate *Rhizoctonia* spp.) in container-grown, azalea nursery stock. Nelson and Campbell (1993) used DCA in an analysis of the spatial patterns of leaf spot and virus diseases on white clover.

7.5 Applications of Spatial Analysis in Root Disease Epidemiology and Management

The analysis and understanding of the spatial dimension of root diseases provide an opportunity to refine studies aimed at developing better management programs for root diseases and will, ultimately, provide an essential component to the conceptual foundation for root disease management. An understanding of the mechanisms of pathogen dispersal and the resultant spatial pattern of the propagules and disease are also keys to the successful design and sampling of field experiments. Additionally, a knowledge of spatial pattern and the ways in which diseases progress is requisite for accurate estimation of crop losses due to root diseases.

7.5.1 Determination and Interpretation of Temporal and Spatial Scales

The need to consider temporal and spatial scales and their interactions in ecological and epidemiological hypotheses has been identified (O'Neill et al. 1986; Peters 1988; Roughgarden et al. 1989; Wiens 1989; Onstad 1992) but has been largely ignored in botanical epidemiology (Madden 1989). The temporal and spatial scales for root epidemics may be defined more easily than for many foliar epidemics because the pathogens are resident in the soil during most phases of the epidemic and because of the more limited dispersal of soilborne pathogens within a given time period. The consistent definition of temporally and spatially appropriate scales for root epidemics will, however, necessitate a shift in thinking and planning of studies from a field and row crop-oriented focus to a more ecologically oriented focus which depends on the ecology and biology of the pathogen-host interaction.

Spatial and temporal scales define the "ecological neighborhood" of the epidemic. The neighborhood is the region within which a pathogen is active or has some influence during the period of time that is of interest (Addicot et al. 1987). Neighborhood sizes can be estimated by examining the cumulative effects and activities of the pathogens (including dispersal) and the influence of the pathogen as a function of increasingly large spatial units during the time period of interest. The scale of the neighborhood must be large enough to encompass all normal movement by the average individual of the pathogen population during the time period of interest (Wiens 1976; Addicot et al. 1987; Onstad 1992).

In most of the spatial analyses performed for root disease epidemics the size of study units (e.g., quadrats) and study areas has been arbitrary rather than being based upon a scale that is ecologically appropriate to the organisms involved and the questions to be answered in the study. Agronomic fields often have served as the study area without regard to specific characteristics of pathogen dispersal during or among growing seasons. Several potential problems may arise from such a situation (Addicot et al. 1987).

One problem is that it will be difficult to compare observational or experimental results within or among species in either the same or different systems due to the fact that the study units may represent different scales of neighborhood types for each of the species. An example of this problem is found among studies on the spatial pattern of propagules of *Macrophomina phaseolina* in soil. Propagules had a random pattern within corn rhizospheres (quadrats 0.02–0.03 m^2, extent 0.58 m^2) (Olanya and Campbell 1989), a random to aggregated pattern in soil prior to and after disking (quadrats 1 m^2, extent 64 m^2) (Olanya and Campbell 1988), an aggregated pattern in cultivated soil (quadrats $0.3\,m^2$, extent $697\,m^2$) (Mihail and Alcorn 1987), and an aggregated pattern in the presence of a highly susceptible crop (quadrats $4.5\,m^2$, extent $465\,m^2$) (Mihail 1989). The neighborhood represented by the corn rhizosphere is much different than that in the other studies. In two of the studies (Mihail 1989; Olanya and Campbell 1989) host plants were present which would provide very different ecological conditions in the soil than in the other two studies (Mihail and Alcorn 1987; Olanya and Campbell 1988) in which host plants were not present. In all four studies the rationale for the study area was probably more one of convenience (i.e., numbers of quadrats placed in an arbitrarily chosen study area so that the area was covered with a manageable number of samples) than one based upon knowledge of the dispersal or movement of the pathogen, thus specific comparisons among the studies are difficult.

A second problem arises when a patch of pathogen or disease occurrence is viewed as a discontinuity in an environmental characteristic that matters to the organism (Wiens 1976). For example, patches may be wet areas that occur in fields and may favor pathogen occurrence. The specified patch size, however, may not correspond to the appropriate ecological neighborhood for that organism, especially if dispersal of the organism occurs over long distances when soil is saturated or flooded following rain.

Finally, different processes in the same system may occur at different scales with the result that it may not be sufficient to examine a particular system at only one temporal or spatial scale. Most often, plant epidemiologists have based the size of sample unit for spatial studies upon the area that needs to be included in the study (e.g., the size of the field) and the number of quadrats in which soil samples can be taken and processed or in which disease can be assessed within cost constraints. This strategy is practical but suffers from several cryptic hazards. One hazard is that the spatial scale at which a study can be conveniently conducted may not correspond closely to an ecological neighborhood. For example, in a study on the aggregation of plant parasitic nematodes with Taylor's b used as a measure of aggregation, Boag and Topham (1984) found that when samples were taken adjacent to one another or at very small intervals, the nematodes tended to have a random pattern, i.e., b values were near 1. As the distance between samples increased so did the values of b for *Longidorus elongatus*, *L. goodeyi*, and *Rotylenchus goodeyi* which suggested that the nematodes were becoming more aggregated. However, differences were observed in the degree of aggregation between species such that *L. elongatus* and *L. goodeyi* had b values of nearly 2 at a sample interval distance of 50–57 cm, whereas *R. goodeyi* had a similar b value at 10–18 cm. Also, the distance between sampling points at which a nearly constant b value was obtained varied between experimental sites and may have been influenced by soil type. Thus, for these nematodes the definition of an ecological neighborhood, and thus the appropriate spatial scale at which to conduct studies, may be dependent on species and soil type.

The data needed to determine neighborhood size for processes such as dispersal and survival of inoculum may be difficult to obtain and may vary with the geographic area of interest. Preliminary spatial and temporal studies need to be conducted in order to determine the appropriate ecological size of sample units. Even with such preliminary studies, however, the criteria for establishing neighborhood size may be necessarily arbitrary and may depend on whether pathogen or host processes are taken as the focus for scale determinations. Also, as spatial scale changes from a regional focus to a local one, the environmental factors that affect the host-pathogen interaction may differ (Burdon et al. 1989).

7.5.2 Design and Sampling of Experiments

Aggregation of inoculum of root pathogens in soil and the resultant diseases poses several challenges, but also provides several opportunities in studies on soilborne pathogens and root diseases. The challenges are to design studies that take spatial patterns of pathogens and diseases into account and to sample such studies in such a way and with sufficient sample numbers such that reliable estimates of inoculum density and disease are obtained. The opportunities, which are dependent on a prior knowledge of the spatial characteristics of root epidemics, arise from an investigator's ability to manipulate the location of field plots in ways that maximize the benefits from spatially aggregated inoculum and disease.

In the design of field experiments a prior knowledge of the actual spatial pattern of inoculum or disease or of the general spatial characteristics of the pathosystem allows one to use such knowledge to improve the efficiency of field plot designs. Noe and Campbell (1985) analyzed the spatial patterns of several nematodes in field soils and found that, although cluster size for species was a field-specific characteristic, the clusters of nematodes were generally ellipsoidal with the long axis oriented in the direction of plant rows. Noe and Campbell (1985) suggested that if maximum variation among field plots was desired, as for crop loss experiments, then plots in a subsequent year should be oriented along crop rows that were present in the previous growing season. If variation among plots was to be minimized, as in a variety trial, then plots should be oriented across previous rows.

For the fungus *Phymatotrichum omnivorum* ellipsoidal clusters of inoculum are also expected to occur because the fungus generally grows more along plant rows than across rows (Jeger et al. 1987). Gerik et al. (1985) used knowledge of spatial factors in cotton root rot epidemics to optimize plot size for studies on cotton root rot caused by *P. omnivorum*. In order to miminize error variance in field studies, they proposed that large plots (at least $33\,m^2$) should be used and plot sizes should be increased by increasing the number of adjacent rows as compared to increasing row length within plots. With the expected ellipsoidal clusters of the pathogen, the increase in the number of adjacent rows would tend to "average out" the inoculum density between plots.

Empirical data on inoculum density and pattern were used by Culbreath et al. (1990) to design experiments on disease progress of Cylindrocladium black rot on seven peanut genotypes. Previous work (Hau et al. 1982) established that microsclerotia of *C. crotalariae* were aggregated spatially in soil. Culbreath et al. (1990) established inoculum density within each intended plot by soil assay for the pathogen (Phipps et al. 1976) and used that spatial pattern information to assign each genotype within each block of the experiments to plots with high, medium, and low propagule densities.

Sampling of plots or fields in which spatial aggregation of inoculum or disease occurs requires a larger sample size to gain the same relative degree of reliability that can be obtained with spatial randomness of propagules or disease and a smaller sample size. The reason for this lies in the fact that with spatial aggregation the variance (s^2) among sample units is relatively greater than with spatial randomness and the determination of sample size is usually based upon some measure of relative variance. For example, if sampling reliability is defined by a coefficient of variability (CV), where CV $= s/\bar{x}$, with s the standard deviation and $\bar{x}$ the mean, then the number of samples (n) to be taken is defined as $n = s^2/(\bar{x}^2 CV^2)$. From this formula it is apparent that sample size increases directly in proportion to the variance and as degree of spatial aggregation increases, the variance also increases.

Sample pattern has also been investigated in relation to spatial aggregation of inoculum for soilborne pathogens. Based upon the simulation studies of Lin et al. (1979), Hau et al. (1982) examined several different simulated sampling patterns on actual propagule count data for *C. crotalariae* and concluded that a W-shaped or diamond-shaped sampling path provided better estimates of inoculum density than did random or single-transect sampling schemes. Mihail and Alcorn (1987) confirmed in their studies on *M. phaseolina* that systematic sampling strategies were the most useful for estimating mean propagule density.

7.5.3 Establishing Crop Loss Estimates

Spatial heterogeneity of inoculum and disease may result in patchy crop injury and losses. Most crop loss models relate the mean density of inoculum or intensity of disease to the amount of loss and assume implicitly that the mean values used for inoculum or disease are derived from a spatially homogeneous population. The result is that, depending on the shape of the curve relating crop loss to inoculum or disease, crop losses may be overestimated or underestimated if simple arithmetic means are used for inoculum density or disease intensity when inoculum or disease is spatially heterogeneous.

The potential for overestimation of crop losses due to nematodes has been identified clearly (Seinhorst 1973; Perry 1983; Noe and Barker 1985) when spatial homogeneity of nematodes is assumed in place of the more realistic assumption of spatial heterogeneity or aggregation of nematodes. Seinhorst (1973) stated clearly that: "...the relation between average population density of nematodes and yield depends on the frequency distribution of nematode density in the field." The practical implications of this statement were amplified by Perry (1983), who suggested that rather than using a mean value for nematode density in a crop loss model, each sample density should be used as a model input to determine expected crop damage. Noe and Barker (1985) reemphasized the need for caution in using mean values of nematode density in crop loss models and showed that using such mean density values could result in overestimation of yield losses by as much as 300% when populations were highly aggregated.

The actual effects of using a mean value from a spatially aggregated population in a crop loss assessment model which assumes a spatially homogeneous population of pathogens or disease were generalized by Hughes (1988). Through simulation studies he showed that whether over- or underestimation of crop losses occurred through use of a mean value for an aggregated population, relative to that for a spatially homogeneous or random population, depended on the shape of the curve relating crop loss and pathogen density or disease intensity. The examples provided for plant-parasitic nematodes (Seinhorst 1973; Perry 1983; Noe and Barker 1985) correctly predicted overestimation of crop losses derived from a single mean value because the curve relating yield and nematode density was convex and allowed for the possibility of compensatory growth by host plants following attacks by nematodes. Hughes (1988) indicated that if the yield-damage curve was concave, suggesting an intraspecific competition for resources with fewer sites actually occupied with spatially aggregated inoculum, then use of a simple mean value for a spatially aggregated population would tend to underestimate crop loss relative to that expected for a spatially homogeneous pathogen population.

References

Abawi GS, Grogan RG (1975) Source of primary inoculum and effects of temperature and moisture on infection of beans by *Whetzelinia sclerotiorum*. Phytopathology 65:300–309

Abawi GS, Grogan RG (1979) Epidemiology of diseases caused by *Sclerotinia* species. Phytopathology 69:899–904

Adams PB (1981) Forecasting onion white rot disease. Phytopathology 71:1178–1181

Adams PB (1986) Production of sclerotia of *Sclerotinia minor* on lettuce in the field and their distribution in soil after disking. Plant Dis 70:1043–1046

Addicott JF, Aho JM, Antolin MF, Padilla DK, Richardson JS, Soluk DA (1987) Ecological neighborhoods: scaling environmental patterns. Oikos 49:340–346
Alby T, Ferris JM, Ferris VR (1983) Dispersion and distribution of *Pratylenchus scribneri* and *Hoplolaimus galeatus* in soybean fields. J Nematol 15:418–426
Allen RN, Newhook FJ (1973) Chemotaxis of zoospores of *Phytophthora cinnamomi* to ethanol in capillaries of soil pore dimensions. Trans Br Mycol Soc 61:287–302
Aylor DE (1990) The role of intermittent wind in the dispersal of fungal pathogens. Annu Rev Phytopathol 28:73–92
Aylor DE, Ferrandino FJ (1990) Initial spread of bean rust close to an inoculated bean leaf. Phytopathology 80:1469–1476
Belmar SB, Jones RK, Starr JL (1987) Influence of crop rotation on inoculum density of *Rhizoctonia solani* and sheath blight incidence in rice. Phytopathology 77:1138–1143
Benson DM (1990) Movement of Phytophthora zoospores through columns of container media. Phytopathology 80:1025 (Abstr)
Benson DM, Campbell CL (1985) Spatial pattern of Phytophthora root rot and dieback of azalea in container-grown nursery stock. Plant Dis 69:1049–1054
Boag B, Brown DJF, Topham PB (1987) Vertical and horizontal distribution of virus-vector nematodes and implications for sampling procedures. Nematology 33:83–96
Boag B, Topham PB (1984) Aggregation of plant parasitic nematodes and Taylor's power law. Nematologica 30:348–357
Brooks DH (1965) Root infection by ascospores of *Ophiobolus graminis* as a factor in epidemiology of the take-all disease. Trans Br Mycol Soc 48:237–248
Bruton BD, Reuveni R (1985) Vertical distribution of microsclerotia of *Macrophomina phaseolina* under various soil types and host crops. Agric Ecosyst Environ 12:165–169
Burdon JJ, Jarosz AM, Kirby GC (1989) Pattern and patchiness in plant-pathogen interactions – causes and consequences. Annu Rev Ecol Syst 20:119–136
Campbell CL, Madden LV (1990) Introduction to plant disease epidemiology. Wiley & Sons, New York, 532 pp
Campbell CL, Jacobi WR, Powell NT, Main CE (1984) Analysis of disease progression and the randomness of occurrence of infected plants during tobacco black shank epidemics. Phytopathology 74:230–235
Campbell CL, Noe JP (1985) The spatial analysis of soilborne pathogens and root diseases. Annu Rev Phytopathol 23:127–148
Campbell CL, Pennypacker SP (1980) Distribution of hypocotyl rot caused in snapbean by *Rhizoctonia solani*. Phytopathology 70:521–525
Campbell CL, Van der Gaag DJ (1993) Temporal and spatial dynamics of microsclerotia of *Macrophomina phaseolina* in three fields in North Carolina over four to five years. Phytopathology 83:1434–1440
Cappaert MR, Powelson ML, Franc GD, Harrison MD (1988) Irrigation water as a source of inoculum of soft rot erwinias for aerial stem rot of potatoes. Phytopathology 78:1668–1672
Chellemi DO, Rohrbach KG, Yost RS, Sonoda RM (1988) Analysis of the spatial pattern of plant pathogens and diseased plants using geostatistics. Phytopathology 78:221–226
Clark WC (1985) Scales of climate impacts. Clim Change 7:5–27
Cliff AD, Ord JK (1973) Spatial autocorrelation. Pion, London
Cliff AD, Ord JK (1981) Spatial processes: models and aplications. Pion, London
Cochran WG (1936) The statistical analysis of field counts of diseased plants. JR Statist Soc Suppl 3:49–67
Converse RH, Seely J, Martin LW (1979) Evidence for random local spread of aphidborne mild yellow-edge virus in strawberries. Phytopathology 69:142–144

Culbreath AK, Beute MK, Wynne JC (1990) Use of spatial patterns and density of inoculum of *Cylindrocladium crotalariae* during field evaluation of partially resistant peanut genotypes. Phytopatholopgy 80:1395–1400
Davis JM (1987) Modeling the long-range transport of plant pathogens in the atmosphere. Annu Rev Phytopathol 25:169–188
Delcourt HR, Delcourt PA, Webb T (1983) Dynamic plant ecology: the spectrum of vegetation change in space and time. Quat Sci Rev 1:153–175
Dillard HR, Grogan RG (1985) Relationship between sclerotial spatial pattern and density of *Sclerotinia minor* and the incidence of lettuce drop. Phytopathology 75:90–94
Dukes PD, Jenkins SF, Thompson SS Jr (1977) Detection and some observations of *Phytophthora parasitica* var. *nicotianae* in ponds used for irrigation of tobacco. Tob Sci 21:97–100
Duniway JM (1976) Movement of zoospores of *Phytophthora cryptogea* in soils of various textures and matric potentials. Phytopathology 66:877–882
Duniway JM, McKeen CD (1987) Dispersal of *Phytophthora cryptogea* zoospores in soils and glass microbeads by water flow. Phytopathology 77:1744 (Abstr)
Easton GD, Nagle ME, Bailey DL (1969) A method for estimating *Verticillium albo-atrum* propagules in field soil and irrigation waste water. Phytopathology 59:1171–1172
Ferrin DM, Mitchell DJ (1986) Influence of initial density sitribution of inoculum on the epidemiology of tobacco black shank. Phytopathology 76:1153–1158
Ferris H, Mullens TA, Foord KE (1990) Stability and characteristics of spatial description parameters for nematode populations. J Nematol 22:427–439
Finney DJ (1947) The significance of associations in a square-point lattice. JR Statistic Soc Suppl 9:99–103
Fitt BDL, McCartney HA (1986) Spore dispersal in relation to epidemic models. In: Leonard KJ, Fry WE (eds) Plant disease epidemiology: population dynamics and management, vol 1. Macmillan, New York, pp 311–345
Franc GD, Harrison MD, Powelson ML (1985) The dispersal of phytopathogenic bacteria. In: MacKenzie DR, Barfield CS, Kennedy GG, Berger RD (eds) The movement and dispersal of agriculturally important biotic agents. Claitor's Publ Div, Claitor's Books, Baton Rouge, pp 37–49
Francl LJ (1986) Spatial analysis of *Heterodera glycines* populations in field plots. J Nematol 18:183–189
Freeman GH (1953) Spread of diseases in a rectangular plantation with vacancies. Biometrika 40:287–296
Frisina TA, Benson DM (1989) Occurrence of binucleate *Rhizocotnia* spp. on azalea and spatial analysis of web blight in container-grown nursery stock. Plant Dis 73:249–254
Garrett SD (1939) Soil conditions and the take-all disease of wheat. IV. Factors limiting infection by ascospores of *Ophiobolus graminis*. Ann Appl Biol 26:47–55
Garrett SD (1970) Pathogenic root-infecting fungi. Cambridge Univ Press, Cambridge
Gates CE, Ethridge FG (1972) A genralized set of discrete frequency distributions with FORTRAN program. Int Assoc Math Geo 4:1–24
Gäumann E (1946) Pflanzliche Infektionslehre. Birkhäuser, Basel
Gerik TJ, Rush CM, Jeger MJ (1985) Optimizing plot size for field studies of Phymatotrichum root rot of cotton. Phytopathology 75:240–243
Gerlagh M (1986) Introduction of *Ophiobolus graminis* into new polders and its decline. Meded Lab Phytopath Wageningen, No 241
Gill DL (1970) Pathogenic *Pythium* from irrigation ponds. Plant Dis Rep 54:1077–1079

Gilligan CA (1988) Analysis of the spatial pattern of soilborne pathogens. In: Kranz J, Rotem J (eds) Experimental techniques in plant disease epidemiology. Springer, Berlin Heidelberg New York, pp 85–98
Goodell P, Ferris H (1980) Plant-parasitic nematode distribution in an alfalfa field. J Nematol 12:136–141
Gottwald TR, Richie SM, Campbell CL (1992) LCOR2 – spatial correlation analysis software for the personal computer. Plant Dis 76:213–215
Graham DC, Harrison MD (1975) Potential spread of *Erwinia* spp. in aerosols. Phytopathology 65:739–741
Graham DC, Wuinn CE, Bradley LF (1977) Quantitative studies on the generation of aerosols of *Erwinia carotovora* var. *atroseptica* by simulated raindrop impaction on blackleg-infected potato stems. J Appl Bacteriol 43:413–424
Gray SM, Moyer JW, Bloomfield P (1986a) Two dimensional distance class models for quantitative description of virus-infected plant distribution lattices. Phytopathology 76:243–248
Gray SM, Moyer JW, Kennedy GG, Campbell CL (1986b) Virus-suppression and aphid resistance effects on spatial and temporal spread of watermelon mosaic virus 2. Phytopathology 76:1254–1259
Gregory PH (1973) The microbiology of the atmosphere, 2nd edn. Leonard Hill, London
Greig-Smith P (1952) The use of random and contiguous quadrats in the study of the structure of plant communities. Ann Bot New Ser 16:293–316
Griffin GJ, Garren HK, Taylor HD (1981) Influence of crop rotation and minimum tillage on the population of *Aspergillus flavus* group in peanut field soil. Plant Dis 65:898–900
Griffin GJ, Tomimatsu GS (1983) Root infection pattern, infection efficiency, and infection density-disease incidence relationships of *Cylindrocladium crotalariae* on peanut in field soil. Can J Plant Path 5:81–88
Harrison MD, Franc GD, Maddox DA (1987) Presence of *Erwinia carotovora* in surface water in North America. J Appl Bacteriol 62:565–570
Hau FC, Campbell CL, Beute MK (1982) Inoculum distribution and sampling methods for *Cylindrocladium crotalariae* in a peanut field soil. Plant Dis 66:568–571
Heesterbeek JAP, Zadoks JC (1987) Modelling pandemics of quarantine pests and diseases: problems and perspectives. Crop Prot 6:211–221
Hill MO (1973) The intensity of spatial pattern in plant communities. J Ecol 61:225–235
Hirst JM (1965) Dispersal of microorganisms. In: Baker KF, Snyder WC (eds) Ecology of soil-borne plant pathogens. Univ California Press, Berkeley, pp 69–81
Hughes G (1988) Spatial heterogeneity in crop loss assessment models. Phytopathology 78:883–884
Hughes G (1990) Characterizing crop responses to patchy pathogen attack. Plant Pathol 39:2–4
Isaac I (1957) Wilt of lucerne caused by species of *Verticillium*. Ann Appl Biol 45:550–558
Iwao S (1968) A new regression method for analyzaing the aggregation pattern of animal populations. Res Popul Ecol 10:1–20
Iwao S (1977) The $\overset{*}{m}$-m statistics as a comprehensive method for analyzing spatial pattern of biological populations and its application to sampling problems. In: Morisita M (ed) Studies on methods of estimating population density, biomass, and productivity in terrestrial animals. Univ of Tokyo Press, Tokyo, pp 21–46
Iwao S (1979) The $\overset{*}{m}$-m method for analyzing the distribution patterns of single and mixed species populations. In: Patil GP, Rosenzweig M (eds) Contemporary

quantitative ecology and related econometrics. International Cooperative Publishing House, Fairlane, MD, pp 215–228
Jarvis WR, Hawthorne BT (1972) *Sclerotinia minor* on lettuce: progress of an epidemic. Ann Appl Biol 70:207–214
Jeger MJ (1990) Mathematical analysis and modeling of spatial aspects of plant disease epidemics. In: Kranz J (ed) Epidemics of plant diseases: mathematical analysis and modeling, 2nd edn. Springer, Berlin Heidelberg New York, pp 53–95
Jeger MJ, Kenerley CM, Gerik TJ, Koch DO (1987) Spatial dynamics of Phymatotrichum root rot in row crops in the blackland region of north central Texas. Phytopathology 77:1647–1656
Johnson KB, Apple JD, Powelson ML (1988) Spatial patterns of *Verticillium dahliae* propagules in potato field soils of Oregon's Columbia Basin. Plant Dis 72:484–488
Jumars PA, Thistle D, Jones ML (1977) Detecting two-dimensional spatial structure in biological data. Oecologia 28:109–123
Kareiva P, Andersen M (1988) Spatial aspects of species interactions: the wedding of models and experiments. In: Hastings A (ed) Community ecology. Springer, Berlin Heidelberg New York, pp 38–54
Kliejunas JT, Ko WH (1976) Dispersal of *Phytophthora cinnamomi* on the island of Hawaii, Phytopathology 66:457–460
Klotz LJ, Wong Po-Ping, DeWolfe TA (1959) Survey of irrigation water for the presence of *Phytophthora* spp. pathogenic to citrus. Plant Dis Rep 43:830–832
Krishna Iyer PV (1949) The first and second moments of some probability distributions arising from points on a lattice and their application. Biometrika 36:135–141
Krishna Iyer PV (1950) The theory of probability distributions on points on a lattice. Ann Math Statist 21:198–217
Legg JT (1953) The natural spread of nettlehead, a virus disease of hops. Rep E Malling Res Stn for 1952, pp 12–127
Levin SA (1989) Challenges in the development of a theory of ecosystem structure and function. In: Roughgarden J, May RM, Levin SA (eds) Perspectives in ecological theory. Princeton University Press, Princeton, pp 242–255
Lin CS, Poushinsky G, Mauer M (1979) An examination of five sampling methods under random and clustered disease distribution using simulation. Can J Plant Sci 59:121–130
Linderman RG (1974) Ascospore discharge from perithecia of *Calonectria theae*, *C. crotalariae*, and *C. kyotensis*. Phytopathology 64:567–569
Ludwig JA (1979) A test of different quadrat variance methods for the analysis of spatial pattern. In: Cormack RM, Ord JK (eds) Spatial and temporal analysis in ecology. International Cooperative Publishing House, Fairlane, MD, pp 289–304
MacDonald JD, Duniway JM (1978) Influence of soil texture and temperature on motility of *Phytophthora cryptogea* and *P. megasperma* zoospores. Phytopathology 68:1627–1630
MacGuidwin AE, Stanger BA (1991) Changes in vertical distribution of *Pratylenchus scribneri* under potato and corn. J Nematol 23:73–81
MacNish GC (1985) Mapping *Rhizoctonia* patch in consecutive cereal crops in Western Australia. Plant Path 34:165–174
Madden LV (1989) Dynamic nature of within-field disease and pathogen distributions. In: Jeger MJ (ed) Spatial components of plant disease epidemics. Prentice-Hall, Englewood Cliffs, pp 96–126
Madden LV, Louie R, Abt JJ, Knoke JK (1982) Evaluation of tests for randomness of infected plants. Phytopathology 72:195–198

Marois JJ, Adams PB (1985) Frequency distribution analyses of lettuce crop caused by *Sclerotinia minor*. Phytopathology 75:957–961
Martin SB, Campbell CL, Lucas LT (1983) Horizontal distribution and characterization of *Rhizoctonia* spp. in tall fescue turf. Phytopathology 73:1064–1068
McCarter-Zorner NJ, Franc GD, Harrison MD, Michaud JE, Quinn CE, Sells IA, Graham DC (1984) Soft rot *Erwinia* bacteria in surface and underground waters in southern Scotland and in Colorado, United States. J Appl Bacteriol 57:95–105
McCartney HA, Fitt BDL (1985) Construction of dispersal models. In: Gilligan CA (ed) Mathematical modelling of crop disease. Adv Plant Pathol 3:107–143
McIntosh DL (1966) The occurrence of *Phytophthora* spp. in irrigation water in British Columbia. Can J Bot 44:1591–1596
McSorley R, Dickson DW (1991) Determining consistency of spatial disperson of nematodes in small plots. J Nematol 23:65–72
Mihail JD (1989) *Macrophomina phaseolina*: spatio-temporal dynamics of inoculum and disease in a highly susceptible crop. Phytopathology 79:848–855
Mihail JD, Alcorn SM (1982) Quantitative recovery of *Macrophomina phaseolina* from soil. Plant Dis 66:662–663
Mihail JD, Alcorn SM (1987) *Macrophomina phaseolina*: spatial patterns in a cultivated soil and sampling strategies. Phytopathology 77:1126–1131
Mircetich SM, Browne GT, Krueger W, Schreader W (1985) *Phytophthora* spp. isolated from surface-water irrigation sources in California. Phytopathology 75:1346–1347 (Abstr)
Modjeska JS, Rawlings JO (1983) Spatial correlation analysis of uniformity data. Biometrics 39:373–384
Moran PAP (1948) The interpretation of statistical maps. J R Statist Soc B 10:243–251
Morisita M (1959) Measuring the dispersion of individuals and analysis of distributional patterns. Mem Fac Sci Kyushu Univ Ser E (Biol) 2:215–235
Nakai T, Ui T (1977) Population and distribution of sclerotia of *Rhizoctonia solani* Kühn in sugar beet field soil. Soil Biol Biochem 9:377–381
Neher D, Duniway JM (1992) Dispersal of *Phytophthora parasitica* in tomato fields by furrow irrigation. Plant Dis 76:582–586
Nelson SC, Marsh PL, Campbell CL (1992) 2D CLASS – a two-dimensional distance class analysis software for the personal computer. Plant Dis 76:213–215
Nelson SC, Campbell CL (1993) Comparative spatial analysis of foliar epidemics on white clover caued by viruses, fungi, and a bacterium. Phytopathology 83: 288–301
Newhook FJ, Young BR, Allen SD, Allen RN (1981) Zoospore motility of *Phytophthora cinnamomi* in particulate substrates. Phytopathol Z 101:202–209
Nicot PC, Rouse DI, Yandell BS (1984) Comparison of statistical methods for studying spatial patterns of soilborne plant pathogens in the field. Phytopathology 74:1399–1402
Noe JP, Barker KR (1985) Overestimation of yield loss of tobacco caused by the aggregated spatial pattern of *Meloidogyne incognita*. J Nematol 17:245–251
Noe JP, Campbell CL (1985) Spatial pattern analysis of plant-parasitic nematodes. J Nematol 17:86–93
Olanya OM, Campbell CL (1988) Effects of tillage on the spatial pattern of *Macrophomina phaseolina*. Phytopathology 78:217–221
Olanya OM, Campbell CL (1989) Density and spatial pattern of propagules of *Macrophomina phaseolina* in corn rhizospheres. Phytopathology 79:1119–1123
O'Neill RV, DeAngelis EL, Waide JB, Allen TFH (1986) A hierarchical concept of ecosystems. Princeton University Press, Princeton
Onstad DA (1992) Temporal and spatial scales in epidemiological concepts. J Theor Biol 158:495–515

Pasquill F (1974) Development of practical systems for calculating the concentration pattern from a single source. In: Pasquill F (ed) Atmospheric diffusion. Ellis Horwood, Chichester, pp 365–380
Pedgley DE (1986) Long distance transport of spores. In: Leonard KJ, Fry WE (eds) Plant disease epidemiology: population dynamics and management, vol 1. Macmillan, New York, pp 346–365
Pennacchio F, D'Errico FP, Tremblay E (1985) Spatial distribution pattern and sequential sampling plan for *Paratylenchus dianthus*. Nematol Medit 13:137–146
Perombelon MCM (1978) Contamination of potato crops by air-borne *Erwinia carotovora*. Proc 4th Int Conf Plant Pathol Bacteriol, Angers, Stn Pathol Veg Phytobacteriol, INRA, pp 563–565
Perombelon MCM, Fox RA, Lowe R (1979) Dispersal of *Erwinia carotovora* in aerosols produced by the pulverization of potato haulm prior to harvest. J Phytopathol 94:249–260
Perombelon MCM, Kelman A (1980) Ecology of the soft rot erwinias. Annu Rev Phytopathol 18:361–387
Perry JN (1983) Effects of spatial heterogeneity on Jones' model for cyst-nematode population dynamics and crop root damage. J Appl Ecol 20:849–856
Peters RH (1988) Some general problems for ecology illustrated by food web theory. Ecology 69:1673–1676
Phelps K, White JG, Henn AJ (1991) Studies on the frequency distribution of *Pythium*-induced cavity spots of carrots. Ann Appl Biol 119:21–30
Phipps PM, Beute MK, Barker KR (1976) An elutriation method for quantitative isolation of *Cylndrocladium crotalariae* microsclerotia from peanut field soil. Phytopathology 66:1256–1259
Pielou EC (1977) Mathematical ecology, 2nd edn. Wiley, New York
Punja ZK, Smith VL, Campbell CL, Jenkins SF (1985) Sampling and extraction procedures to estimate numbers, spatial pattern, and temporal distribution of sclerotia of *Sclerotium rolfsii* in soil. Plant Dis 69:469–474
Rittenhouse CM, Griffin GJ (1985) Pattern of *Thielaviopsis basicola* in tobacco field soil. Can J Plant Pathol 7:377–381
Roach SA (1968) The theory of random clumping. Methuen, London
Roth DA, Griffin GJ (1981) Cylindrocladium root rot of black walnut seedlings and inoculum pattern in nursery soil. Can J Plant Pathol 3:1–5
Roughgarden G, May RM, Levin SA (1989) Perspectives in ecological theory. Princeton Univ Press, Princeton
Rowe RC, Beute MK (1975) Ascospore formation and discharge by *Calonectria crotalariae*. Phytopathology 65:393–398
Rowe RC, Johnston SA, Beute MK (1974) Formation and dispersal of *Cylindrocladium crotalariae* microsclerotia in infected peanut roots. Phytopathology 64:1294–1297
Rowe RC, Farley JD, Coplin DL (1977) Airborne spore dispersal and recolonization of steamed soil by *Fusarium oxysporum* in tomato greenhouses. Phytopathology 67:1513–1517
Ruddick SM, Williams ST (1972) Studies on the ecology of actinomycetes in soil. V. Some factors influencing the dispersal and adsorption of spores in soil. Soil Biol Biochem 4:93–103
Sakai AK, Oden NL (1983) Spatial pattern of sex expression in silver maple (*Acer saccharinum* L.): Morisita's index and spatial autocorrelation. Am Nat 122: 489–508
Schuh W, Frederiksen RA, Jeger MJ (1986) Analysis of spatial patterns in sorghum downy mildew with Morisita's index of dispersion. Phytopathology 76:446–450
Seinhorst JW (1973) The relation between nematode distribution in a field and loss in yield at different average nematode densities. Nematologica 19:421–427

Shew BB, Beute MK, Campbell CL (1984) Spatial pattern of southern stem rot caused by *Sclerotium rolfsii* in six North Carolina peanut fields. Phytopathology 74:730–735

Shew HD, Main CE (1985) Rhizoctonia leaf spot of flue-cured tobacco in North Carolina. Plant Dis 69:901–903

Shokes FM, McCarter SM (1979) Occurrence, dissemination, and survival of plant pathogens in surface irrigation ponds in southern Georgia. Phytopathology 69:510–516

Smith VL, Rowe RC (1984) Characteristics and distribution of propagules of *Verticillium dahliae* in Ohio potato field soils and assessment of two assay methods. Phytopathology 74:553–556

Sokal RR (1979) Ecological parameters inferred from spatial correlograms. In: Patil GP, Rosenzweig M (eds) Contemporary quantitative ecology and related ecometrics. International Cooperative Publishing House, Fairlane, MD, pp 167–196

Sokal RR, Oden NL (1978a) Spatial autocorrelation in biology 1. Methodology. Biol J Linn Soc 10:199–228

Sokal RR, Oden NL (1978b) Spatial autocorrelation in biology 2. Some biological implications and four applications of evolutionary and ecological interest. Biol J Linn Soc 10:229–249

Sokal RR, Thomson JD (1987) Applications of spatial autocorrelation in ecology. In: Legendre P, Legendre I (eds) Developments in numerical ecology. NATO ASI Ser G14. Springer, Berlin Heidelberg New York, pp 431–466

Southwood TRE (1978) Ecological methods, 2nd edn. Wiley, New York

Stanghellini ME, von Bretzel P, Kronland WC, Jenkins AD (1982) Inoculum densities of *Pythium aphanidermatum* in soils of irrigated sugar beet fields in Arizona. Phytopathology 72:1481–1485

Steadman JR, Maier CE, Schwartz HF, Kerr ED (1975) Pollution of surface irrigation waters by plant pathogenic organisms. Water Resour Bull 11:796–804

Swiecki TJ, MacDonald JD (1991) Soil salinity enhances Phytophthora root rot of tomato but hinders asexual reproduction by *Phytophthora parasitica*. J Am Soc Hortic Sci 116:471–477

Taylor JD, Griffin GJ, Garren KH (1981) Inoculum pattern, inoculum density-disease incidence relationships, and population fluctuations of *Cylindrocladium crotalariae* microsclerotia in peanut field soil. Phytopathology 71:1297–1302

Taylor LR (1961) Aggregation, variance and the mean. Nature 189:732–735

Taylor LR (1971) Aggregation as a species characteristic. In: Patil GP, Pielou EC, Waters WE (eds) Statistical ecology. Pennsylvania State Univ Press, University Park, pp 357–377

Taylor LR (1984) Assessing and interpreting the spatial distribution of insect populations. Annu Rev Entomol 29:321–357

Taylor LR, Woiwood IP, Perry JN (1978) The density-dependence of spatial behaviour and the rarity of randomness. J Anim Ecol 47:383–406

Thomson SV, Allen RM (1974) Occurrence of *Phytophthora* species and other potential plant pathogens in recycled irrigation water. Plant Dis Rep 58:945–949

Thomson SV, Allen RM (1976) Mechanisms of survival of zoospores of *Phytophthora parasitica* in irrigation water. Phytopathology 66:1198–1202

Todd H (1940) Note on random associations in a square point lattice. J R Statist Soc Suppl 7:78–82

Tomasino SF, Conway KE (1987) Spatial patern, inoculum density-disease incidence relationship, and population dynamics of *Sclerotium rolfsii* on apple rootstock. Plant Dis 71:719–724

Upton G, Fingleton B (1985) Spatial data analysis by example, vol 1. Wiley, New York

Vanderplank JE (1946) A method for estimating the number of random groups of adjacent diseased plants in a homogeneous field. Trans R Soc S Afr 31:269
Vanderplank JE (1963) Plant diseases: epidemics and control. Academic Press, New York
Vanderplank JE (1968) Disease resistance in plants. Academic Press, New York
Venette JR, Kennedy BW (1975) Naturally produced aerosols of *Pseudomonas glycinea*. Phytopathology 65:737–738
Von Broembsen SL (1984) Distribution of *Phytophthora cinnamomi* in rivers of the South Western Cape Province of South Africa. Phytophylactica 16:227–229
Weste G, Marks GC (1987) The biology of *Phytophthora cinnamomi* in Australasian forests. Annu Rev Phytopathol 25:207–229
Wheeler T, Kenerley EM, Jeger MJ, Starr JL (1987) The effects of quadrat and core size on determining the spatial pattern of *Criconemella sphaerocephalus*. J Nematol 19:413–419
Wiens JA (1976) Population responses to patchy environments. Annu Rev Ecol Syst 7:81–120
Wiens JA (1989) Spatial scaling in ecology. Funct Ecol 3:385–397
Yang XB, Snow JP, Berggren GT (1991) Patterns of Rhizoctonia foliar blight on soybean and effect of aggregation on disease development. Phytopathology 81:287–293

8 Host Resistance

H.D. Shew and B.B. Shew

8.1 Introduction

Host resistance is the most economical and potentially the most effective tool for the management of root diseases and soilborne pathogens. It requires no specialized equipment to use, reduces or eliminates the need for application of synthetic chemicals, and may shorten rotation times. Resistance to root diseases and soilborne pathogens has played a major role, and in the future, will most likely play an even greater role in the sustained production of agricultural crops.

The first major advances in the development of resistance to soilborne pathogens were the pioneering studies of Orton (1909). He initiated studies on the Fusarium wilt diseases of cotton, watermelon, and cowpea around 1900 in the United States, and developed resistant germplasm and varieties using a variety of methods that included hybridization and selection of surviving plants grown in highly infested soil. Similar successes were reported before 1915 for the Fusarium wilt diseases of flax, tomato, and cabbage (reviewed by Walker 1969). Rapid progress was made in the first half of the twentieth century in the introduction of resistance to soilborne pathogens into a range of crop species (Walker 1965). Today, most of our crop plants have had resistance to one or more soilborne pathogens incorporated into high yielding cultivars.

Plants have evolved a wide array of defense mechanisms to pathogens. Parlevliet (1981) classifies these mechanisms as avoidance, tolerance, and resistance. Identification and incorporation of these mechanisms into agronomically acceptable crops have been a major goal of plant breeders and plant pathologists. The goal of this chapter is to enhance our understanding of resistance to soilborne pathogens and root diseases, by exploring how we currently identify, characterize, develop, and use resistance. The examples and discussions used will center on resistance to fungal pathogens. The reader is referred to several excellent reviews on this topic for nematodes (Boerma and Hussey 1992; Roberts 1992; Williamson et al. 1992; Young 1992; Barker 1993) and bacteria (Hayward 1991; Whalen 1991). Mechanisms of pathogenesis and the molecular biology of the resistance response are also beyond the scope of this chapter and have been reviewed elsewhere (e.g., Keen 1992; Lindgren et al. 1992).

C.L. Campbell/D.M. Benson (Eds.)
Epidemiology and Management of Root Diseases

8.2 Terminology

A clear understanding of terminology is always important, but especially so in any discussion of host resistance and related areas. The following terms are defined to prevent confusion arising from differing interpretations of the terminology used.

Disease avoidance refers to the lack of disease on a potentially susceptible host that results from the decreased frequency of contact between inoculum and infection court. It differs from *disease escape* in that avoidance is the result of genetically controlled mechanisms in the host, whereas escape is merely a spatial phenomenon. Some types of *passive resistance* and *preformed resistance* are avoidance mechanisms. This phenomenon has received limited attention in the study of epidemics of root diseases.

As discussed by Clarke (1986) the term *tolerance* has been used loosely to refer to the ability of a host plant to yield well even when infected and diseased. Much confusion has arisen concerning the use of the term tolerance, primarily because some researchers consider tolerance a form of resistance and because it is difficult to differentiate from some types of partial resistance. Tolerance refers to the ability of a host plant to endure or sustain levels of a pathogen or disease that would cause greater impairment of growth or yield in other hosts of the same or similar species. Tolerance should not be used to refer to any host property that delays or reduces development of the pathogen or disease (Clarke 1986).

Resistance is the ability of a host plant to reduce, delay, or prevent development of a pathogen and/or disease. Any discussion of disease resistance invariably points to the wide array of terms that have been introduced in an attempt to subdivide types of disease resistance into unambiguous groups. The advantages and disadvantages of various terms have been presented in several reviews and will not be repeated here (e.g., Vanderplank 1968, 1984; Harris and Frederiksen 1984; Burdon 1992). None of the terms proposed have gained universal usage, illustrating that significant differences still exist among researchers in preferred terminology.

Proposed resistance terminology may reflect the genetic basis of resistance, the epidemiological consequences of resistance, or describe the effect of resistance without reference to its genetic basis or consequences. Problems and confusion arise when terms confound more than one of these areas or when definitions are amended as new information is obtained. In this chapter, we primarily will use the terms *partial resistance* and *complete resistance* to describe resistance to soilborne pathogens and root diseases.

Complete resistance prevents growth and reproduction of the pathogen. This resistance results in a sharp decline in the pathogen population in the absence of a saprophytic phase, reproduction on susceptible weed hosts, or in the rhizosphere and roots of symptomless hosts. *Nonhost resistance* is resistance exhibited by plant species not considered to be a host of a particular organism. It is often thought of as immunity but it may have many responses in common with complete resistance (Heath 1981; Hadwiger et al. 1990). *Partial resistance* simply means resistance that is not complete. It is not synonymous with *horizontal resistance* (see Vanderplank 1968). Parlevliet (1989, 1992) defines partial resistance as quantitatively expressed resistance, but partial resistance is not necessarily quantitatively inherited.

Resistance may be to the pathogen, disease, or both. Pathologists, especially those who work with root diseases, frequently do not distinguish between these forms of resistance because the type or mechanism of resistance is not known. Complete resistance is almost always resistance to the pathogen, whereas partial resistance most likely incorporates mechanisms to the pathogen and the disease.

Pathogenicity is the ability to cause disease. *Virulence* (specific virulence) is pathogenicity characterized by specific genetic interactions between genotypes of the host and the pathogen. The occurrence of races of the pathogen is the common result of these interactions. *Aggressiveness* (general virulence) refers to the level of disease caused by an isolate of the pathogen on a host; specific interactions with host genotypes are not observed.

8.3 Components of Partial Resistance to Root Pathogens

Parlevliet (1979) introduced the concept that partial resistance could be subdivided into several components. This concept fostered an awareness of the potential for using partial resistance for management of foliar pathogens. Pathologists and breeders have since characterized resistance components and studied their inheritance in several pathosystems (Parlevliet 1992).

Polycyclic infection resulting in logistic increase in plant disease provided the conceptual framework for understanding the importance of resistance components such as latent period, infectious period, and spore production (Parlevliet 1979). These components have received less attention in studies of root disease resistance, partly because they seem less important in pathosystems that traditionally (and sometimes, erroneously) have been viewed as monocyclic, and partly because they are difficult to measure for root pathogens. Nevertheless, partial resistance often is the best or only option for management of root pathogens. A systematic approach to identifying resistance components should allow breeders to focus on the traits that have the most potential for enhancing partial resistance to root pathogens.

The most important components in expression of resistance vary depending on the population dynamics of the root pathogen. Population dynamics can be characterized as polycyclic, monocyclic, or heterogeneous, based on production and dispersal of secondary inoculum during a host growth cycle. Root pathogens that are polycyclic include root-knot nematodes, *Phytophthora* spp., *Pythium* spp., and *Aphanomyces euteiches*. No secondary inoculum is dispersed from hosts infected by monocyclic root pathogens; the epidemic is governed by the amount of primary inoculum. The vascular wilt pathogens *Fusarium oxysporum* and *Verticillium* spp. are classic examples, but they are exceptional because they cause systemic infection. Root rotting pathogens such as *Cylindrocladium crotalariae* and *Macrophomina phaseolina* provide different examples of monocyclic disease dynamics. Pathogens with heterogeneous population dynamics produce a few cycles of secondary infection, but primary inoculum still influences the course of the epidemic (Gilligan 1990). Examples include *Sclerotium cepivorum*, *Thielaviopsis basicola*, *Phymatotrichum omnivorum*, and *Gaeumannomyces graminis*.

Population dynamics cannot be inferred from statistical fit of theoretical models to disease progress curves (Huisman 1982; Pfender 1982). Direct observation of sporulation on root surfaces (Pulliam 1986; Mauk and Hine 1988), soil assays documenting changes in inoculum density (Ferrin and Mitchell 1986), observation of secondary infection through root contacts or mycelial extension (Crowe and Hall 1980; Percy and Rush 1985), and presence of secondary infections after point source inoculation in noninfested fields (Shew 1987) are examples of evidence for (or, if absent, against) pathogen reproduction and dispersal in a growing season.

Theoretical characteristics of root disease epidemics have been most thoroughly studied for monocyclic diseases (Jeger 1987; Gilligan 1990). These studies have extended the monomolecular model of disease increase by linking differential equations that describe dynamics of rates of infection, root growth, and lesion expansion. Jeger (1987) showed that the shape of the disease progress curve, the final amount of diseased area, and the final amount of root growth could be predicted entirely from these three rates. His models rest on several assumptions which may not be valid under many conditions, but their predictions provide valuable insight into disease dynamics and suggest certain components that may contribute to resistance to root pathogens.

Root disease dynamics are simplest when there is a fixed pool of primary inoculum, no root growth, no lesion expansion, and no production of secondary inoculum. These conditions partially satisfy the assumptions implicit in the use of the monomolecular model as a biological model, and also provide a starting point for many other models (Baker 1971; Gilligan 1985; Jeger 1987). Under these conditions, expression of host resistance will be determined entirely by infection efficiency, lesion size, and carrying capacity. Of these, infection efficiency has received the most attention and can be defined as the number of infections produced on a unit of host tissue per unit inoculum. Carrying capacity refers to the maximum (asymptotic) level of disease possible on a particular host.

The size of the root system and the rate of root growth also determine dynamics of root disease, because the probability of contact between infection court and inoculum increases with increasing root length (Gilligan 1985). Root growth is one of the many reasons that disease progress curves for monocyclic root diseases often appear S-shaped (Huisman 1982). Jeger (1987) theorized that for monomolecular pathosystems, the balance between the rate of root growth and the rate of infection was critical to root disease dynamics. He found that *slower* rates of root growth, relative to the infection rate, can lead to *lower* levels of disease in the absence of lesion expansion. This phenomenon, which might be referred to as Jeger's paradox, occurs because of inoculum depletion in the pathozone. Limited evidence suggests that slow rates of root growth may contribute to apparent "resistance" under broader conditions than those outlined by Jeger (Jones 1990; Brune and Beute 1993). Root growth rates or rooting patterns that result in less disease are avoidance mechanisms, as discussed in Section 8.5.

The rate of lesion area expansion (and to a lesser extent, lesion area) profoundly affects dynamics of monocyclic root disease. If lesions do not expand, the absolute area of diseased root levels off, whereas root growth continues. Even in the absence of host growth (perhaps for crown or hypocotyl diseases) the portion of root area diseased approaches an asymptote of less than 100% when lesions do not expand. Infection efficiency and lesion size determine the carrying capacity; a lower asymptotic level of disease is a measure of resistance. With lesion expansion, however, the portion of root diseased approaches 100% in the absence of root growth, or when roots grow more slowly than lesions expand. The ability of the host to restrict lesion size or slow lesion expansion is critical to expression of resistance to monocyclic root disease.

Root diseases caused by polycyclic pathogens have components of resistance in common with foliar pathogens because in either case disease suppression depends on reducing the number of pathogen generations per growing season and the amount of inoculum produced per generation. Longer latent periods, shorter infectious periods, and lower rates of sporulation will result in higher levels of partial resistance. Nevertheless, straightforward application of foliar models and resistance concepts to polycyclic root diseases overlooks important differences between the pathosystems. For example, root growth influences disease progress not only by varying the amount of host tissue available for infection (as with foliar diseases), but also by increasing the probability of contacting inoculum, as discussed for monocyclic pathogens. In a simulation model of Phytophthora root rot on Fraser fir (a polycyclic disease), Reynolds et al. (1986) found that root growth contributed greatly to secondary infection by initiating root-to-root contacts. Other impacts of root growth on polycyclic disease await further investigation.

Brassett and Gilligan (1988) and Gilligan (1990; Chap. 6, this Vol.) developed an analytical model for root pathogens with heterogeneous population dynamics. These pathogens produce secondary inoculum, but primary inoculum continues to contribute to the epidemic rather than being ephemeral, as is assumed with truly polycyclic pathogens. The models describing disease dynamics generated curves that ranged from monomolecular to asymmetric sigmoid (Brassett and Gilligan 1988); high rates of secondary infection resulted in the most rapid disease progress. As the coefficient of secondary infection decreased, the coefficient of primary infection had greater influence on the epidemic. Thus, for heterogeneous pathosystems, breeders should first seek to limit secondary infection through longer latent periods and lower levels of inoculum production, and then strive to reduce infection efficiency. Lesion size and rate of lesion expansion probably impact both phases of the epidemic through increasing root-to-root spread and sporulating tissue in the polycyclic phase and by determining asymptotic levels of disease in the primary infection phase.

8.4 Measurement of Resistance Components

Infection efficiency probably is the component of resistance to root disease that has been measured most often. For example, classic inoculum density-disease incidence curves (Baker 1971) measure infection efficiency indirectly. Usually, however, the host was not exposed to a known absolute amount of viable inoculum because inoculum density was expressed per gram of soil. Correction for multiple infections may be necessary (Gregory 1948; Waggoner 1981) and timing of inoculation and disease assessment also is critical to avoid confounding inoculum efficiency with other components of resistance. Reynolds et al. (1985a) and Tomimatsu and Griffin (1982) are among the few authors who have attempted to account explicitly for these confounding factors, but their calculations required the measurement of several parameters. Direct measurement of infection efficiency through precise placement of inoculum on roots has been accomplished and is simpler conceptually (Harris and Beute 1982; Jones 1990). Plants should be grown in soil that is pathogen-free, disease-conducive, and if possible, natural; propagules should be as similar as possible in ability to cause infection. Culture-produced inoculum often is inferior or performs differently than natural inoculum and should be avoided in favor of inoculum produced on natural substrate when possible. Inoculum levels must be selected carefully to avoid overwhelming partial resistance. Probability theory has been used to determine realistic amounts of inoculum (Harris and Beute 1982).

An interesting advantage in assessment of infection efficiency with root pathogens is the availability of selective media for their culture. Thus, several authors (e.g., Tomimatsu and Griffin 1982; Huisman 1988; Jones 1990) have assayed directly for infection rather than lesion formation as is common with foliar pathogens. Root sections plated for assay of infection efficiency should be as long as possible to avoid fragmenting of single infection sites (Tomimatsu and Griffin 1982; Huisman 1988). Researchers may want to consider resistance to infection and resistance to lesion formation as separate components of resistance; in either case, they need to specify clearly the component measured.

Measurement of latent period, defined as the time from inoculation to inoculum production, requires careful experimentation. The time of contact between infection court and inoculum is not known precisely if the host is planted in bulk, infested soil. Time of inoculation can be controlled by placing inoculum directly on roots (Harris and Beute 1982; Jones 1990; Milholland and Daykin 1993), by allowing roots to grow from noninfested into infested soil (Mitchell 1975), or by planting the host in a soil infested with a zoospore-producing pathogen and then imposing a flooding period (Duniway 1983; Shew 1983; Kenerley et al. 1984). Likewise, observation of time to inoculum production requires root observation chambers, in vitro incubation, and observation of inoculated roots (Reynolds et al. 1985a), or destructive sampling at frequent intervals after inoculation. Time from inoculation to spore production will vary among a population of root lesions, influencing the precision and accuracy of latent period estimation (Shaner 1980). Estimates of latent period will be most reliable near the population mean, but the first lesions that sporulate have more impact on generation time and thus final population level.

Lesion size and rate of lesion area expansion probably are critical components of resistance to root diseases, but seldom are measured explicitly. The most commonly measured variables that relate to lesion size are root rot indices. Selection for resistance based on root rot indices has been relatively successful (e.g., Pataky et al. 1983; Singleton et al. 1992; Shew and Shoemaker 1993). If infection efficiency, carrying capacity of the host, and root growth factors are equal, root rot indices that estimate the necrotic root area would relate directly to lesion size. Otherwise these factors are confounded in root rot indices and more precise measurement may be needed.

Rate of lesion expansion can be determined by point inoculation followed by plating root sections in sequence from the site of inoculation at various times after inoculation (Reynolds et al. 1985b). Jones (1990) observed expansion of lesions on roots of tobacco caused by *Phytophthora parasitica* var. *nicotianae*. Roots were point inoculated through root observation plates that were installed beside plants at transplanting. Soil was removed daily from observation plates and lesion size measured on undisturbed roots. Measurement of the rate of lesion expansion also is possible in root observation boxes (Jones 1990) and on detached tissues in vitro (Brenneman et al. 1988; Chappell 1992).

Inoculum production can be expressed per lesion, per lesion (or necrotic) area, per area (or gram) of root, or per gram of soil. Each of these expressions may reflect different mechanisms of resistance, and more than one of them probably is necessary to fully characterize resistance. Again, careful experimentation is needed to assure that the roots and lesions examined are of uniform size and age. Dispersal of inoculum is required for secondary infection to occur, so only inoculum which is capable of contacting and infecting healthy tissue should be quantified when the purpose is to measure resistance that limits polycyclic pathogen increase. Resistance also may influence inoculum quality, that is, the capacity to initiate infection or survive in the absence of the host. Host resistance could influence inoculum size (Black and Beute 1984c), type, nutrient status, or germinability.

8.5 Disease Avoidance and Disease Escape

Disease avoidance and disease escape are terms used to describe the absence of disease on an otherwise susceptible host due to the presence of temporal, spatial, or physical barriers between the pathogen and its host (Parlevliet 1981; Burdon 1992). These terms should not be considered synonymous. In fact, disease escape may be the result of disease avoidance mechanisms. Avoidance mechanisms or traits are genetically controlled, whereas escape results from a fortuitous set of circumstances that prevents host infection; disease escape would therefore more appropriately be termed infection escape (Agrios 1980). Agrios (1980) discusses escape in a broad sense. He includes numerous factors that operate outside the host as well as many host and pathogen factors as potentially important in disease escape. In this chapter, host factors that are heritable and that result in escape are considered avoidance mechanisms for reasons discussed below. Disease escape can nonetheless play a very significant role in the management of root diseases through the use of delayed planting (e.g., Sidebottom and Beute 1989a), irrigation management (Duniway 1983), and other cultural practices (Sumner, Chap. 11).

The effects of disease escape are readily observed in the early stages of development of root disease epidemics due to the clumped distribution of initial inoculum typical of soilborne pathogens (Campbell and Noe 1985). Escape in these situations can be especially troublesome or disastrous for plant breeders, who select plants based on their ability to survive in a disease nursery during early generations of testing. To prevent escapes, disease nurseries are often either very highly infested locations or they are infested each year with the pathogen. In either case, moderate levels of partial resistance and disease avoidance may be overlooked if unrealistic inoculum densities are used.

Disease avoidance has received limited attention in the study of root diseases and soilborne pathogens, except for diseases caused by *Sclerotinia* and *Sclerotium* spp. Canopy characteristics create favorable or unfavorable microclimates for these fungi before and during pathogenesis (e.g., Schwartz et al. 1978; Shew and Beute 1984; Chappell 1992). Less attention has been given to the role of avoidance mechanisms in the development of root diseases because it is difficult to quantify root growth dynamics, especially under field conditions. In addition, avoidance mechanisms generally are not important in greenhouse experiments due to restricted root development in pots, inoculum levels used, and inoculum distribution. Unlike the classical examples of avoidance exhibited by the closed-flower habit of cereals (Caldwell 1966), where little or no disease occurs, avoidance mechanisms for root diseases probably never result in the total absence of disease. Thus, they are difficult to separate from components of partial resistance.

Avoidance operates prior to host-pathogen contact (inoculation). Factors that affect inoculation of roots include root number (fewer roots mean fewer pathogen contacts), root distribution (spatial and temporal dynamics, see Chaps. 6 and 7, this Vol.), rate of root growth, and the quality and quantity of root exudates. Although reduced pathogen attraction via altered root exudation is generally considered a resistance mechanism, resistance implies a response by the host plant to the presence of a pathogen or pathogen product. Plant roots or germinating seeds that attract fewer pathogen propagules (e.g., Keeling 1974; Milholland 1975) in field soil may not appear as resistant in controlled inoculations or in field trials where artificially high inoculum levels are used.

Root growth dynamics play an important role in the epidemics of root pathogens. Avoidance of pathogen propagules has been documented or hypothesized in several root diseases and may be common in cultivars that exhibit apparent partial resistance to soilborne pathogens. These cultivars may actually have smaller (usually less branched) root systems than susceptible cultivars. This observation has been documented in such diverse systems as *P. p. nicotianae* on tobacco (Jones 1990), *C. crotalariae* on peanut (Brune and Beute 1993), *Streptomyces ipomoea* on sweet potato (Ristaino 1993), and *Striga* spp. on several hosts. Huisman and Gerik (1989) discuss the role that the rate of root elongation may have in avoiding infection. Fungi that respond slowly to root exudates may be passed by rapidly growing root tips.

Can breeders take advantage of avoidance mechanisms? The primary reason for characterizing avoidance mechanisms apart from disease escape is to broaden our thinking about plant traits that enhance host performance. By broadly classifying heritable traits as escape traits (Agrios 1980), researchers may overlook potentially important mechanisms of resistance for some groups of pathogens. Because avoidance mechanisms are most likely quantitatively inherited, the use of techniques to identify quantitative trait loci (QTLs) (e.g., Stuber 1990; Edwards 1992) may provide the opportunity to identify these traits, allowing their use in a breeding program.

Although avoidance of pathogens may lead to fewer root infections, smaller root systems may have detrimental effects on other plant traits. Genotypes resistant to root pathogens frequently have lower yields and do not tolerate environmental stresses, for example, drought stress, as well as susceptible cultivars. Are smaller root systems a pleiotropic effect of resistance genes or have smaller root systems been selected inadvertently because they survive better in field disease evaluations? Are they present only in cultivars with partial resistance or is this trait also common with monogenic complete resistance? Plant pathologists must address these questions if we are to understand how avoidance mechanisms can best be used in management of root diseases and before breeders will readily incorporate smaller root systems into new cultivars.

8.6 Tolerance

There are currently two general views of tolerance to pathogens and disease in plants. The broad-sense view is presented by Mussell (1980). In the broad sense, tolerance refers to the ability of a cell, whole plant, or population of plants to provide a habitat for pathogen development but still perform acceptably. Tolerance is not distinguished from partial resistance in this conceptual approach. The broad-sense concept of tolerance has been used by some researchers studying the Phytophthora root rot of soybean (e.g., Schmitthenner 1985; McBlain et al. 1991a,b). In this system, genotype reactions are initially based on a cotyledon inoculation test designed to identify complete resistance. Unfortunately, but not surprisingly, results of the cotyledon test and field results do not always agree. This has led to the use of the broad-sense tolerance concept, wherein tolerance is used to refer to any mechanism other than complete resistance which reduces disease severity. Tolerance mechanisms include race nonspecific endurance, rate-reducing resistance (Tooley and Grau 1982), and root resistance (McBlain et al. 1991a). The high heritabilities observed for broad-sense tolerance to *Phytophthora megasperma* f. sp. *glycinea* (Walker and Schmitthenner 1984) probably results from the selection for multiple components of resistance and possibly tolerance. A variety of inoculation techniques have been used to identify different "components of tolerance" (McBlain et al. 1991b). The use of tolerance in the broad sense is at best confusing and should be avoided.

A narrow, but more realistic, view of tolerance is presented by Clarke (1986), Cook and Evans (1987), and Schafer (1971). Clarke (1986) indicates that a plant may tolerate the parasite, the disease, or both. Tolerance is not a type of resistance; it may occur in plants along with resistance and it may segregate independently of resistance (Trudgill 1991). This is another good reason for separating resistance and tolerance mechanisms.

Identification and selection for tolerance require its measurement or quantification. This is accomplished with methods that measure pathogen biomass, the amount of disease, and the extent of growth or yield suppression (Clarke 1986). Root diseases present unique problems for the study of tolerance for the same reasons as discussed for partial resistance. Accurate assessment of disease severity and pathogen biomass is difficult. This does not mean that tolerance should be ignored or combined with resistance for convenience. Crute (1986) describes methods used to demonstrate tolerance to club root and *Plasmodiophora brassicae*.

Just as resistance-susceptibility represents a continuum in plants, tolerance-intolerance may be considered as a continuum. This concept has received much attention in nematology literature (Dropkin and Nelson 1960; Barker 1993), but very little with other root pathogens. Some susceptible cultivars may more appropriately be termed intolerant. For example, Shew and Shoemaker (1993) found that some burley tobacco cultivars suffer unusually high levels of yield suppression at similar or lower root rot severities than other cultivars.

8.7 Examples of Host-Pathogen Systems

Host resistance plays an important role in disease management in many root pathosystems. Readers may want to consult Crute (1986), Law and Milholland (1991, 1992), and Tooley and Grau (1982, 1984a,b) for discussions of resistance in specific pathosystems, and Singleton et al. (1992) for a review of methodology for working with soilborne pathogens. We have selected four pathosystems: black shank of tobacco, a polycyclic root and stem rot disease; black root rot of tobacco, a heterogeneous root rot disease; Cylindrocladium black rot of peanut, a monocyclic root rot; and Verticillium wilt of tomato, a monocyclic vascular wilt, to illustrate the concepts of root disease resistance presented in this chapter. These pathosystems were selected because they differ in epidemiology, level of specificity in the host-pathogen interaction, and in the mechanisms, components, and inheritance of resistance. All of the diseases are of great economic importance and the same or closely related pathogens cause diseases on many hosts. Although there is a wealth of information about resistance in these pathosystems, they have not been studied with all of the concepts of resistance presented in this chapter specifically in mind. Therefore, some information about epidemiology or resistance may be incomplete. We hope these examples will serve to illustrate the components of resistance common to root pathosystems as well as some components that may have been overlooked in the past.

8.7.1 Black Shank of Tobacco

Black shank is a destructive root and crown rot disease of tobacco (*Nicotiana tabacum*) that occurs worldwide (Lucas 1975). This polycyclic disease is caused by the soilborne fungus *Phytophthora parasitica nicotianae* (syn. *P. nicotianae nicotianae*). Infection of roots and stems typically results in the production of characteristic aboveground symptoms (Shew and Lucas 1991) which makes possible the quantitative assessment of the effects of host resistance or other treatments on epidemic development (Campbell and Powell 1980; Campbell et al. 1984; Ferrin and Mitchell 1986). Tobacco is considered the only natural host of tobacco isolates of *P. p. nicotianae*, even though other genera can be infected by artificial inoculation (Lucas 1975; Holmes and Benson 1993; H.D. Shew, unpubl.). Host specificity of tobacco isolates is determined by the absence of the elicitor parasiticein (Ricci et al. 1992). Variation in virulence and aggressiveness of isolates of *P. p. nicotianae* has been reported. Four races of the pathogen have been described (Apple 1962; Prinsloo and Pauer 1973; McIntyre and Taylor 1978). The common race is designated race 0 and is found worldwide (Lucas 1975). Aggressiveness varies among isolates of race 0 (Apple 1957, 1961; Dukes and Apple 1962) and race 1 (H.D. Shew, unpubl.); aggressiveness of isolates of other races has received less attention.

Identification and measurement of resistance components require techniques for quantitative assay of the pathogen and for disease assessment. Semiselective agar media (reviewed by Mitchell and Kannwischer-Mitchell 1992) are frequently used in the quantification of pathogen propagules in soil (e.g., Kannwischer and Mitchell 1981; Ferrin and Mitchell 1986; Jones 1990). An ELISA technique has been described for quantification of the pathogen in root tissue (Jones and Shew 1989).

Management of black shank is based on the use of resistant cultivars in most areas of tobacco production. Several sources of resistance to black shank have been identified (Lucas 1975; Wernsman and Rufty 1987). Most cultivars of flue-cured tobacco have partial resistance from the cigar-wrapper cultivar Florida 301. Resistance from Florida 301 is effective against all races of the pathogen. Levels of resistance range from low to high in currently available cultivars. The resistance is controlled by a partially dominant gene but is modified by the genetic background of susceptible parents (Lucas 1975). Resistance to *P. p. nicotianae* also has been introduced into *N. tabacum* from *N. longiflora* and *N. plumbaginifolia* (reviewed by Wernsman et al. 1974). The monogenic resistance from *N. longiflora* provides complete resistance to race 0 of *P. p. nicotianae* but no resistance to race 1. Two types of resistance have been identified in *N. plumbaginifolia*; one type provides partial resistance to races 0 and 1 and the other type provides complete resistance to race 0 and no resistance to race 1. Race 2, which occurs only in South Africa, is based on the ability to attack the single-gene resistance from Delcrest 202. Race 3, found only in Connecticut, can attack plants with resistance from *N. plumbaginifolia* but not *N. longiflora*.

8.7.1.1 Components of Resistance

The complete resistance to race 0 of *P. p. nicotianae* obtained from *N. longiflora* and *N. plumbaginifolia* does not prevent root infection, but prevents colonization of the root beyond the epidermis (Powell 1956; Hanchey and Wheeler 1970). A similar response was observed in tobacco callus of monogenic resistant lines inoculated with zoospores of the pathogen (de Zoeten et al. 1982). Race 1 of *P. p. nicotianae* emerged very quickly in response to the two sources of resistance to race 0 (Lucas 1975). It is interesting to note that race 1 causes disease on genotypes with either of the resistance genes, but they are not the same gene, because race 3 is characterized by a differential response to the two genes. Only one cultivar with monogenic resistance, the burley cultivar Kentucky 14 × L-8 with resistance from *N. plumbaginifolia*, is currently grown in the United States. No cultivar with the *N. longiflora* resistance gene is currently in use because of the occurrence of race 1, some undesirable traits associated with the resistance gene, and the high level of partial resistance available.

Several studies have attempted to identify and quantify components of partial resistance associated with the Florida 301 source. Partial resistance does not affect levels of infection in tobacco (Pulliam 1986; Jones 1990), but it does decrease the number of infections that develop into visible lesions (Jones 1990). Thus, the distinction between infection and lesion formation is important.

Partial resistance does affect lesion expansion (Jones 1990). Cultivars that ranged in level of resistance from susceptible to highly resistant were inoculated at specific sites on the root system 5–6 weeks after transplanting into uninfested field soil. Rows were covered with polyethylene to prevent soil from becoming saturated or flooded, which suppressed or eliminated secondary infections caused by zoospores. Disease was evaluated every week based on development of aboveground symptoms, which occurred as a direct result of lesion expansion into the stem. Resistance delayed symptom development and reduced final disease incidence. In susceptible cultivars and in cultivars with low levels of partial resistance, disease progress curves often had the classic S-shape characteristic of logistic disease increase. Incubation period (time from inoculation to aboveground symptom development), more correctly the rate of lesion expansion, thus may affect the shape of the disease progress curve.

Secondary inoculum production is inhibited by partial resistance (Shew 1985; Ferrin and Mitchell 1986). Resistant cultivars not only suppress inoculum production by inhibiting lesion expansion, but also by suppressing inoculum production per unit area of lesion (Shew and Jones, unpubl.). The combined effects of partial resistance on lesion expansion and secondary inoculum production result in rate-reducing resistance that delays epidemic development and reduces or eliminates spread of the pathogen during the growing season (Campbell et al. 1984; Shew 1987). These observations help explain why level of initial inoculum is important in predicting disease severity for a resistant cultivar where secondary inoculum is limited, but is not a good predictor for disease incidence for a susceptible cultivar where secondary inoculum production is abundant (Ferrin and Mitchell 1986; Shew 1987).

Soon after the release of cultivars with high levels of partial resistance, Apple (1961) observed that isolates of *P. p. nicotianae* obtained from these cultivars were more aggressive on tobacco than isolates from susceptible cultivars. Passage of isolates through resistant cultivars also increased aggressiveness (Dukes and Apple 1961). This increased aggressiveness may partially explain the apparent "loss" or reduction in level of resistance often observed for resistant cultivars over a period of years. Specific adaptation (increased level of specific virulence) by *P. p. nicotianae* to tobacco cultivars with partial resistance has not been demonstrated, but warrants investigation.

8.7.1.2 Avoidance and Tolerance

Zoospore attraction is similar in susceptible and resistant tobacco genotypes (Dukes and Apple 1961; Pulliam 1986; Jones 1990). Similar results have been observed on other hosts with other members of the genus *Phytophthora* (Hickman 1970). Interestingly, attraction occurs all along tobacco roots in soil (Jones 1990), not just at wounds and the area behind the root tip. Inoculum efficiency is greater on primary roots than secondary or tertiary roots (English and Mitchell 1989).

Evidence indicates that avoidance mechanisms occur in tobacco cultivars rated as having moderate to high levels of resistance to black shank. Cultivars rated as highly resistant have smaller root systems than susceptible cultivars. Jones (1990) quantified numbers and lengths of roots of tobacco cultivars with a range of resistance levels to *P. p. nicotianae* and found that numbers of roots decreased as resistance rating increased. Unfortunately, a smaller root system may be detrimental to some important plant attributes such as yield and drought tolerance, and in tobacco perhaps is related to premature flowering and susceptibility to wind damage. It is not known if root system size and architecture are linked to resistance factors or if they have been inadvertently selected during screening in disease nurseries. The practice of selecting survivors, and any trait associated with their survival, in highly infested soil has been used extensively in screening for resistant germplasm in tobacco and many other crops.

8.7.2 Black Root Rot of Tobacco

Black root rot is primarily a cortical rot caused by *Thielaviopsis basicola* (synanamorph: *Chalara elegans*). The disease occurs worldwide on tobacco (Lucas 1975) and on at least 137 other plant species (Otani 1962). This disease rarely kills plants and cannot be diagnosed accurately without observing roots for the presence of characteristic black lesions and perhaps sporulation by the fungus (Shew and Lucas 1991). Chlamydospores and endoconidia are produced on the root surface, but because their movement in soil is limited, they probably are effective only in reinfecting other roots on the same plant. The production of secondary inoculum with limited dispersal results in heterogeneous disease dynamics. The primary aboveground symptom of the disease is stunting, which results in uneven stands of plants. Because black root rot does not cause distinctive aboveground symptoms, the disease does not readily lend itself to classical epidemiological investigations.

Host specialization has been reported in *T. basicola* (e.g., Lucas 1975; Tabachnik 1979; Corbaz 1985) but isolates from a given host species are not limited to that species (Tabachnik et al. 1979). Corbaz (1985) placed isolates from tobacco and legumes into the same subgroup. Variation in aggressiveness in isolates of *T. basicola* also has been observed (Clayton 1969; Lucas 1975; Tabachnik et al. 1979; Meyer and Shew 1991b). Stover (1950a) described brown and gray isolates of *T. basicola*; brown isolates were highly aggressive and gray isolates were less aggressive on tobacco. Color designation is not used to group isolates since this initial work. No races of *T. basicola* have been characterized, although Stover (1950b) placed isolates in four physiological races based primarily on level of aggressiveness and origin of the isolate.

Assay techniques have been developed for the qualitative and quantitative recovery of *T. basicola* from soil (reviewed by Shew and Meyer 1991). The semiselective agar medium of Specht and Griffin (1985), coupled with simplified assay procedures (Shew and Meyer 1992), make highly selective recovery of *T. basicola* from soil possible.

Management of black root rot on tobacco and other crops depends primarily on the use of resistance and cultural practices. There are several types of resistance to black root rot found in tobacco. The first type is complete resistance (Clayton 1969). This resistance is controlled by a single dominant gene transferred from *N. debneyi*. Early problems with late maturity, low yields, and small leaf size associated with this resistance have been overcome in recent years. Resistance obtained from oriental tobacco TI 87 provides moderate to high levels of partial resistance (see Wilkinson et al. 1991). A second source of partial resistance from oriental tobacco in TI 89 has not been used due to tight linkages with undesirable traits (Clayton 1969). A third source of partial resistance is found in cultivars of *N. tabacum*. A group of recessive genes controls inheritance of this low to moderate partial resistance (Clayton 1969).

8.7.2.1 Components of Resistance

The mechanism of resistance to *T. basicola* found in cultivars with the monogenic resistance from *N. debneyi* has not been determined. The effect on infection is not known. Lesions are occasionally observed but isolates of *T. basicola* obtained from these lesions are not able to infect resistant cultivars upon reinoculation. Inoculum densities to *T. basicola* decrease when cultivars with this resistance are planted, indicating no secondary inoculum is produced on roots (Shew and Shoemaker 1993). No known races (isolates) of the pathogen can attack cultivars with this resistance.

Is the absence of race development due to the "strength" of the resistance gene or the biology of *T. basicola*? First, the mechanism of resistance to *T. basicola* from *N. debneyi* should be determined. This may help explain why this resistance has been durable and may aid in development of resistance to *T. basicola* in other hosts. Perhaps this resistance can be transferred via molecular techniques to genera of plants where no genes for resistance to *T. basicola* have been found. Absence of race development might also be explained by the biology of *T. basicola*. The broad host range and high saprophytic ability of *T. basicola* (Lucas 1975) may eliminate the selection pressure that a strong resistance gene places on a specialized pathogen.

Mechanisms of partial resistance have been investigated. An early report by Conant (1927) describes the events in pathogenesis on cultivars with a range of resistance. Resistance was temperature-dependent, with susceptible cultivars becoming resistant when soil temperatures reached 25–30 °C. The primary resistance mechanism was walling-off of the pathogen by development of cork layers in the host. Resistance did not alter penetration, and was based on the rapidity of cork layer formation. Interestingly, Conant also concluded that mass action of hyphae could overcome this type of resistance, indicating that initial inoculum level is very important in expression of resistance. In contrast, research by Jewett (1938) supported a chemical nature of resistance.

The importance of initial inoculum level on severity of black root rot in tobacco has been reported (Specht and Griffin 1988; Meyer et al. 1989; Meyer and Shew 1991a). However, as reported in tobacco black shank, initial inoculum density is more predictive of disease severity in a resistant cultivar than in a susceptible cultivar (Meyer et al. 1989, Meyer and Shew 1991a). Assuming pathogen presence, soil chemical factors were better predictors of disease severity than initial inoculum density when cultivars with low levels of partial resistance were grown (Meyer et al. 1989). Partial resistance also may reduce secondary inoculum production (Shew and Shoemaker 1993) and in some cultivars lesion development (Specht and Griffin 1988; Meyer and Shew 1991a; Shew and Shoemaker 1993). The burley cultivar Ky 14 × L8 frequently has high root rot ratings but suppresses secondary inoculum production.

No studies have been completed on the effects of partial resistance on levels of aggressiveness in the pathogen, although variability in aggressiveness has been documented (Meyer and Shew 1991b). Levels of disease and rankings of resistant cultivars often vary with location but whether this is due to environmental factors or adaptation of isolates to a cultivar (specific virulence factors) has not been determined.

Development of resistance to black root rot has a long history (reviewed by Clayton 1969). The incorporation of complete resistance from *N. debneyi* appears to have provided a durable source of resistance, but breeding for enhanced levels of partial resistance (e.g., Wilkinson et al. 1991) will prove valuable if a race of the pathogen develops that can attack the *N. debneyi* source of resistance. Combination of high levels of partial resistance from *N. tabacum* and monogenic resistance should prove to be very durable for this disease.

8.7.2.2 Avoidance and Tolerance

In some cultivars of burley tobacco disease severity and secondary inoculum production are not good indicators of overall plant performance. For example, cultivars rated as having moderate levels of partial resistance, based on yield in the presence of the pathogen, may have similar or even higher root rot severity ratings and result in greater final inoculum densities of *T. basicola* than some cultivars rated as having low levels of resistance (Wilkinson et al. 1991; Shew and Shoemaker 1993). This indicates the occurrence of tolerance-intolerance mechanisms in cultivars of burley tobacco to *T. basicola* and black root rot. Mechanisms of partial resistance apparently have been confounded with mechanisms of tolerance (sensu Clarke 1986) to the pathogen or disease (Shew and Shoemaker 1993).

Avoidance mechanisms have not been investigated with this disease.

8.7.3 Cylindrocladium Black Rot of Peanut

Cylindrocladium crotalariae causes a destructive root, peg, and pod rot of peanut (*Arachis hypogaea*). The disease is most severe in North Carolina and Virginia, but occurs sporadically in other peanut-producing regions. Root infection results in chlorosis, wilting, and death of shoots. Disease incidence in excess of 80% and yield losses exceeding 50% are common when susceptible cultivars are planted in fields with infested soil and conducive environment. Wilting and presence of red-orange perithecia on tap roots, crowns, and stems of infected plants are diagnostic for Cylindrocladium black rot, which allows epidemiological studies to be based on aboveground symptomatology (Culbreath et al. 1991). Ascospores produced by the perfect stage (*Calonectria crotalariae*) are ephemeral and apparently play no role in the epidemiology of the disease (Rowe and Beute 1974). Cylindrocladium black rot is thought to be monocyclic because infections occur from microsclerotia, which are produced embedded in root tissue (Rowe et al. 1974b). Root contacts potentially could cause secondary infections, but high soil temperatures probably inhibit infection during much of the growing season (Sidebottom and Beute 1989a).

C. crotalariae has a broad host range and there is no evidence for specialization on hosts commonly used in peanut rotations (Rowe and Beute 1973; Hadley et al. 1979a). Host species can, however, influence propagule size and the ability to induce root rot (Black and Beute 1984a,c).

Techniques are available for quantitative assay of the pathogen. Microsclerotia free in soil and in organic debris are recovered by elutriation (Phipps et al. 1976) or wet sieving (Griffin 1977) and plated on a semiselective medium (Phipps et al. 1976; Griffin 1977). Methods for inoculum production and quantitative inoculation in greenhouse (Phipps and Beute 1977; Black et al. 1984) and microplot (Phipps and Beute 1979) trials have been published.

The search for sources of resistance to Cylindrocladium black rot began soon after the disease became established in North Carolina and Virginia in the 1970s. Cylindrocladium black rot had the potential to devastate peanut production and early attempts at chemical control were unsuccessful or uneconomical (Rowe et al. 1974a). Breeding for resistance to the disease has focused on the virginia market-type peanut, because it is grown in areas where the disease is most severe. The virginia market type is similar to the virginia botanical types (*A. hypogaea* subsp. *hypogaea* var. *hypogaea*), but modern market types generally have originated from crosses that incorporate genetic material from more than one botanical group. Among botanical groups of peanut, resistance to *C. crotalariae* was highest and most prevalent in spanish type (*A. hypogaea* subsp. *fastigiata* var. *vulgaris*); valencia types (*A. hypogaea* subsp. *fastigiata* var. *fastigiata*) were highly susceptible (Wynne et al. 1975; Coffelt and Garren 1982). Intermediate resistance was found among virginia types. High levels of resistance also may be available in wild species of *Arachis* (Coffelt and Garren 1982; H.T. Stalker, 1993 pers. comm.).

All known resistance to *C. crotalariae* is partial and appears to be inherited quantitatively (Hadley et al. 1979b). Heritability of resistance is moderate, ranging from 0.43 to 0.73, depending on the material evaluated and the methods used for calculations (Hadley et al. 1979b; Green et al. 1983). A resistant germplasm, NC 3033, was released in 1976 (Beute et al. 1976). NC 3033 is a virginia-type peanut although it has spanish types included in its parentage. The poor agronomic qualities of NC 3033 make it unsuitable for commercial production, but it often is used as a resistant control in genotype evaluations. The first cultivar with resistance to CBR was NC 8C, which was released in 1983 (Wynne and Beute 1983).

8.7.3.1 Components of Resistance

Harris and Beute (1982) showed in histological studies that both resistant and susceptible genotypes are infected by *C. crotalariae*. After penetration, resistant genotypes were more effective at walling off infections with wound periderms than were susceptible genotypes. Periderms in resistant genotypes were less likely to be breached by the advancing fungus, which limits lesion expansion. The mechanism triggering this differential response is not known. Histological evidence suggests that toxic substances also contribute to pathogenesis, but their identity or mode of action has not been elucidated (Johnson and Beute 1974; Harris and Beute 1982).

Expression of resistance to Cylindrocladium black rot depends on the level of inoculum present. Inoculum densities of 0.5, 5, and 50 microsclerotia per g soil routinely are used to characterize the resistance of genotypes in the greenhouse (Phipps and Beute 1977). Susceptible genotypes become severely diseased at inoculum densities as low as 0.5 microsclerotia per g of soil (about one microsclerotium per mm root; Harris and Beute 1982), whereas NC 3033 expresses some resistance at inoculum densities of 50 microsclerotia/g or higher (Phipps and Beute 1977). Genotypes usually are ranked in the field based on incidence of dead and wilted plants and in the greenhouse by a 0–5 root rot index. The root rot index is based on the percentage of the root system with symptoms of the disease and potentially confounds differences in infection frequency with differences in lesion development due to lesion coalescence (Black and Beute 1984c). Root growth factors probably are not confused with resistance in greenhouse screenings because roots are confined to a small column of soil, which they colonize completely during a typical 5- to 10-week evaluation period (Black and Beute 1984c). Root rot ratings made in greenhouse evaluations generally are predictive of resistance in the field (Hadley and Beute 1979b; Pataky et al. 1983).

Inoculum density also affects resistance evaluations in the field. Hau et al. (1982) found that inoculum of *C. crotalariae* was clustered, with densities ranging from 0 to 46 microsclerotia/g soil in one field. Even the highly resistant genotype NC 3033 will become diseased at the high range of inoculum density, whereas even susceptible genotypes will escape infection if planted in plots with no inoculum. Pataky et al. (1983) showed that significant errors in classifying genotype response were possible because of spatial pattern of inoculum. Analysis of covariance has been used to correct for differences in inoculum density (Sidebottom and Beute 1989b). Culbreath et al. (1991) were able to assign genotypes to plots in different inoculum classes by sampling soil and determining inoculum density before planting. This method allowed very precise separation of genotypes based on the number of dead and wilted plants.

It is unclear whether resistance suppresses production of microsclerotia in roots. Numbers of microsclerotia produced per gram of root tissue collected from field-grown peanuts were highly variable and were not good predictors of resistance (Green et al. 1983; Pataky et al. 1983). Occasionally, the largest numbers of microsclerotia were recovered from resistant genotypes. Highly susceptible genotypes may have become severely diseased too early to support high levels of pathogen reproduction, or roots containing microsclerotia may not have been recovered from highly susceptible genotypes because of extensive rotting. Nevertheless, soil populations of microsclerotia generally are smaller following culture of resistant genotypes (Black and Beute 1984a,b).

The potential for populations of *Cylindrocladium crotalariae* to adapt to resistant genotypes has been investigated (Hadley et al. 1979a; Black et al. 1984). Isolates from the resistant genotype NC 3033 had more general virulence (aggressiveness) and specific virulence than isolates from the susceptible Florigiant when they were evaluated on NC 3033 and Florigiant in greenhouse experiments (Hadley et al. 1979a). In later experiments in field microplots, NC 3033 and Florigiant were grown in monoculture for 4 years (Black et al. 1984). Isolates originating from NC 3033 had greater general virulence (aggressiveness) after 4 years on both genotypes than isolates from Florigiant, and one isolate had high levels of specific virulence on NC 3033. When the same isolate was composited with 5 or 11 other isolates in the same microplot experiment, however, no increased virulence was detected. Either the isolate was a poor competitor, or genetic recombination among isolates masked the presence of virulence genes in heterokaryons (Black et al. 1984). The threat that resistance could be overcome appears slight, as long as farmers refrain from planting *Cylindrocladium*-resistant genotypes in monoculture. In the 10 years since resistance has been deployed in areas of high disease incidence, there is no evidence that it has been overcome. The low levels of partial resistance in NC 8C and the current cultivar, NC 10C, may not exert much selection pressure on the pathogen, especially when used in concert with rotation and metam sodium fumigation, as currently is recommended. If more highly resistant cultivars are released in the future, potential for specific adaptation should be evaluated. Techniques employing the polymerase chain reaction might allow better monitoring of isolates than was possible in the past.

8.7.3.2 Avoidance and Tolerance

Peanuts may avoid infection by *C. crotalariae*. Some evidence suggests that exudates from susceptible genotypes stimulate microsclerotial germination more than exudates from resistant genotypes (Krigsvold et al. 1982). Root growth dynamics may also play a role in disease avoidance. Brune and Beute (1993) quantified root length, depth of rooting, and rate of root growth in genotypes susceptible and partially resistant to Cylindrocladium black rot. They found that rate of root growth and total root length were less for resistant than susceptible genotypes. Earlier, Pataky et al. (1983) found that genotypes with apparently equal levels of partial resistance in greenhouse evaluations had subtle differences in disease when grown in the field. Root growth dynamics may explain these differences, because they would not enter into greenhouse evaluations of resistance. Rooting patterns may partially explain why resistant genotypes respond better to fumigation with metam sodium than susceptible genotypes. After treatment with metam sodium, populations of *Cylindrocladium crotalariae* are lowest in the middle of the plant bed. Infections may be delayed on genotypes that take longer to grow out of the region of lower inoculum density (Cline and Beute 1986).

8.7.4 Verticillium Wilt of Tomato

Verticillium wilt of tomato (*Lycopersicon esculentum*) is a monocyclic, vascular wilt disease caused by the soilborne fungi *Verticillium dahliae* and *V. albo-atrum*. *V. dahliae*, which produces microsclerotia, is the primary pathogen in most tomato-producing areas (Pohronezny 1991). Presence and severity of Verticillium wilt are based on subjective ratings of stunting, wilting, chlorosis, and vascular browning (Visser and Hattingh 1980). Across a range of initial inoculum densities, wilting provides the most accurate assessment of disease (Visser and Hattingh 1980). Yield suppression also has been used to assess the impact of Verticillium wilt on resistant and susceptible genotypes (Grogan et al. 1979; Bender and Shoemaker 1984).

V. dahliae has a very wide host range, but host specialization has been reported among isolates obtained from various hosts (Horiuchi et al. 1990; Okoli et al. 1993). Using a set of differential hosts, Horiuchi et al. (1990) found that isolates from tomato also caused disease on other hosts, but isolates from crucifers, sweet pepper, and eggplant infected but did not cause disease on tomato. Ten percent of *V. dahliae* isolates obtained from tomato field soils in California did not cause disease on tomato; pathogenicity of the isolates on other hosts was not determined (Grogan et al. 1979). Two races of the fungus have been described based on the ability to cause disease on tomato cultivars that contain the dominant *Ve* resistance gene obtained from *L. esculentum* var. *cerasiforme* 'Peru Wild'. Isolates within each race of *V. dahliae* vary significantly in aggressiveness (Grogan et al. 1979).

Qualitative and quantitative assay techniques have been developed for recovery of *V. dahliae* from soil (reviewed by Harris et al. 1993) and host tissue (Davis et al. 1983). A specialized selective medium also has been developed for the quantification of root infections by *V. dahliae* on cotton (Huisman 1988). An immunoenzymatic staining technique was developed for detecting *V. dahliae* in roots (Gerik et al. 1988), and molecular techniques such as DNA probes or the polymerase chain reaction should prove useful in the detection and quantification from host tissue.

Management of Verticillium wilt of tomato has relied heavily on the use of the *Ve* resistance gene. The very long-lived microsclerotia and the expense of fumigation have further enhanced the importance of host resistance. Partial resistance to *V. dahliae* has been described (Jones and Crill 1975; Okie and Gardner 1982; Bender and Shoemaker 1984) and probably was important in the slow spread of race 2 of the pathogen into some tomato areas (Bender and Shoemaker 1984).

8.7.4.1 Components of Resistance

Resistance provided by the single dominant *Ve* gene is not complete (Armen 1985). This type of resistance has been called "vertical partial resistance" by Vanderplank (1984). Race 1 can be isolated from root, stem, and petiole tissues of cultivars that contain the *Ve* gene following controlled inoculations (Armen 1985) and from field plants grown in infested soil (Blackhurst and Wood 1963; Grogan et al. 1979; Bender and Shoemaker 1984). Infection of race 1 resistant cultivars by isolates of race 1 causes an increase in root growth (Blackhurst and Wood 1963; Sinha and Wood 1967), but yield suppression may also result in some years and locations (Bender and Shoemaker 1984). Resistant cultivars suppress fungal growth throughout the plant and fewer xylem vessels, especially in petioles, contain fungal hyphae and tyloses (Tjamos et al. 1975; Armen 1985). This reduced colonization suppresses wilting and associated symptoms of the disease. Cultivars with the *Ve* gene are apparently resistant to toxins produced by race 1 of *V. dahliae* (Orentein et al. 1989). Inoculum production by the pathogen was not quantified in these plants.

Race 2 of *V. dahliae* was observed soon after the *Ve* gene was deployed in tomato cultivars and is thought to be a component of field populations of *V. dahliae*. Bender and Shoemaker (1984) found race 2 isolates in fields never planted with cultivars having race 1 resistance; race 2 does not totally disappear from a field if race 1 susceptible cultivars are planted. Race 2 is now known to occur worldwide. Isolates of race 1, in general, cause greater losses than race 2 isolates on tomato cultivars that are susceptible to both races (Grogan et al. 1979; Bender and Shoemaker 1984). This would imply a loss in aggressiveness in race 2 and a direct cost of added virulence to the pathogen. Apparently not all isolates have reduced aggressiveness, because Armen (1985) found no difference in yield suppression by race 1 and race 2 on susceptible tomato. Development of race 2 isolates in race 1 resistant cultivars was similar to that of race 1 in susceptible cultivars (Armen 1985). Inoculum production by race 2 on cultivars with or without the *Ve* gene has not been measured. This is unfortunate, because along with yield, inoculum production may be the most important indicator of the effectiveness of resistance. Inoculum production in previous crops also is important because inoculum density determines the behavior of the epidemic in later crops.

Mechanisms of partial resistance have not been quantified. Partially resistant cultivars become infected and support growth of the pathogen such that intermediate levels of yield suppression are observed (Okie and Gardner 1982). Bender and Shoemaker (1984) suggest that the indeterminate growth habit, compared to a determinate growth habit, may be involved in partial resistance. Indeterminant cultivars continue to grow throughout the season and the added root growth may compensate for infections by *Verticillium*. The effect of partial resistance on inoculum has not been quantified, but suppression of inoculum production could play a role in the durability of race 1 resistance by suppressing initial inoculum densities in subsequent crops.

8.7.4.2 Avoidance and Tolerance

Avoidance or tolerance mechanisms have not been identified in this pathosystem.

8.8 Summary

The examples discussed illustrate some of the great diversity in defense mechanisms that plants use against root pathogens. The primary defense in each pathosystem was resistance, whether complete, partial, or both. Complete resistance has been introduced into many crops, but with few exceptions race development quickly nullified its usefulness. Pathosystems where pathogen races have not overcome resistance include Fusarium wilt of cabbage (Walker 1965) and black root rot of tobacco. What can we learn from pathosystems where monogenic resistance has been durable? We should spend more time investigating pathosystems where resistance genes have not failed.

For most root diseases and soilborne pathogens, partial resistance is the best or only resistance available. Although we have emphasized components of partial resistance in this chapter, the examples illustrate that useful resistance can be identified and incorporated into new cultivars without explicit selection for components of resistance. Indeed, such selection often would be prohibitively time-consuming, expensive, and inefficient (Parlevliet 1992). Our purpose in discussing components of resistance were: first, to identify these components for root diseases; second, to determine if certain components are common to many root pathosystems; and third, to determine if plant pathologists and breeders need to reexamine how they evaluate resistance to root pathogens.

Inoculum density relationships were important in the polycyclic, heterogenneous, and monocyclic pathosystems examined; expression of resistance depended on inoculum density. The use of methods that account for this dynamic relationship is of utmost importance in resistance evaluations.

The theoretical models and practical examples discussed suggest that lesion size and/or rate of lesion expansion are critical in determining final disease severity. Direct selection for these components, or indirect selection through the use of necrosis indices, has been used successfully in enhancing resistance to root diseases, and should be emphasized in breeding programs. In contrast, resistance to infection is not a critical component of resistance in any of the examples discussed. Better distinctions between infection per se and subsequent lesion formation may be necessary to allow full understanding of the importance of this component.

Resistance that inhibits inoculum production is not well characterized. Often, data are highly variable and measurements are confounded by other components of resistance such as lesion expansion. The increasing recognition of the importance of secondary inoculum in some root pathosystems should lead to more quantitative studies of its role in resistance.

The role of avoidance in root disease suppression has received limited attention. However, the black shank and Cylindrocladium black rot examples verify the theoretical importance of root growth parameters in disease dynamics. Root growth dynamics must be easily measured and highly heritable if they are to be considered specifically in breeding programs. This area of research may be amenable to the use of techniques for the identification of quantitative trait loci. Altered root growth also should not compromise plant performance. Identification of partial resistance in careful greenhouse, microplot, and field experiments can minimize inadvertent selection of undesirable rooting characteristics in "resistant" genotypes.

Tolerance is frequently discussed but infrequently studied. A clearer understanding of how tolerance is identified and inherited will encourage investigations into its possible role in disease management. Unfortunately, there are very few examples currently in the literature that researchers can use as model systems. This should change as more research is conducted on components of partial resistance in plants because researchers will identify systems where these components do not explain field observations.

Resistance, avoidance, and tolerance are vital to the sustainable management of root diseases and soilborne pathogens. Although these defense mechanisms can occur in the same plant, it is not known whether they segregate independently. Further study is needed to identify and characterize resistance, avoidance, and tolerance so that they can be used most efficiently in cultivar development.

Acknowledgments. This paper was partially funded by USAID peanut CRSP grant DAN-40480G-SS-2065-00. Recommendations do not represent an official position or policy of USAID.

References

Agrios GN (1980) Escape from disease. In: Horsfall JG, Cowling EB (eds) Plant disease, vol V. Academic Press, New York, pp 17–37

Apple JL (1957) Pathogenic, cultural, and physiological variation within *Phytophthora parasitica* var. *nicotianae*. Phytopathology 47:733–740

Apple JL (1961) Influence of resistant varieties on virulence level within natural populations of *Phytophthora parasitica* var. *nicotianae*. Plant Dis Rep 45:968–971
Apple JL (1962) Physiological specialization within *Phytophthora parasitica* var. *nicotianae*. Phytopathology 52:351–354
Armen J (1985) Histopathology and the relation of symptom expression to crop yield of *Verticillium dahliae* races 1 and 2 on susceptible and resistant tomato cultivars. MS Thesis, North Carolina State University at Raleigh
Baker R (1971) Analyses involving inoculum density of soilborne pathogens in epidemiology. Phytopathology 61:1280–1292
Barker KR (1993) Resistance/tolerance and related concepts/terminology in plant nematology. Plant Dis 77:111–113
Bender CG, Shoemaker, PB (1984) Prevalence of Verticillium wilt of tomato and virulence of *Verticillium dahliae* race 1 and race 2 isolates in western North Carolina. Plant Dis 68:305–309
Beute MK, Wynne JC, Emery, DA (1976) Registration of NC 3033 peanut germplasm. Crop Sci 16:887
Black MC, Beute MK (1984a) Effects of rotations with susceptible and resistant peanuts, soybeans, and corn on inoculum efficiency of *Cylindrocladium crotalariae* on peanuts. Plant Dis 68:401–405
Black MC, Beute MK (1984b) Different ratios of general: specific virulence variance among isolates of *Cylindrocladium crotalariae* from different peanut genotypes. Phytopathology 74:941–945
Black MC, Beute MK (1984c) Relationships among inoculum density, microsclerotium size, and inoculum efficiency of *Cylindrocladium crotalariae* causing root rot on peanuts. Phytopathology 74:1128–1132
Black MC, Beute MK, Leonard KJ (1984) Effects of monoculture with susceptible and resistant peanuts on the virulence of *Cylindrocladium crotalariae*. Phytopathology 74:945–950
Blackurst FM, Wood RKS (1963) Resistance of tomato plants to *Verticillium albo-atrum*. Trans Br Mycol Soc 46:385–392
Boerma HR, Hussey RS (1992) Breeding plants for resistance to nematodes. J Nematol 24:242–252
Brassett PR, Gilligan CA (1988) A model of primary and secondary infection in botanical epidemics. Z Pflanzenkr Pflanzenschutz 95:352–360
Brenneman TB, Phipps PM, Stipes RJ (1988) A rapid method to assess pathogenicity of *Sclerotinia minor* on peanut. Phytopathology 78:815
Brune PD, Beute MK (1993) Root growth dynamics as a factor in resistance of peanut to Cylindrocladium root rot. Proc Am Peanut Res Ed Soc 25:20
Burdon JJ (1992) Diseases and plant population biology. Cambridge Univ Press, Cambridge, pp 49–72
Caldwell RM (1966) Advances and challenges in the control of plant diseases through breeding. US Dep Agric, Agri Res Ser, ARS 33–110, pp 117–125
Campbell CL, Noe JP (1985) The spatial distribution of soilborne pathogens and root diseases. Annu Rev Phytopathol 23:129–148
Campbell CL, Powell NT (1980) Progression of disease induced by soilborne pathogens: tobacco black shank. Prot Ecol 2:177–182
Campbell CL, Jacobi WR, Powell NT, Main CE (1984) Analysis of disease progress and the randomness of occurrence of infected plants during tobacco black shank epidemics. Phytopathology 74:230–235
Chappell GF (1992) Evaluation of peanut genotype resistance to *Sclerotinia minor*. MS Thesis, North Carolina State University at Raleigh
Clarke DD (1986) Tolerance of parasites and disease in plants and its significance in host-parasite interactions. In: Ingram DS, Williams PH (eds) Advances in plant pathology, vol 5. Academic Press, New York, pp 161–197

Clayton EE (1969) The study of resistance to the black root rot disease of tobacco. Tob Sci 13:30–37

Cline WO, Beute MK (1986) Effect of metam sodium, peanut genotype and inoculum density on incidence of Cylindrocladium black rot. Peanut Sci 13:41–45

Coffelt TA, Garren KH (1982) Screening for resistance to Cylindrocladium black rot in peanuts (*Arachis hypogaea* L.). Peanut Sci 9:1–5

Conant GH (1927) Histological studies of resistance in tobacco to *Thielavia basicola*. Am J Bot 14:457–480

Cook R, Evans R (1987) Resistance and tolerance. In: Brown RH, Kerry BR (eds) Principles and practice of nematode control in crops. Academic Press, London, pp 179–231

Corbaz R (1985) Pathotypes et variations du pouvoir pathogene chez *Chalara elegans* Nag Raj et Kendrick (=*Thielaviopsis basicola*). Phytopath Z 113:289–299

Crowe FJ, Hall DH (1980) Vertical distribution of sclerotia of *Sclerotium cepivorum* and host root systems relative to white rot of onion and garlic. Phytopathology 70:70–73

Crute IR (1986) The relationship between *Plasmodiophora brassicae* and its hosts: the application of concepts relating to variation in inter-organismal associations. In: Ingram DS, Williams PH (eds) Advances in plant pathology, vol 5. Academic Press, New York, pp 1–52

Culbreath AK, Beute MK, Campbell CL (1991) Spatial and temporal aspects of epidemics of Cylindrocladium black rot in resistant and susceptible peanut genotypes. Phytopathology 81:144–150

Davis JR, Pavek JJ, Corsini DL (1983) A sensitive method for quantifying *Verticillium dahliae* colonization in plant tissue and evaluating resistance among potato genotypes. Phytopathology 73:1009–1014

deZoeten GA, Gaard G, Haberlach GT, Helgeson JP (1982) Infection of tobacco callus by *Phytophthora parasitica* var. *nicotianae*. Phytopathology 72:743–746

Dropkin VH, Nelson PE (1960) The histopathology of root-knot nematode infections in soybeans. Phytopathology 50:442–449

Dukes PD, Apple JL (1961) Chemotaxis of zoospores of *Phytophthora parasitica* var. *nicotianae* by plant roots and certain chemical solutions. Phytopathology 51:195–197

Dukes PD, Apple JL (1961) Influence of host passage on virulence of *Phytophthora parasitica* var. *nicotianae*. Plant Dis Rep 45:362–365

Dukes PD, Apple JL (1962) Relationship of zoospore production potential and zoospore motility with virulence in *Phytophthora parasitica* var. *nicotianae*. Phytopathology 52:191–193

Duniway JM (1983) Role of physical factors in the development of *Phytophthora* diseases. In: Erwin DC, Bartnicki-Garcia S, Tsao PH (eds) *Phytophthora*: its biology, taxonomy, ecology, and pathology. Am Phytopathol Soc, St Paul, pp 175–187

Edwards M (1992) Use of molecular markers in the evaluation and introgression of genetic diversity for quantitative traits. Field Crops Res 29:241–260

English JT, Mitchell DJ (1989) Use of morphometric analysis for characterization of tobacco root growth in relation to infection by *Phytophthora parasitica* var. *nicotianae*. Plant Soil 113:243–249

Ferrin DM, Mitchell DJ (1986) Influence of initial density and distribution of inoculum on the epidemiology of tobacco black shank. Phytopathology 76:1153–1158

Gerik JS, Huisman OC (1988) Study of field-grown cotton roots with *Verticillium dahliae* using an immunoenzymatic staining technique. Phytopathology 78:1174–1178

Gilligan CA (1985) Construction of temporal models. III. Disease progress of soilborne pathogens. In: Gilligan CA (ed) Mathematical modelling of crop disease. Advances in plant pathology, vol 3. Academic Press, London, pp 67–105
Gilligan CA (1990) Mathematical modelling and analysis of soilborne pathogens. In: Kranz J (ed) Epidemics of plant diseases: mathematical modelling and analysis, 2nd edn. Springer, Berlin Heidellberg New York, pp 96–142
Green CC, Wynne JC, Beute MK (1983) Genetic variability and heritability estimates based on the F_2 generation from crosses of large-seeded Virginia-type peanuts with lines resistant to Cylindrocladium black rot. Peanut Sci 10:47–51
Gregory PH (1948) The multiple-infection transformation. Ann Appl Biol 35:412–417
Griffin GJ (1977) Improved selective medium for isolating *Cylindrocladium crotalariae* microsclerotia from naturally infested soils. Can J Micro 23:680–683
Grogan RG, Ioannou N, Schneider RW, Sall MA, Kimble KA (1979) Verticillium wilt on resistant and tomato cultivars in California: virulence of isolates from plants and soil and relationship of inoculum density to disease incidence. Phytopathology 69:1176–1180
Hadley BA, Beute MK, Leonard KJ (1979a) Variability of *Cylindrocladium crotalariae* response to resistant host plant selection pressure in peanut. Phytopathology 69:1112–1114
Hadley BA, Beute MK, Wynne JC (1979b) Heritability of Cylindrocladium black rot resistance in peanut. Peanut Sci 6:51–54
Hadwiger LA, Chiang CC, Pettinger A, Chang MM (1990) Strategy to improve disease resistance by transferring "non-host" disease resistance genes from peas to potatoes. In: Bills DD, Kung S (eds) Biotechnology and food safety. Butterworth-Heinemann, Boston, pp 241–253
Hanchey P, Wheeler H (1970) Pathological changes in ultrastructure: tobacco roots infected with *Phytophthora parasitica* var. *nicotianae*. Phytopathology 60:33–39
Harris NE, Beute MK (1982) Histological responses of peanut germplasm resistant and susceptible to *Cylindrocladium crotalariae* in relationship to inoculum density. Phytopathology 72:1250–1256
Harris MK, Frederiksen RA (1984) Concepts and methods regarding host plant resistance to arthropods and pathogens. Annu Rev Phytopathol 22:247–272
Harris DC, Yang JR, Ridout MS (1993) The detection and estimation of *Verticillium dahliae* in naturally infested soil. Plant Pathol 42:238–250
Hau FC, Campbell CL, Beute MK (1982) Inoculum distribution and sampling methods for *Cylindrocladium crotalariae* in a peanut field. Plant Dis 66:568–571
Hayward AC (1991) Biology and epidemiology of bacterial wilt caused by *Pseudomonas solanacearum*. Annu Rev Phytopathol 29:65–87
Heath MC (1981) A generalized concept of host-parasite specificity. Phytopathology 71:1121–1123
Hickman CJ (1970) Biology of *Phytophthora* zoospores. Phytopathology 60:1128–1135
Holmes KA, Benson DM (1994) Evaluation of *Phytophthora parasitica* var. *nicotianae* for biocontrol of *Phytophthora parasitica*, a causal agent of pre-emergence damping off of *Catharanthus roseus*. Plant Dis 77:193–199
Horiuchi S, Hagiwara H, Takeuchi S (1990) Host specificity of isolates of *Verticillium dahliae* towards cruciferous and solanaceous plants. In: Hornby D (ed) Biological control of soil-borne plant pathogens. CAB International, Wallingford, Oxon, pp 285–297
Huisman OC (1982) Interrelations of root growth dynamics to epidemiology of root-invading fungi. Annu Rev Phytopathol 20:303–327
Huisman OC (1988) Seasonal colonization of roots of field-grown cotton by *Verticillium dahliae* and *V. tricorpus*. Phytopathology 78:708–716

Huisman OC, Gerik JS (1989) Dynamics of colonization of plant roots by *Verticillium dahliae* and other fungi. In: Tjamos EC, Beckman CH (eds) Vascular wilt diseases of plants, basic studies and control. Springer, Berlin Heidelberg New York, pp 1–17
Jeger MJ (1987) The influence of root growth and inoculum density on the dynamics of root disease epidemics: theoretical analysis. New Phytol 107:359–378
Jewett FL (1938) Relation of soil temperature and nutrition to the resistance of tobacco to *Thielavia basicola*. Botan Gaz 100:276–297
Johnson SA, Beute MK (1974) Histopathology of Cylindrocladium black rot of peanut. Phytopathology 64:649–653
Jones JP, Crill P (1975) Reaction of resistant, tolerant, and susceptible tomato varieties to Verticillium wilt. Plant Dis Rep 59:3–6
Jones KJ (1990) Components of resistance in *Nicotiana tabacum* to *Phytophthora parasitica* var. *nicotianae*. PhD Thesis, North Carolina State University at Raleigh
Jones KJ, Shew HD (1989) Components of resistance of tobacco to *Phytophthora parasitica* var. *nicotianae*. Phytopathology 79:1183
Kannwischer ME, Mitchell DJ (1981) Relationships of numbers of spores of *Phytophthora parasitica* var. *nicotianae* to infection and mortality of tobacco. Phytopathology 71:69–73
Keeling BL (1974) Soybean seed rot and the relation of seed exudate to host susceptibility. Phytopathology 64:1445–1447
Keen NT (1992) The molecular biology of disease resistance. Plant Mol Biol 19:109–122
Kenerley CM, Papke K, Bruck RI (1984) Effect of flooding on development of Phytophthora root rot in Fraser fir seedlings. Phytopathology 74:401–404
Krigsvold DT, Griffin GJ, Hale MG (1982) Microsclerotial germination of *Cylindrocladium crotalariae* in the rhizospheres of susceptible and resistant peanuts. Phytopathology 72:859–864
Law TF, Milholland RD (1991) Production of sporangia and oospores of *Phytophthora fragariae* in roots of strawberry plants. Plant Dis 75:475–478
Law TF, Milholland RD (1992) Susceptibility of strawberry genotypes to infection and colonization by races of *Phytophthora fragariae* and the growth responses of inoculated genotypes. Plant Dis 76:335–339
Lindgren PB, Jakobek JL, Smith JA (1992) Molecular analysis of plant defence responses to plant pathogens. J Nematol 24:330–337
Lucas GB (1975) Diseases of tobacco, 3rd edn. Biological Consulting Associates, Raleigh
Mauk PA, Hine RB (1988) Infection, colonization of *Gossypium hirsutum* and *G. barbadense*, and development of black root rot caused by *Thielaviopsis basicola*. Phytopathology 78:1662–1667
McBlain BA, Zimmerly MM, Schmitthenner AF, Hacker JK (1991a) Tolerance to Phytophthora rot in soybean. I. Studies of the cross "Ripley" × "Harper". Crop Sci 31:1405–1411
McBlain BA, Hacker JK, Zimmerly MM, Schmitthenner AF (1991b) Tolerance to Phytophthora rot in soybean. II. Evaluation of three tolerance screening methods. Crop Sci 31:1412–1417
McIntyre JL, Taylor GS (1978) Race 3 of *Phytophthora parasitica* var. *nicotianae*. Phytopathology 68:35–38
Meyer JR, Shew HD (1991a) Development of black root rot on burley tobacco as influenced by inoculum density of *Thielaviopsis basicola*, host resistance, and soil chemistry. Plant Dis 75:601–605
Meyer JR, Shew HD (1991b) Soils suppressive to black root rot of burley tobacco, caused by *Thielaviopsis basicola*. Phytopathology 81:946–954

Meyer JR, Shew HD, Shoemaker PB (1989) Populations of *Thielaviopsis basicola* and the occurrence of black root rot on burley tobacco in western North Carolina. Plant Dis 73:239–242

Milholland RD (1975) Pathogenicity and histopathology of *Phytophthora cinnamomi* on highbush and rabbiteye blueberry. Phytopathology 65:789–793

Milholland RD, Daykin ME (1993) Colonization of roots of strawberry cultivars with different levels of susceptibility to *Phytophthora fragariae*. Phytopathology 83: 538–542

Mitchell DJ (1975) Density of *Pythium myriotylum* oospores in soil in relation to infection of rye. Phytopathology 65:570–575

Mitchell DJ, Kannwischer-Mitchell ME (1992) *Phytophthora*. In: Singleton LL, Mihail JD, Rush CM (eds) Methods for research on soilborne phytopathogenic fungi. APS Press, St Paul, pp 31–38

Mussell H (1980) Tolerance to disease. In: Horsfall JG, Cowling EB (eds) Plant disease an advanced treatise, vol V. How plants defend themselves. Academic Press, New York, pp 39–52

Okie WR, Gardner RG (1982) Screening tomato seedlings for resistance to *Verticillium dahliae* races 1 and 2. Plant Dis 66:34–37

Okoli CAN, Carder JH, Barbara DJ (1993) Molecular variation and sub-specific groupings within *Verticillium dahliae*. Mycol Res 97:233–239

Orenstein J, Nachmias A, Lips H (1989) Inhibition of tomato root growth in culture by a *Verticillium dahliae* phytotoxin. Phytoparasitica 17:269–278

Orton WA (1909) The development of farm crops resistant to disease. US Dep Agric Yearb 1908:453–464

Otani Y (1962) Studies on the black root rot disease caused by *Thielaviopsis basicola* (Berk and Br) Ferraris. Bull Okayama Tob Res Stn 23, pp 1–118

Parlevliet JE (1979) Components of resistance that reduce the rate of epidemic development. Annu Rev Phytopathol 17:203–222

Parlevliet JE (1981) Race non-specific resistance. In: Jenkyn JF, Plumb RT (eds) Strategies for the control of cereal disease. Blackwell Scientific, Oxford, pp 47–54

Parlevliet JE (1989) Identification and evaluation of quantitative resistance. In: Leonard KJ, Fry WE (eds) Plant disease epidemiology, vol 2. Genetics resistance and management. McGraw-Hill, New York, pp 215–248

Parlevliet JE (1992) Selecting components of partial resistance. In: Stalker HT, Murphy JP (eds) Plant breeding in the 1990s. CAB International, Wallingford, Oxon

Pataky JK, Black MC, Beute MK, Wynne JC (1983) Comparative analysis of Cylindrocladium black rot resistance in peanut: greenhouse, microplot, and field testing procedures. Phytopathology 73:1615–1620

Percy RG, Rush CM (1985) Evaluation of four upland cotton genotypes for a rate-limiting resistance to Phymatotrichum root rot. Phytopathology 75:463–466

Pfender WF (1982) Monocyclic and polycyclic root diseases: distinguishing between the nature of the disease cycle and the shape of the disease progress curve. Phytopathology 72:31–32

Phipps PM, Beute MK (1977) Sensitivity of susceptible and resistant peanut cultivars to inoculum densities of *Cylindrocladium crotalariae* microsclerotia in soil. Plant Dis Rep 61:300–303

Phipps PM, Beute MK (1979) Population dynamics of *Cylindrocladium crotalariae* microsclerotia in naturally-infested soil. Phytopathology 69:240–243

Phipps PM, Beute MK, Barker KR (1976) An elutriation method for quantitative isolation of *Cylindrocladium crotalariae* microsclerotia from peanut field soil. Phytopathology 66:1255–1259

Pohronezny KL (1991) Verticillium wilt. In: Jones JB, Jones JP, Stall RE, Zitter TA (eds) Compendium of tomato diseases. APS Press St Paul, pp 23–24

Powell NT (1956) Host parasite relations of *Phytophthora parasitica* var. *nicotianae* in roots of resistant and susceptible tobacco varieties, *Nicotiana longiflora* and *Nicotiana plumbaginifolia*. MS Thesis, North Carolina State University at Raleigh

Prinsloo GC, Pauer GDC (1973) Die identifikasie van rasse van *Phytophthora nicotianae* (B de Haan) *nicotianae* wat in Suid Afrika voorkom. Phytophylactica 6:217–220

Pulliam RJ (1986) Response of susceptible and resistant tobacco cultivars to infection by *Phytophthora parasitica* var. *nicotianae*. MS Thesis, North Carolina State University at Raleigh

Reynolds KM, Benson DM, Bruck RI (1985a) Epidemiology of Phytophthora root rot of Fraser fir: estimates of rhizosphere width and inoculum efficiency. Phytopathology 75:1010–1014

Reynolds KM, Benson DM, Bruck RI (1985b) Epidemiology of Phytophthora root rot of Fraser fir: root colonization and inoculum production. Phytopathology 75:1004–1009

Reynolds KM, Gold HJ, Bruck RI, Benson DM, Campbell CL (1986) Simulation of the spread of *Phytophthora cinnamomi* causing a root rot of Fraser fir in nursery beds. Phytopathology 76:1190–1201

Ricci P, Trentin F, Bonnet P, Venard P, Mouton-Perronnet F, Bruneteau M (1992) Differential production of parasiticein, an elicitor of necrosis and resistance in tobacco, by isolates of *Phytophthora parasitica*. Plant Pathol 41:298–307

Ristaino JB (1993) Effect of resistance to *Streptomyces ipomoea* on disease, yield, and dry matter partitioning in sweetpotato. Plant Dis 77:193–196

Roberts PA (1992) Current status of the availability, development, and use of host plant resistance to nematodes. J Nematol 24:213–227

Rowe RC, Beute MK (1973) Susceptibility of peanut rotational crops (tobacco, cotton and corn) to *Cylindrocladium crotalariae*. Plant Dis Rep 57:1035–1039

Rowe RC, Beute MK (1974) Ascospore formation and discharge by *Calonectria crotalariae*. Phytopathology 64:393–398

Rowe RC, Beute MK, Wells JC, Wynne JC (1974a) Incidence and control of Cylindrocladium black rot of peanuts in North Carolina during 1973. Plant Dis Rep 58:348–352

Rowe RC, Johnson SA, Beute MK (1974b) Formation and dispersal of *Cylindrocladium crotalariae* microsclerotia in infected peanut roots. Phytopathology 64:1294–1297

Schafer JF (1971) Tolerance to plant disease. Annu Rev Phytopathol 9:235–252

Schmitthenner AF (1985) Problems and progress in control of Phytophthora root rot of soybean. Plant Dis 69:362–368

Schwartz HF, Steadman JR, Coyne DD (1978) Influence of *Phaseolus vulgaris* blossoming characteristics and canopy structure upon reaction to *Sclerotinia sclerotiorum*. Phytopathology 68:465–470

Shaner G (1980) Probits for analyzing latent period data in studies of slow rusting resistance. Phytopathology 70:1179–1182

Shew BB, Beute MK (1984) Effects of crop management on the epidemiology of southern stem rot of peanut. Phytopathology 74:530–535

Shew HD (1983) Effects of soil matric potential on infection of tobacco by *Phytophthora parasitica* var. *nicotianae*. Phytopathology 73:1160–1163

Shew HD (1985) Inoculum production by *Phytophthora parasitica* var. *nicotianae* in roots of flue-cured tobacco. Phytopathology 75:1347

Shew HD (1987) Effect of host resistance on spread of *Phytophthora parasitica* var. *nicotianae* and subsequent development of tobacco black shank under field conditions. Phytopathology 77:1090–1093

Shew HD, Lucas GB (1991) Compendium of tobacco diseases. APS Press, St Paul 90pp

Shew HD, Meyer JR (1992) *Thielaviopsis*. In: Singleton LL, Mihail JD, Rush CM (eds) Methods for research on soilborne phytopathogenic fungi. APS Press, St Paul, pp 171–174
Shew HD, Shoemaker PB (1993) Effects of host resistance and soil fumigation on populations of *Thielaviopsis basicola* and development of black root rot on burley tobacco. Plant Dis 77:1035–1039
Sidebottom JR, Beute MK (1989a) Control of Cylindrocladium black rot of peanut with cultural practices that modify soil temperature. Plant Dis 73:672–676
Sidebottom JR, Beute MK (1989b) Inducing soil suppression to Cylindrocladium black rot of peanut through crop rotations with soybean. Plant Dis 73:679–685
Singleton LL, Mihail JD, Rush CM (1992) Methods for research on soilborne phytopathogenic fungi. APS Press, St Paul
Sinha AK, Wood RKS (1967) An analysis of responses of resistant and of susceptible tomato plants to *Verticillium* infection. Ann Appl Biol 59:143–154
Specht LP, Griffin GJ (1985) A selective medium for enumerating low populations of *Thielaviopsis basicola* in tobacco field soils. Can J Plant Pathol 7:438–441
Specht LP, Griffin GJ (1988) Relation of inoculum density of *Thielaviopsis basicola* to the severity of black root rot and the growth of tobacco in naturally infested soil. Can J Plant Pathol 10:15–22
Stover RH (1950a) The black root rot disease of tobacco. I. Studies on the causal organism *Thielaviopsis basicola*. Can J Res C 28:445–470
Stover RH (1950b) The black root rot disease of tobacco. II. Physiological specialization of *Thielaviopsis basicola* on *Nicotiana tabacum*. Can J Res C 28:726–738
Stuber CW (1990) Molecular markers in the manipulation of quantitative characters. In: Brown AHD, Clegg MT, Kahler AL, Weir BS (eds) Plant population genetics, breeding, and genetic resources. Sinauer Associates, Sunderland, pp 334–350
Tabachnik M, DeVay JE, Garber RH, Wakeman RJ (1979) Influence of soil inoculum concentrations on host range and disease reactions caused by isolates of *Thielaviopsis basicola* and comparison of soil assay methods. Phytopathology 69:974–977
Tjamos EC, Smith IM (1975) The expression of resistance to *Verticillium albo-atrum* in monogenically resistant tomato varieties. Physiol Plant Pathol 6:215–225
Tomimatsu GS, Griffin GJ (1982) Inoculum potential of *Cylindrocladium crotalariae*: infection rates and microsclerotial density-root infection relationships on peanut. Phytopathology 72:511–517
Tooley PW, Grau CR (1982) Identification and quantitative characterization of rate-reducing resistance to *Phytophthora megasperma* f. sp. *glycinea* in soybeans seedlings. Phytopathology 72:727–733
Tooley PW, Grau CR (1984a) Field characterization of rate-reducing resistance to *Phytophthora megasperma* f. sp. *glycinea* in soybean. Phytopathology 74:1201–1208
Tooley PW, Grau CR (1984b) The relationship between rate-reducing resistance to *Phytophthora megasperma* f. sp. *glycinea* and yield of soybean. Phytopathology 74:1209–1216
Trudgill DL (1991) Resistance to and tolerance of plant parasitic nematodes in plants. Annu Rev Phytopathol 29:167–192
Vanderplank JE (1968) Disease resistance in plants. Academic Press, New York
Vanderplank JE (1984) Disease resistance in plants, 2nd edn. Academic Press, Orlando
Visser S, Hattingh MJ (1980) Criteria for quantal host response of tomato plants to infection by *Verticillium dahliae*. Plant Dis 64:207–208
Waggoner PE (1981) Models of plant diseases. Bioscience 31:315–319
Walker JC (1965) Host resistance as it relates to root pathogens and soil microorganisms. In: Baker KF, Snyder WC (eds) Ecology of soil-borne plant pathogens:

prelude to biological control. University of California Press, Berkeley, pp 314–320
Walker JC (1969) Plant pathology. McGraw-Hill, New York, pp 786–802
Walker AK, Schmitthenner AF (1984) Heritability of tolerance to Phytophthora rot of soybean. Crop Sci 24:490–491
Wernsman EA, Rufty RC (1987) Tobacco. In: Fehr WR (ed) Principles of cultivar development, vol 2. Crop species. Macmillan, New York, pp 669–698
Wernsman EA, Matzinger DF, Powell NT (1974) Genetic investigations of intraspecific and interspecific sources of black shank resistance in tobacco. Tob Sci 18:15–18
Whalen MC (1991) Advances in breeding for resistance to bacterial pathogens. In: Murray DR (ed) Advanced methods in plant breeding and biotechnology. CAB International, Wallingford, Oxon, pp 299–318
Wilkinson CA, Rufty RC, Shew HD (1991) Inheritance of partial resistance to black root rot in burley tobacco. Plant Dis 75:889–892
Williamson VM, Ho J-Y, Ma HM (1992) Molecular transfer of nematode resistance genes. J Nematol 24:234–241
Wynne JC, Beute MK (1983) Registration of NC 8C peanut. Crop Sci 23:184
Wynne JC, Rowe RC, Beute MK (1975) Resistance of peanut genotypes to *Cylindrocladium crotalariae*. Peanut Sci 2:54–56
Young LD (1992) Problems and strategies associated with long-term use of nematode resistant cultivars. J Nematol 24:228–233

9 Chemical Management

H.V. Morton

9.1 Strategies for Use of Chemical Pesticides

Soilborne pathogens can infect a plant at any time during its growth and development, starting within 6–12 h of germination as shown by Nelson et al. (1986). Thus, the choice of a control measure for such damping-off diseases has to be made prior to planting. The controls available for perennials, and those diseases including the vascular wilts and sclerotial soilborne pathogens, are more involved. Key factors such as crop safety, environmental conditions, cultural practices, the complexity of the soil (e.g., absorption capacity), and the life cycle of the pathogen all interact to make chemical management of disease a challenging task. The above-mentioned factors also limit the options of chemicals available. Chemical control of pathogens in soil can be considered one of the last frontiers of "uncontrolled" diseases.

Certainly, the reports of crop losses due to root disease can attest to this scenario. Disease loss estimates by the Cotton Disease Council over the past 50 years show that seedling diseases have contributed most to disease losses (2–3% per annum). Black shank (*Phytophthora nicotianae*), Granville wilt (*Pseudomonas solanacearum*), and root-knot nematodes (*Meloidogyne* spp.) are among the top five diseases recorded annually on tobacco in the United States. The problem of replanting peaches would be another major loss factor when considering diseases of perennials. These are just three examples out of a long list. The following is a discussion on the chemical control practices available.

9.1.1 Soil Fumigation

By nature, most fumigants are highly volatile biocides with a broad spectrum of activity. Fumigants can eliminate soilborne bacteria, fungi, nematodes, insects, mites, and weeds creating a pest-free environment. Fumigation also depletes populations of many beneficial and disease-suppressive factors. Typically, this reduction in microflora and fauna is short-lived and is followed by a resurgence of activity (e.g., bacteria, *Trichoderma* spp., oomycetes,

C.L. Campbell/D.M. Benson (Eds.)
Epidemiology and Management of Root Diseases

and nematodes such as *Paratrichodorus* spp.) after the fumigant volatilizes. Fumigants are commonly used in greenhouse and plant bed operations where applications are made in air-tight containers, or are confined by a plastic cover. In the field, the use of fumigants is limited to high-value crops in a broadcast, row, or bed pattern with applications made by gravity flow or under pressure from an inert gas through chisels or knives. The closer the setting of the chisels, the more uniform the distribution of the fumigant (Goring 1957).

The value of fumigants, which are expensive and labor intensive to apply, lies in their ability to eliminate microorganisms that are causal agents of otherwise hard to control diseases (e.g., fungi that produce sclerotial resting bodies; also fungal and nematode diseases associated with replanting problems in orchards and vineyards).

One product that can be applied in a different manner than most soil-injected fumigants is metam sodium (Vapam) which is a water-soluble liquid; thus, metam sodium can be applied via irrigation both in garden soils via a drench or to crops via overhead or furrow irrigation. Best results are obtained if a tarpaulin or cover is used for 2 days; seeding should be delayed if a tarpaulin is used.

A recent strategy being extensively researched in Israel is the use of low rates of a fumigant followed by the introduction of a biocontrol agent which has a high level of tolerance to the fumigant, e.g., *Trichoderma* spp. With a view to environmental fate, this may be a means to sustain use of some valuable fumigant products in the future.

Besides being expensive, fumigants have several environmental and usage drawbacks (e.g., air and groundwater contamination). There may be an essential waiting period before planting of up to 2 weeks at a critical time in the spring in northern areas and in the fall in southern states. They cannot be applied to no-till or minimum tillage due to the crop refuse and lack of seed bed tilth. The environmental drawbacks include inherent toxicity to all forms of life and possible groundwater contamination.

9.1.2 At-Plant Application of Fungicides

The at-plant application of fungicides and nematicides is favored by growers as it can be included in the planting operations, thereby reducing the number of trips across the field.

Basically, at-plant applications fall into one of four categories, viz.: (1) broadcast sprays: these may or may not be incorporated into the soil. Frequently, such applications are made in conjunction with a herbicide, e.g., metalaxyl on tobacco. Whether or not the product is incorporated depends upon the physical characteristics of the chemical. Such features as water solubility, volatility, and photodegradation more often than not require that the product be incorporated. While broadcast spray applications are the most efficacious, it should be recognized that they are also the most expensive. Consequently, this approach is usually limited to high-value crops such as ornamentals, strawberries, tobacco, and vegetables (especially transplanted seed beds). (2) Band applications over the row: band applications use less chemicals than the broadcast method. Most band applications are made in front of the planter press wheel with either a spray or granule in a 7 to 14 in. (18–36 cm) band. The press wheel affords a means of shallow incorporation. (3) In-furrow application: this is used primarily for seedling damping-off diseases. The aim is to apply the fungicide into the planting furrow as well as into the covering soil (termed a bow-wave application). Again, the product may be applied via spray or as a granule. The relatively small amounts of chemical required, and the accuracy of placement have made these applications popular. (4) Potting mix incorporation: the final at-plant method pertains to container-grown ornamentals and vegetable transplants. Frequently, fungicides are incorporated into the potting mix of container-grown crops at the time the mix is prepared. A second method is via drench. For a soil depth of less than 4 in. (10 cm), 1 pint (473 ml) of solution is applied per square foot (929 cm^2) of greenhouse or shadehouse bench. When there is more than 4 in. (10 cm) of soil, typically 1.5–2 pints (709–946 ml) of solution is applied.

9.1.3 Seed Dressings

The dividends of treating seed is well recognized. The degree to which seed is treated depends upon the value of the seed and the value of the crop. Virtually all vegetable seed is treated, whereas only about one-third of all wheat seed is treated in the USA. Seed treatment fungicides are usually applied as a liquid or slurry. In almost all cases, the old practice of dusting seed has been replaced due to improved coverage afforded by new seed treating equipment and concern for potential worker exposure during the seed treatment process. Currently, the only extensive use of seed dusting is on potatoes, as the liquid application favors soft rot caused by *Erwinia carotovora*.

Seed treatment applications are commonly made by the company producing the seed (e.g., corn or cotton), the elevator where the seed is stored (e.g., peanuts, soybeans, or wheat), or by the farmer using hopper box treatment. On-farm treatment may increase in popularity because seed companies have trouble disposing chemically treated seed; it must be burned in an incinerator. In the mid-1970s, following the banning of the mercurials, the predominant seed applications were made with broad-spectrum contact fungicides (e.g., captan, thiram, and PCNB-ETMT). This has changed with the introduction of systemic compounds which are more efficacious against specific pathogens (e.g., carboxin, metalaxyl, tridimenol). These systemic compounds must often be tank-mixed with other fungicides to broaden the spectrum of activity.

With the increasing concern for the environment, the use of seed as a carrier of fungicides to control diseases caused by soilborne and foliar pathogens has become more attractive. The basic manufacturers have thus placed more emphasis in this market segment, which currently represents 14% of the fungicide sales in the USA. The fruits of these efforts are now becoming available in Europe (e.g., phenylpyrroles, triazoles). Similarly, many of those working with biocontrols are investigating seed treatments as the ideal vehicle for fungicides.

As the products used in seed treatments have been improved, so has the equipment used to apply them. Much of the credit for this goes to Gustafson's efforts in the USA (e.g., Gustafson's Triple-Treater). Similarly, improvements have been made in the on-farm treaters which are commonly affixed to the back of a truck carrying the seed. One older method still in practice (mainly because of its simplicity) is that of planter-box applications. In recent years, seed companies have greatly improved the formulations used in planter-box treatments. Planter-box applications are applied in four steps: (1) a seed hopper is half-filled, and half of the desired treatment is applied over the seed; (2) the seed and treatment are mixed by stirring the material with an ax handle; (3) the remaining seed and treatment are added to fill the box; and (4) step 2 is repeated. The seed which is partially coated falls into the seed furrow along with a portion of the fungicide. The major drawbacks of this application method are (1) uneven seed coverage; the smaller the seed, the more difficult it becomes to obtain the necessary coverage of the seed coat; and (2) the time it takes to treat the seed during the planting process.

Another method of treating seed on the farm is that of seed-piece dips. This application method is used when vegetative parts are used for propagation (e.g., sugarcane, flower bulbs). Those planting water-seeded rice also use seed dipping to apply fungicides. This method is limited by restrictions on the disposal of the chemical solution utilized in treatments.

A lot of progress is being made in the field of seed coatings, which usually employ a water-soluble polymer. These coatings greatly improve the distribution and adhesion of active ingredients at low rates on the seed. Maude (1990) compared liquid and dry materials and found that liquid seed treatments had a less uniform distribution but better adhesion than did seed treated with dry materials.

Seed treatments must be safe for the seed and they should not diminish germination during storage. In the case of legumes, seed treatments should also not interfere with Bradyrhizobial inoculants applied at planting.

9.1.4 Postplant Treatments

Postplant treatments are usually made to achieve disease control that is not afforded with preplant treatments. This may be especially important if the edible portion of a long-season crop is a root crop (e.g., potato, peanuts). For diseases that develop later in the crop cycle, postplant treatments may provide better control; e.g., *Sclerotium rolfsii* appears on peanuts at pegging and *Sclerotinia minor* at flowering. Another use of postplant treatments is a "rescue treatment". Unfortunately, there are several restrictions placed upon postplant treatments.

1. They are commonly integrated with other cultural practices, for example, lay-by fertilization or cultivation (e.g., tobacco, potato, vegetables) to limit plant injury and soil compaction.
2. The growers' equipment must be able to go over the top of the crop without damaging the crop, thereby making the correct placement of the treatment difficult.
3. For best results, activation of chemicals often requires rainfall or adequate soil moisture. For best results, the chemical should be taken up by the roots, and preferably possess curative properties.

Today, many growers have the opportunity to apply postplant treatments via irrigation. Such applications require special injector equipment, a metering device, and special valves to prevent backflow into a well or impoundment of surface water. Because of the capital expenditure required for the installation of irrigation equipment, most growers produce high-value crops with limited crop rotation under irrigation. While irrigation may improve the performance of some chemicals, growers often provide excess water to crops under irrigation. The abundance of moisture over extended periods frequently increases disease losses by soilborne pathogens, creating a greater need for postplant treatments.

9.2 Types of Chemical Pesticides

9.2.1 Nematicides

Nematodes are surprisingly resistant to most chemicals, due in part to the impermeability of their cuticle and protective egg covering. Table 9.1 gives an overview of the currently registered nematicides in the USA. The first seven materials listed are soil fumigants; the remainder are either organophosphates or carbamates. The fumigants all require a waiting period prior to planting and many must be confined with a plastic cover over the treated soil. Thus, these materials require a large investment before the crop is planted. The primary advantage of a fumigant is the quick reduction in nematode numbers within days of application.

It is worth noting that no new products have been registered since 1974, and that several materials are no longer available. From the grower's perspective, the ideal soil fumigant was ethylene dibromide (EDB); it was

Table 9.1. Status of nematicides in the USA

Common Name	Market introduction	Remarks
Chloropicrin	1920	*Pro:* Broad-spectrum biocide including plant-pathogenic soil fungi, bacteria, and soil insects and weeds (used in mixture with other fumigants). **Con:** Five- to 21-day delay after treating before planting. Hazardous to workers, but abnoxious vapors encourage safety.
Methyl bromide	1942	*Pro:* Broad-spectrum biocide except bacteria. Highly efficacious, used on seed beds in mixtures with chloropicrin. One- to 4-day delay in planting. **Con:** Hazardous, bromine in soil.
1,3-Dichloropropene 1,2-Dichloropropane (D,D)	1943	*Pro:* Highly effective against cyst and root knot nematodes. **Con:** Hazardous, unacceptable toxicology, detected in groundwater. Removed from market 1980s.
1,3-Dichloropropene	1946	*Pro:* Highly effective, broad-spectrum nematicide, some soil insects. **Con:** Three weeks or longer wait to plant. Groundwater contamination potential, air pollutant in California, temporarily withdrawn in California.
Ethylene dibromide (EDB)	1945	*Pro:* Effective, economical, works best against ectoparasitic nematodes. Low volume rate per acre. **Con:** Toxicology. Removed from market in 1980s.
Dibromochloropropane (DBCP)	1954	*Pro:* Safety to crop, can be used around living plants. **Con:** Slow acting, groundwater contamination potential. Removed from marketplace in 1980's.
Metam sodium	1957	*Pro:* Broad-spectrum, can be applied in irrigation water, low mammalian toxicity, not restricted for use in USA. **Con:** Limited efficacy, needs water to activate, long waiting period after treating.
Methylisothiocyanate (breakdown product of metam sodium)	1963	*Pro:* Broad-spectrum, can be applied in cold weather, if soil tillable, tarpaulins unnecessary. **Con:** Somewhat limited efficacy. Not reregistered in USA beyond 1992, still available outside USA.

Table 9.1. *Continued*

Common Name	Market introduction	Remarks
Ethoprop	1967	*Pro:* Nematicidal and insecticidal. **Con:** Limited efficacy, especially with abundant rainfall in sandy soils. Phytotoxic.
Carbofuran	1967	*Pro:* Systemic, some phytotoxicity, no waiting period. **Con:** Avian toxicity has resulted in removal of granular form from market. May be decomposed by soil microflora with continuous usage.
Aldicarb	1970	*Pro:* Nematicidal and insecticidal, broad-spectrum, good efficacy, systemic, no waiting period. **Con:** Acute toxicity, high groundwater contamination potential.
Fenamiphos	1973	*Pro:* Broad labeling, high efficacy, low potential for leaching. **Con:** May be decomposed by soil microflora with continuous usage.
Oxamyl	1974	*Pro:* Systemic, low phytotoxicity, can be used to treat living plants. **Con:** Limited efficacy, especially in sandy soils.

highly efficacious and reasonably priced. This product could be applied pre- or postplant because of its crop safety.

Fumigants having broad-spectrum biocidal activity are expected to have limited value in the future because of reregistration requirements under today's Environmental Protection Agency (EPA) regulatory climate. As a result of a reregistration law enacted in 1988, all products registered under FIFRA (Federal Insecticide, Fungicide, and Rodenticide Act) must be reregistered; many of the fumigants are being subjected to much more rigid regulatory guidelines than they were at the time of the initial registration. There is only one new fumigant currently in development, Enzone (tetrathiocarbamate salts which are a reformulation of carbon bisulfide), from Unical (Sacramento, CA). Enzone currently has an experimental use permit on citrus, grapes, and potatoes to control nematodes and phylloxera.

Similarly, there is one new organophosphate nematicide, fosthiazate (Nemathorin) from Ishihara (ISK Biotech, Mentor, OH) which has an experimental use permit on tobacco and tomatoes.

Table 9.2. Examples of chemical control of major soilborne diseases

Pathogen	State of control	Product
Pythium, Phytophthora	Good	Captan, metalaxyl, fosetyl-Al
Rhizoctonia, Pseudocercosporella	Fair	PCNB, benomyl
Fusarium, Verticillium	Weak	Fumigants
Phymatotrichum, Armilaria	None	–
Gaeumannomyces, Plasmodiophora	None	–
Sclerotium rolfsii, Sclerotinia	Fair-good	PCNB, iprodione

9.2.2 Soil Fungicides

Progress in the development of soil fungicides has been limited by the fact that many chemical moieties are (1) metabolized quickly into nonfungicidal degradation products in the soil or (2) bound chemically/physically in soil, especially those high in organic matter or clay. Once bound, chemicals cannot provide the required efficacy to control soilborne diseases. Coupled with this is the fact that many of the microorganisms in soil have very resilient resting bodies (e.g., sclerotia), thus the task of soilborne disease management becomes all the more difficult. Schwinn (1982) provided an overview of the state of this art (Table 9.2); not much has changed since this review as so few new fungicides have been registered (see Table 9.3).

Control of soilborne pathogens by chemicals is accomplished by reducing the initial inoculum in the soil and/or the rate of infection. The amount of infection can be reduced by protecting the infection court, impeding the pathogen from reaching the root, and by chemical induction of host plant resistance. There are several new chemical moieties in development which have an impact on soilborne diseases.

9.2.2.1 Soil-Applied Fungicides Under Development

There are several new fungicide moieties which have been tested by university cooperators.

1. *Oomycetes*
a) Oxadixyl (Sandofan/Anchor) – This is a phenylamide similar to metalaxyl, but not as active in the soil. It is being developed by Sandoz Crop Protection.
b) Dimethomorph (CME-151) – A morpholine with activity on foliar oomycetes; it is currently being evaluated as a soil fungicide by Shell. Dimethomorph has no activity against members of Pythiaceae.

Table 9.3. EPA-approved seed and soil fungicides

Common name	Trade mark(s)	Seed	Soil	Foliar	Remarks
A For use against *Pythium* and *Phytophthora* spp.					
• Metalaxyl	Apron Ridomil Subdue	X	X	–	Pro: Highly efficacious, systemic, used at low rates. Nononcogenic. Multicrop. Con: Subject to fungal resistance.
• Propamocarb	Banol	–	X	–	Pro: Fungistat which is systemic, limited soil mobility. Con: Usage limited to turf and ornamentals.
• Fosetyl-Al	Aliette	–	X	X	Pro: Moves basipetally; can also be used as a soil drench. Con: Spectrum limited to *Phytophthora* spp. Should not be used with Cu-sprays.
• Captan	Captan	X	–	–	Pro: Broad-spectrum seed treatment. Multicrop. Con: Residual efficacy is limited as it is a protectant.
• Etridiazole	Koban Terrazole Truban	X	X	–	Pro: Used in combination with Terrachlor. Con: Uses limited to cotton, onamentals, turf.
B For use against *Rhizoctonia solani*					
• Carboxin	Vitavax	X	X	–	Pro: Broad-spectrum systemic seed treatment. Multicrop. Also active on *S. rolfsii*. Con: Limited to seed and mid-season peanut soil uses.
• PCNB	Terrachlor	X	X	–	Pro: Broad-spectrum protectant. Con: High rates, limited labeling.
• Chloroneb	Demosan	X	X	–	Pro: Highly effective in control of postemergence damping-off. Con: High rates, limited labeling.
• Thiram	Thiram	X	(X)[a]	–	Pro: Multicrop seed protectant. Con: Limited use in soil.

Table 9.3. *Continued*

Common name	Trade mark(s)	Seed	Soil	Foliar	Remarks
C For use against "other" diseases					
Fusarium					
• Thiabendazole	Mertect	X	–	–	Pro: Control of *Fusarium*.
thiophanate-methyl	Topsin-M	X	–	–	Con: Use limited to bulbs, corns, tubers.
• Benomyl	Benlate	–	(X)	–	Pro: Provides control of numerous diseases in greenhouses by drench. Con: Lacks activity in the field. Manufacturer recently withdrew all greenhouse uses.
Sclerotinia					
• Iprodione	Rovral	–	X	X	Pro: Also controls *Botrytis*, *Rhizoctonia*. Con: Subject to fungal resistance.
• Vinclozolin	Ronilan	–	–	X	Pro: Preventive and curative. Con: Limited number of crops on the label.
Sclerotium rolfsii					
• PCNB	Terraclor	–	X	–	Pro: Safe. Available as spray or granular. Con: Limited to peanuts, high rate.
Seedborne diseases					
• Triadimenol	Baytan	X	–	–	Pro: Controls many seedborne cereal diseases, also foliar diseases. Low rates. Con: Limited crops.

[a] (X) means that it can be applied to soil, but with limited success.

2. *Basidiomycetes*: Kataria and Verma (1990) tested several new compounds with activity against *R. solani*. The authors evaluated the efficacy against isolates AG-2.1 and AG-4 on canola (rapeseed) and found differences in efficacy on the part of some products against the two isolates, both in vitro and in vivo. As might be expected, some compounds performed better on the seed than as soil treatments because the active ingredient can be bound up in the soil by organic matter or clay particles.
   The most consistent new products under development are:

a) Tolclofos-methyl (Rizolex) has been registered in Japan and Europe for several years. Being an organophosphate, tolclofos-methyl has a novel mode of action compared to most fungicides. Tolclofos-methyl was shown to be fungicidal to sclerotia of *R. solani* by Barnes et al. (1983). As a soil fungicide, it is used at rates of 0.5–1 lb. ai/A (active ingredients per acre) (0.56–1.12 kg ai/ha). Sandoz Crop Protection has the rights to develop this Sumitomo Ltd. product in the USA.
b) Flutolanil (Moncut) is a benzanilide compound developed by Nihou Nohyaku; the US rights are with AgrEvo (SN-84364). Flutolanil is a systemic fungicide with protective and curative properties. In the soil, it is used at rates of 1–3 lb. ai/A (1.12–3.4 kg ai/ha) banded over the row, while the seed treatment rates are 0.75–1.5 oz. ai/cwt (active ingredients per hundred weight) (21.3–42.5 g). The half-life in soil is 40–60 days. Flutolanil has also performed well against *Sclerotium rolfsii* on peanuts and *Rhizoctonia solani* in ornamentals.
c) Iprodione (Rovral) has been developed and marketed for several uses in the USA, but only recently has Rhone-Poulenc decided to pursue opportunities for this product in the soil to control *R. solani*. In cotton, the rate of Rovral being evaluated is 0.1–0.2 lb. ai/A (0.1–0.22 kg ai/ha) applied in a band over the row. Results show that the liquid spray performs better than granular formulations (Baldwin et al. 1991).
d) Pencycuron (Moncern) has very strong activity towards some AG groups of *R. solani*, but none to others (Kataria and Verma 1990). This product has been tested in the US against rice sheath blight, but has not been as thoroughly investigated in soil.
e) Terbuconazole (Folicur) is a triazole from Bayer/Miles which is being developed as a seed treatment and provides control of limb rot of peanuts caused by *R. solani* following foliar applications to control leaf spot.

3. *Ascomycetes*

a) Fluazinam (ASC-66825) is a new class of fungicide with a unique spectrum of activity which includes *Sclerotinia minor*, *Sclerotium rolfsii*, and *Rhizoctonia solani* (respective ED_{50} values in vitro were 0.0025, 0.035, and 0.19 μg/ml, according to Smith et al. 1992).

4. *Deuteromycetes*

a) Phenylpyrroles (CGA-142705, CGA-173506) are a new class of fungicide which is active against *Fusarium* spp. and *R. solani* (ED_{50} values of 0.18 and 0.04 μg/ml for CGA-173506 on *Fusarium* spp. and *R. solani*, respectively).

9.2.3 Seed Treatments

The major market for US seed treatments is cereals, albeit only ca. one-third of the acreage is being planted with treated seed. Today, fungicide seed treatment of small grain is based on five active substances: captan, carboxin, maneb, pentachloronitrobenzene (PCNB), and thiram. Two other specialty products, Apron (metalaxyl) and Imazalil, are also registered for use on cereals. On many other crops, captan is the most widely used product for damping-off, although the combination of metalaxyl and carboxin has become the product of choice in areas of severe disease development, e.g., on cotton in the Mississippi Delta region in the USA.

9.2.3.1 New Cereal Seed Treatments Under Development

Several new triazole seed treatments hold considerable promise on cereals. These include triadimenol (Baytan) which was registered for use by EPA in 1990, and two new products with registration packages currently in review, viz., tebuconazole (Raxil) and difenoconazole (Dividend). By virtue of its systemic properties, triadimenol has the unique property of controlling overwintering powdery mildew infections on cereals. However, both tebuconazole and difenoconazole have advantages in crop safety over triadimenol. In addition, difenoconazole is the first fungicide to offer complete control of the seed and soilborne phases of dwarf bunt (*Tilletia controversa*; Table 9.4).

Another class of seed-dressing material in development includes the phenylpyrrole fungicides, an example of which is fenpicolanil (Beret). The phenylpyrrole fungicides are based on the antibiotic pyrrolnitrin produced by *Pseudomonas fluorescens* (Arima et al. 1965). This group has a broad spectrum of activity which includes *Fusarium* spp. and *Rhizoctonia solani*. Beret is rapidly establishing itself as the standard for control of *Fusarium nivalis* (*Gerlachia nivalis*) in winter cereals in Europe.

9.2.4 Fungicides for Actively Growing Plants

The demands made upon fungicides used postplant are considerably greater, as these uses are most commonly rescue-type treatments. The ideal postplant treatment is one which moves basipetally in the plant. Among the

Table 9.4. Comparison of the efficacy of three cereal seed dressings (percent control)

		Disease				
Seed dressing	Rate g ai per 100 kg	Loose smut	Stripe smut	Bunt	Dwarf bunt	Septoria
Vitavax 200	90	66	95	85	0	80
Baytan	30	97	100	100	12	85
Dividend	24	98	100	100	100	100

fungicides, fosetyl-Al possesses this property and provides good control of Phytophthora root rot in this manner. The only nematicide that has shown limited basipetal activity is oxamyl.

Frequently, postplant treatments are made at thinning, lay-by, or hilling (e.g., iprodione for control of Sclerotinia bottom rot of lettuce). Sometimes such treatments are applied through drip irrigation (e.g., metalaxyl for control of Phytophthora root rot in citrus). The key considerations for products used postplant in the soil are (1) high water solubility; (2) availability (i.e., not bound to the soil particulates); and (3) ability to translocate within the plant. Metalaxyl is such a product.

Certain crops are more suited to postplant treatments because of the management of the crop; for example, perennial crops afford opportunities for postplant treatments. Another suitable crop is peanuts; at-pegging treatments are commonly used on peanuts to control *Pythium myriotylum*, *Rhizoctonia solani*, *Sclerotium rolfsii*, and *Sclerotinia minor*.

Above all, postplant fungicide treatments have to be safe for the crop; this further limits the list of available products.

9.3 Chemical Management Approaches

9.3.1 Protective vs. Curative Applications

As the basic chemical companies established in vivo screens for fungicides rather than in vitro screens, we have seen more new products with a systemic mode of action. Unlike protectant products which only affect fungi on the outside of the host, curative products may work on the host surface and/or within the host. The control provided may be fungicidal or fungistatic (nematicidal or nematistatic). Through a better knowledge of disease epidemiology, we are now able to predict the onset of some diseases. With the availability of such fungicides with curative properties, and a better knowledge of disease epidemiology, we have two key ingredients to facilitate the use of chemical controls in integrated pest management (IPM). The value of the crop dictates the risk a grower is willing to take and the benefit a chemical can provide (see Sect. 9.3.6).

9.3.2 In-Furrow and Band vs. Broadcast Application

Most of the applications made to control soilborne pathogens are now made in a band, or in-furrow, at planting. The reason for this is primarily the judicious use of the products (cost/benefit/risk). Similarly, treatments made to established crops are confined to the root zone wherever possible.

One approach to the management of soil diseases is the use of chemicals at planting to provide protection during the period in which the crop is developing resistance or tolerance to disease after reaching a particular stage of growth ("adult" plant resistance). An example of this is the in-furrow use of metalaxyl on soybeans to control *Phytophthora megasperma* var. *sojae*.

9.3.3 Use of Chemical Pesticides in Conjunction with Biologics

A weakness of many biological control agents is their inconsistent performance. What has been a real surprise to many scientists who have combined biocontrols with chemicals is (1) the fact that the rate of chemical application can be reduced (Howell 1991) or (2) the synergism provided by the combination (CIBA-GEIGY in-house trials).

In the opinion of the author, there will be considerable emphasis placed on this type of research during the 1990s, as such combinations afford two other key attributes to chemicals: a fungicidal resistance strategy and a means to offset patent expiration, as such combinations can be patented or remain proprietary due to data protection at EPA.

9.3.4 Slow-Release Formulations

Slow-release formulations of chemicals have become a major factor in the pharmaceutical industry for the release of drugs within animal systems. However, there are very few examples of slow-release formulations used in agriculture. One of the major reasons for the limited use of this technology is the additional expense such systems add to the formulation of active ingredients. Many questions must be resolved before slow-release formulations are accepted as a mechanism of applying a pesticide against soilborne pathogens or root diseases. A major question is: "Do slow-release systems increase biodegradation of the active ingredient?" The utility of slow release varies with (1) the active ingredients (water solubility, mode of action); (2) environmental factors (what triggers the release); and (3) the epidemiology of a given disease. In the past, far more effort has gone into the safe packaging of pesticides than into slow-release systems.

9.3.5 Residual Effects of Fungicides on Pathogens

Attempts to eliminate covered smut (*Tilletia* spp.) and loose smut (*Ustilago* spp.) from cereals through seed treatments are examples of the residual effect of fungicides on diseases of simple interest. To have residual effects on pathogens, it is important for chemical control measures to reduce the levels of overwintering inoculum and/or the initial infection (e.g., haustoria production).

One new approach to controlling diseases (both foliar and soil) is by means of host plant defense mechanisms. Systemic acquired resistance is characterized by the development of a disease-resistant state in plants that have reacted to previous infection with a hypersensitive response, e.g., tobacco to tobacco mosaic virus (Ross 1961). This phenomenon is now receiving attention as a means to improve the residual control of pathogens.

Another interesting area of research is the impact melanin biosynthesis inhibitors has on developing, overwintering microsclerotia of *Verticillium dahliae*. Treated microsclerotia do not develop melanin and the resulting albino microsclerotia are not fit to overwinter.

Due to the complex physical and chemical traits of soils, many chemical treatments never reach their target; this, along with the reproductive capacity of soilborne microorganisms, often limits the residual effects of most chemicals in soil.

9.3.6 Risk vs. Benefit

Loss due to disease can be considered a biological phenomenon, a social problem, or both (Main 1977). The goal of commercial production agriculture is to optimize production and the quality of food, feed, and fiber with minimum risk to man and the environment. Zadoks (1973) distinguished crop destruction by disease into three progressive phases:

1. Injury: any observable deviation from the normal crop; injury may lead to damage.
2. Damage: any decrease in quality or quantity of a product; damage may lead to loss.
3. Loss: any decrease in economic returns from reduced yields and/or higher costs of agricultural activities designed to reduce damage.

When evaluating benefits and risks, there are supporters of sustainable agriculture who argue for a substantial reduction in reliance on chemicals. In contrast, other well-known scientists such as Ames and Gold (1990) claim that plants produce 10000 times more natural pesticides compared to the synthetic pesticides available for registered use. Research and the expenses of pest control incurred by the farmer are leading agriculture more and more to an integrated pest management (IPM) mode which requires expert decisions for crop management. For example, control of soilborne diseases may impose the need for long-term crop rotation, which may only become a reality if nonfood crops (grasslands) are subsidized. IPM requires a higher level of crop management (frequent scouting by a trained observer), which in turn requires a thorough knowledge of the pests, their economic thresholds, and the measures that can be used for their control.

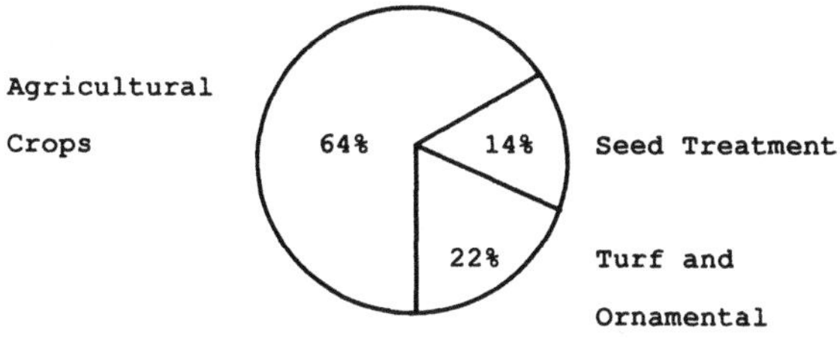

Fig. 9.1. US fungicide markets for agricultural crops, seed treatment, and turf ornamentals

9.4 Summary

Agriculture is the largest industry in the USA. It accounted for 17.5% of our GNP in 1985. North America, with only 5% of the world's population, produces almost 20% of the world's wheat, over 40% of the world's corn, over 50% of the world's soybeans, and 20% of the world's cotton (Barnett 1990). However, the USA only represents 10% of the world's fungicide sales. Figure 9.1 shows a breakdown of the US fungicide market ($520 million in 1990).

Use of fungicides helps the US to contribute nearly 60% (9 million tons) of the food aid around the world. The American farmer today produces enough food for over 70 people, reflecting an efficiency which accounts for 80% more crop output on an acre of land than the previous generation. A study by the National Agricultural Chemicals Association (1989) on the benefits of fungicides showed that US food prices would rise 13% and supplies of fruits, vegetables, peanuts, and other products would drop 32, 21, 68, and 6%, respectively, if protective fungicides were not available. We have a great deal yet to learn about soilborne diseases. It is a fitting challenge as we move into the twenty-first century.

References

Ames BN, Gold LS (1990) Too many rodent carcinogens: mitogenesis increases mutagenesis. Science 249:970–971

Arima K, Imanaka H, Kousaka M, Fukuda A, Tamura G (1965) Studies on pyrrolnitrin, a new antibiotic. J Antibiot Ser A 8:201–219

Baldwin CH Jr, Hope JH, Young HS (1991) Aliette (fosetyl-AL) and Al Rovral (Iprodione): a new fungicide combination for seedling disease control. Proc Beltwide Cotton Conf 1:87

Barnes G, Harris RI, Russell PE (1983) Tolclofos methyl, a new fungicide for the control of *Rhizoctonia solani*. Proc 10th Int Congr Plant Prot 1, Nov. 20–25, 1983, Brighton. Brit Cr Prot Council, Croydon, pp. 460–464

Barnett AG (1990) Agricultural chemicals, their role, the risk and the balance. Seminar at Iowa State University

Goring CAI (1957) Factors influencing diffusion and nematode control by soil fumigants. Agric Chem Dev Bull No 110, Dow Chemical Company
Howell CR (1991) Biological control of Pythium damping-off of cotton with seed-coating preparations of *Gliocladium virens*. Phytopathology 81:738–741
Kataria HR, Verma PR (1990) Efficacy of fungicidal seed treatments against pre-emergence damping-off and post-emergence seedling root rot of growth chamber grown canola caused by *Rhizoctonia solani* AG 2-1 and AG-4. Can J Plant Pathol 12:409–416
Main CE (1977) Crop destruction – the raison d'etre of plant pathology. In: Horsfall JG, Cowling EB (eds) Plant disease, vol 1. Academic Press, New York, pp 55–78
Maude RB (1990) Seed treatment. Pestic Outlook 1:16–22
Nelson EB, Chao WL, Norton JM, Nash GT, Harman GE (1986) Attachment of *Enterobacter cloacae* to hyphae of *Pythium ultimum*: possible role in the biological control of Pythium pre-emergence damping-off. Phytopathology 76:327–335
Ross AF (1961) Systemic acquired resistance induced by localize virus infections in plants. Virology 14:340–358
Schwinn FJ (1982) Chemical control of fungal diseases: importance and problems. In: Dekker J, Georgopoulos SG (eds) Fungicide resistance in crop protection. Centre for Agricultural Publishing and Documentation (PUDOC), Wageningen, pp 7–15
Smith FD, Phipps PM, Stiper RJ (1992) Fluazinam, a new fungicide for control of Sclerotimia blight and other soilborne pathogens. Peanut Sci 19:115–120
Zadoks JC (1973) Schade, vol 4. Gewasbescherming, Wageningen, pp 71–75

10 Biological Management

D.R. Fravel and C.A. Engelkes

10.1 Ecological Opportunities for Biocontrol

Biological control agents add another dimension to the familiar disease triangle. Interactions among the plant, pathogen, biocontrol agent, and environment must be considered in development of management systems for soilborne pathogens. The most successful systems exploit aspects of each member of the tetrad resulting in the maximum benefit to crop production with a minimum expenditure of resources. Knowledge of ecology and epidemiology is particularly helpful in determining when and where the biocontrol agent is most likely to function, when and where the pathogen may be diverted from causing disease, and in rational screening for new biocontrol agents. Timing and placement of the biocontrol agent are often more important to the success of control than the population size of the biocontrol agent. Thus, temporal and spatial relationships among plant, pathogen, and biocontrol agent, as well as how environment modifies these interactions, must be explored.

This chapter discusses ecological approaches to biocontrol. Examples are given to illustrate concepts rather than for an all-inclusive review of the literature. Examples are limited to biocontrol through introduction of beneficial fungi or bacteria and are primarily from agricultural production systems rather than laboratory studies.

10.1.1 Identification of Low Ecological Diversity

Most biocontrol agents introduced onto an agar plate grow and proliferate readily. In contrast, populations of these same organisms may be quickly diminished if they are introduced into raw soil. Although there are chemical and physical differences between agar plates and raw soils, biological differences are most likely to account for the difficulty of establishment of introduced microorganisms in raw soil. In particular, microbial competition for nutrients is considerably more intense in raw soil than on an agar plate because of the greater ecological diversity (greater number of species) in the raw soil. Consider a continuum of increasing ecological diversity with

C.L. Campbell/D.M. Benson (Eds.)
Epidemiology and Management of Root Diseases

the agar plate at one end of the continuum, an undisturbed ecosystem at the other end, and various agricultural systems such as hydroponic, greenhouse crop, row crop, and field crop arranged in between (Marois and English 1989). Chances for successful introduction of the biocontrol agent are increased by introducing it at a time when and/or place where microbial competition is lower. This can be a time and place that occurs naturally during the growing cycle or they can be created for the purpose of introducing the biocontrol agent.

One of the earliest successes in biocontrol was the introduction of *Phlebia gigantea* (*Peniophora gigantea*) to control root rot of forest trees caused by *Heterobasidion annosum* (*Fomes annosus*) (Rishbeth 1975). Spores of *P. gigantea* are applied to freshly cut stumps in liquid suspension or in chain saw oil. *P. gigantea* quickly colonizes the stump, a place of low biological diversity, thus excluding *H. annosum* which otherwise could gain access to healthy trees through root grafts with the cut tree.

Greenhouse systems, which are low in biological diversity, offer another opportunity for introducing biocontrol agents. Lumsden and Locke (1989) successfully exploited this lower diversity to control *Pythium* and *Rhizoctonia* damping-off of zinnia, cotton, and cabbage. Control was accomplished by introducing *Gliocladium virens* into the soilless potting mix in which these transplants are grown.

Horticultural procedures for eggplant were utilized to introduce *Talaromyces flavus* for successful biocontrol of Verticillium wilt (Marois et al. 1982). Eggplant transplants are grown in soilless potting mix in the greenhouse for 6 to 8 weeks before transplanting to the field. Ascospores of *T. flavus* were added during watering while the transplants were in the greenhouse. Thus, the biocontrol agent had 6 to 8 weeks to become established with little competition before being transplanted to a situation of greater biological diversity in the field.

10.1.2 Creation of Ecological Simplicity

In many cases it is possible to create situations of low biological diversity to enhance biocontrol. Methods frequently used include heat or chemical disinfestation of soils. For example, Locke et al. (1985) introduced *Trichoderma viride* into steamed greenhouse soil mix. The biocontrol agent was allowed to colonize the soil mix before planting chrysanthemum cuttings and challenging them with *Fusarium oxysporum* f. sp. *chrysanthemi*, causal agent of Fusarium wilt.

Sublethal treatments with chemicals or heat offer yet another opportunity for successful introduction of biocontrol agents. Although it has been recognized for at least 40 years that weakened pathogen propagules may be more susceptible to biocontrol (Bliss 1951), only recently has this concept been put into practice. When pathogen-infested soil is heated or treated with chemicals at lower than recommended rates, four processes may occur simultaneously: viability of pathogen propagules is reduced, vulnerability of pathogen propagules to potential biocontrol agents is increased, activity of biocontrol agents is affected, and competition may be reduced making it easier to establish some biocontrol agents (modified from Katan 1981).

Combining a reduced rate of Telone C17 with *Talaromyces flavus* resulted in significant reductions in Verticillium wilt incidence on eggplant in the field and an increase in fruit production (Marois et al. 1982). Exposing sclerotia of *Sclerotium rolfsii* to sublethal metham sodium, heating, drying, or heat shock resulted in reduced sclerotia viability due to microbial activity (Smith 1972; Elad et al. 1980; Henis and Papavizas 1983; Lifshitz et al. 1983). Sublethal doses of heat also rendered *Armillaria mellea* and *Rosellinia necartrix* more vulnerable to antagonism by *Trichoderma* spp. (Munnecke et al. 1976; Sztejnberg et al. 1987). Sublethal heat treatments caused a weakening of propagules of *F. oxysporum* f. sp. *melonis* and *F. oxysporum* f. sp. *niveum* resulting in decreased pathogenicity and loss of spore viability beyond the initial mortality due to microbial action (Freeman and Katan 1988). Sublethal solarization has been used to enhance biocontrol of Verticillium wilt of artichoke (Tjamos and Paplomatas 1988) and olive (Tjamos et al. 1991).

10.2 Strategies for Biocontrol

Factors other than ecological diversity must also be considered in approaching biocontrol. In particular, it is important to consider when and where the pathogen is physically most accessible and when and where the pathogen is physiologically most vulnerable. Thus, when and where to apply the biocontrol agent will differ somewhat from system to system. Whether the pathogen is killed or inactivated near the root or far from it, the net effect of most of the biocontrol agents currently in use is to reduce the amount of inoculum which infects the roots. Four main strategies for biocontrol have been employed: (1) protection of the rhizosphere or spermosphere; (2) impediment of the progress of the pathogen toward the root; (3) destruction of pathogen inoculum; and (4) plant-mediated biocontrol. These strategies are not intended as rigid categories. Instead, organizing current knowledge may stimulate thought about how to approach new situations. It should also be recognized that various mechanisms of biocontrol may be deployed with any strategy. For example, a "good competitor" might protect the infection court as well as slow the invasion of soil by the pathogen.

10.2.1 Protection of the Rhizosphere or Spermosphere

Much work has focused on colonization of the rhizosphere and the ability of biocontrol agents applied to seeds or tuber pieces (Burr et al. 1978) to grow and function in the rhizosphere (Windels and Kommedahl 1982; Chao et al. 1986; Ahmad and Baker 1987). First, delivery of the biocontrol agent is greatly simplified. For infection to occur, the pathogen must contact the plant. Thus, instead of trying to locate the pathogen in bulk soil, we can deliver small amounts of the biocontrol agent to the seed or root surface. The rhizosphere and spermosphere are the sites of most microbial activities since nutrients are more abundant in these sites than in bulk soil. Thus, the root or seed may provide nutrients to sustain the biocontrol agent (Roberts 1991). Biocontrol agents which work through (1) antibiosis; (2) competition for nutrients; (3) physical exclusion of the pathogen (Waksman 1952); or (4) a combination of these mechanisms have all been used to protect the rhizosphere or spermosphere.

10.2.1.1 Antibiosis

Antibiosis plays a key role in biological control of crown gall, caused by *Agrobacterium radiobacter* var. *tumifaciens*, by nonpathogenic *A. radiobacter* strain 84 (Kerr 1980). Strain 84 has a plasmid which codes for production of a highly specific nucleotide bacteriocin called agrocin 84 that selectively inhibits most pathogenic agrobacteria. Only agrobacteria strains which produce agrocin control crown gall.

Another system for which antibiosis is well established as the major mechanism is in biocontrol of take-all of wheat, caused by *Gaeumannomyces graminis* var. *tritici*, by *Pseudomonas fluorescens*, and *P. aureofaciens* (Weller and Thomashow 1990). The antibiotic phenazine-1-carboxylic acid has been isolated from rhizospheres of healthy wheat roots colonized by *P. fluorescens* (Thomashow et al. 1990). Phenazine-producing pseudomonads suppress primary infections and limit secondary spread of *G. graminis* var. *tritici* (Bull et al. 1991) compared to mutants defective in phenazine synthesis (Thomashow and Weller 1988) in the laboratory, as well as in the field (Weller and Cook 1983).

Antibiosis is also a mechanism of biocontrol by *Gliocladium virens* strain GL-21 (Lumsden et al. 1991). GL-21 produces the antibiotic gliotoxin that is active at a low concentration in controlling *Pythium*- and *Rhizoctonia*-induced damping-off of greenhouse-grown zinnia seedlings (Lumsden and Locke 1989). This antibiotic was produced abundantly in soilless potting medium and at lower levels in mineral soils amended with GL-21 as a wheat bran-alginate formulation (Lumsden et al. 1992).

Enzymes may be involved indirectly in antibiosis. Extracellular glucose oxidase produced by *Talaromyces flavus* reacts with glucose to produce hydrogen peroxide (Kim et al. 1990). Microsclerotia of *Verticillium dahliae* are killed by hydrogen peroxide. Strains of *T. flavus* deficient in glucose oxidase production do not control Verticillium wilt of eggplant (Fravel and Roberts 1991). Parasitism of microsclerotia of *V. dahliae* is enhanced by the addition of root exudates, suggesting that hydrogen peroxide may weaken microsclerotia rendering them more susceptible to parasitic attack by *T. flavus*.

The relevance of mycoparasitism to biocontrol by *Gliocladium virens* was questioned by Howell (1987) who examined six UV-induced mutants obtained from two strains of *G. virens*. In contrast to their parent strains, these mutants lacked any lytic activity towards *Rhizoctonia solani*. Yet, they did not differ in antagonistic activity in soil from the parent strains.

10.2.1.2 Competition

Competition in a general sense has also been used to achieve biocontrol. Many suppressive soils may effect biocontrol by slowing progress of the pathogen (Schneider 1982). For example, suppressive soil mixed with conducive soil resulted in control of Fusarium wilt (Yuen et al. 1985). In situations where the pathogen reinvades the planting soil for each disease cycle, the infection court may be protected by slowing the advance of the pathogen. This method has been used to protect tomato (Marois and Mitchell 1981; Lemanceau and Alabouvette 1991), chrysanthemum (Locke et al. 1985), and various other hosts (Alabouvette 1990) from *Fusarium* species. Similarly, Ko (1971) protected papaya transplants by surrounding them with raw, pathogen-free soil in fields infested with *Phytophthora palmivora*. Germlings of *Trichoderma hamatum* and *Gliocladium virens* suppressed the saprophytic activity of *Rhizoctonia solani* in the field (Lewis and Papavizas 1991).

The competitive advantage of first possession of a substrate has been recognized for some time (Garrett 1970). Few researchers who have studied soil systems would doubt the importance of competition for limiting nutrients in soil and in the rhizosphere. Yet, competition for nutrients is difficult to study experimentally. Competition for iron, carbon, and nitrogen has been reported.

Competition for iron by plant growth-promoting rhizobacteria (PGPR) (Kloepper and Schroth 1978) has been particularly well documented (Leong 1986). Strains of PGPR, including *Pseudomonas fluorescens*, *P. putida*, and some closely related strains, produce siderophores which have a high affinity for soluble ferric iron (Fe^{+3}). Siderophore-producing PGPR enhanced growth of potato plants under iron-limiting conditions by depriving cyanide-producing pseudomonads of this vital element in the root zone (Bakker and Schippers 1987).

Fusarium species are sensitive to intrageneric competition for nutrients, especially carbon. Alabouvette (1986) demonstrated that greater concentrations of glucose were needed to overcome fungistasis in soils conducive to *F. oxysporum* f. sp. *dianthi*, *F. o. lycopersici*, and *F. o. melonis* compared to soils suppressive to these pathogens. Germination of chlamydospores of *F. oxysporum* f. sp. *melonis* and *F. o. vasinfectum* was significantly reduced in the rhizospheres of melon and cotton, respectively, by *Trichoderma harzianum* (Sivan and Chet 1989). However, this inhibition was negated by the addition of seed exudates, which suggested that competition for nutrients was involved in biocontrol.

Alternatively, the presence of plant exudates may be masked by the biocontrol agent by interception of root signals (Paulitz 1990; Nelson 1991a). The possibility of signal interception is intriguing since signal interception would permit protection of a greater volume than would be expected from the population size of the biocontrol agent.

10.2.1.3 Exclusion

Biocontrol agents may protect the rhizosphere or spermosphere by physically excluding the pathogen (i.e., competition for space). Competition for space may be difficult to distinguish from competition for nutrients since nutrients are nearly always associated with substrates (e.g., organic debris, roots, seeds). Thus, competition for nutrients on the root surface may be observed as physical exclusion of the pathogen. Competition for both nutrients and space is most likely an important factor in several biocontrol systems including the use of ectomycorrhizae to protect pine roots from *Phytophthora cinnamomi* (Marx 1972). Similarly, the ability of *Enterobacter cloacae* to control *Pythium ultimum* is related to its ability to bind to hyphae of seed colonizing *Pythium* spp. and to inhibit further hyphal development and infection of the seed surface (Nelson et al. 1986).

10.2.1.4 Combined Mechanisms

Many, if not most, biocontrol agents use more than one mechanism to protect roots from soilborne pathogens. Use of the various mechanisms may be simultaneous, at different times on the same plant, or one mechanism may operate on one plant while the same biocontrol agent on a different plant may use a different mechanism. Multiple mechanisms of biocontrol can be illustrated with plant growth-promoting rhizobacteria (PGPR) which are discussed more fully in recent reviews (Schippers et al. 1987a,b; Défago and Haas 1990).

PGPR are believed to stimulate plant growth and increase yields indirectly by their aggressive colonization of root systems. *Pseudomonas* spp.

are the primary colonizers of young root tissues because of their flexible metabolisms, short germination times, and mobility. Production of both antibiotics and siderophores has been cited as a factor related to the ability of PGPR to displace or exclude soilborne pathogens and deleterious rhizosphere microorganisms (Suslow and Schroth 1982), which affect plants without causing obvious disease symptoms. In this case, antibiosis and siderophore production may be mechanisms of competition between soilborne pathogens and PGPR (Bruehl 1975; Lockwood 1986).

Production of siderophores and hydrogen cyanide by fluorescent pseudomonads plays varying roles as a biocontrol mechanism at different times and on different plants. Siderophore production is a beneficial attribute of PGPR primarily when iron availability is a limiting factor. *P. fluorescens* produces the siderophore pyoverdine, which contributes to biocontrol of damping-off of cotton caused by *Pythium ultimum* (Loper 1988). However, mutants of *P. putida* strain N1R, deficient in pyoverdine biosynthesis (Pu^-), protected cucumber seedlings against Pythium damping-off as well as the wild-type parent strain N1R in three different agricultural soils (Paulitz and Loper 1991). Similarly, production of hydrogen cyanide by fluorescent pseudomonads had a negative effect on potatoes which are sensitive to hydrogen cyanide (Bakker and Schippers 1987) and a positive effect as a mechanism of biocontrol of black root rot of tobacco caused by *Thielaviopsis basicola* (Voisard et al. 1989).

In many cases, we know that the biocontrol agent colonizes the rhizosphere, although the mechanism(s) of action may not yet be fully elucidated. For example, *Trichoderma* is widely used in biocontrol (e.g., Ahmad and Baker 1987; Sivan and Chet 1989; Sivan and Harman 1991) and *Trichoderma* may have multiple mechanisms (Papavizas 1985).

10.2.2 Plant-Mediated Biocontrol

Strains of PGPR with biocontrol ability may enhance plant growth directly through the induction of plant disease resistance. Van Peer et al. (1991) induced resistance in carnation to Fusarium wilt caused by *F. oxysporum* f. sp. *dianthi* by bacterizing roots with *Pseudomonas* sp. The bacterium was applied to carnation roots 1 week prior to challenging carnation stems with the wilt pathogen. This root application caused accumulation of phytoalexins in the carnation stems contributing to disease control. Wei et al. (1991) induced systemic resistance in cucumber to anthracnose caused by the foliar pathogen *Colletotrichum orbiculare* by treating seeds with six PGPR strains. In both cases, spatial separation of the PGPR strains from the pathogen would make competition inoperable as the mechanism of biocontrol.

10.2.3 Impeding Movement of the Pathogen

Although colonization of the rhizosphere has distinct advantages for biocontrol, other strategies have been successfully employed to reduce the amount of pathogen inoculum reaching the roots. As roots grow through soil they may contact pathogen propagules and the pathogen grows, or in some cases moves, only a very short distance to infect the root. However, in certain situations, the pathogen may grow through relatively larger volumes of soil (or other growing medium) contacting plant roots. Extensive pathogen invasion is especially likely to occur if the soil/growing medium contains few microorganisms (i.e., low diversity) due to heat or chemical treatment, or the conditions of its manufacture. Such situations offer excellent opportunities for biocontrol, especially in greenhouse crops. A variety of biocontrol agents have been used to prevent or retard invasion of pathogenic *Fusarium* spp.

Nonpathogenic *Fusarium oxysporum*, isolated from the suppressive soil from Châteaurenard, effectively controlled Fusarium wilts when introduced into a disinfested wilt-conducive soil (Rouxel et al. 1979). Similarly, strains of nonpathogenic *F. oxysporum* plus fluorescent *Pseudomonas* strains prevented invasion of rock wool by aerially disseminated spores of pathogenic *F. oxysporum* f. sp. *lini* and *F. o. radicis lycopersici* (causal agents of flax wilt and root and crow rot of tomato, respectively) (Lemanceau and Alabouvette 1991).

10.2.4 Destruction of the Pathogen

Pathogen inoculum may be destroyed through mycoparasitism or killed by antibiosis. At first thought, it may seem an impossible task to locate and destroy the overwintering propagules of a pathogen distributed throughout the soil. However, this sort of thinking ignores a number of critical features of plant pathogens and biocontrol agents. Two of the more obvious ones are (1) that pathogens are not uniformly or randomly distributed in soil. They occur in an aggregate distribution and we have a priori knowledge of where to expect these aggregates, i.e., in association with the crop plants or plant residues. (2) Unlike chemical pesticides, biocontrol agents are living organisms and, hence, are capable of reproduction, growth, and spread.

Sporidesmium sclerotivorum behaves in nature as an obligate parasite of *Sclerotinia* species. This biocontrol system takes advantage of several aspects of the pathogen, the biocontrol agent, and the host plant. Sclerotia of *S. minor* are produced on lettuce plants at the end of the growing season. When this crop residue is plowed under at the end of the growing season, the sclerotia are distributed in aggregates in the top 8 cm of soil, with more than half in the top 2 cm (Adams 1986). Disking does little to disperse these aggregates. *Sporidesmium* can grow from an infected sclerotium up to 3 cm to infect a new sclerotium. Thus, if even one sclerotium in an aggregate is infected by the biocontrol agent, the whole aggregate will eventually be destroyed. In this system, the biocontrol agent is sprayed onto lettuce plants at the end of the growing season and the crop is plowed under. Although it may take several seasons to work, *S. sclerotivorum* sprayed on the plants can give complete control of Sclerotinia lettuce drop in the field at rates as low as 0.2 kg/ha (Adams and Fravel 1990). One of the key elements in this system is the accessibility of the sclerotia while they are still above ground.

10.3 Successful Implementation of Biocontrol

In 1990, the USDA celebrated 100 years of research on biocontrol of insects. In contrast, concentrated research on biocontrol of plant diseases has been growing only over approximately the past 25 years. Research in biocontrol of soilborne plant pathogens has laid a foundation of knowledge and concepts upon which we may build to begin using biocontrol agents in agricultural and horticultural production.

10.3.1 Constraints to Biocontrol

Some of the barriers to biocontrol are inherent to working with living organisms, while other barriers may exist in our minds only (Baker and Cook 1974; Schroth and Hancock 1981; Scher and Castagno 1986; Nelson 1991b). Most of us are accustomed to working with chemical pesticides and have inadvertently internalized specific expectations. Thus, we automatically expect to apply a small amount of a compound and find the pathogen dead soon after. We need to be aware that alternative methods may also result in reduction of losses to plant diseases.

Information on environmental effects on biocontrol agents, interactions among microbes, microbes and plants, and how the environment modifies these interactions is needed. During all phases of production, formulation, delivery, and functioning in the field, biocontrol agents are more sensitive to temperature and moisture extremes than are chemical pesticides. For example, differences in efficacy of biocontrol agents have been reported between fall and spring crops (Lemanceau and Alabouvette 1991; Fravel et al. 1992), while similar differences are not generally reported for pesticides.

Studies on microbial interactions are shedding light on inconsistencies in biocontrol of plant diseases. Weller and Cook (1983) found that a combination of two pseudomonad strains with different mechanisms of biocontrol reduced severity of take-all of wheat more than either strain alone. Similarly, combinations of *Pseudomonas* strains and nonpathogenic fusaria acted synergistically to control Fusarium wilt of flax and Fusarium crown and root rot of tomato (Lemanceau and Alabouvette 1991). Application of only *Pseudomonas* strains had no effect on the diseases.

Knowledge of production, formulations, and delivery is also critical to the widespread use of biocontrol agents and currently much information is lacking in this area. The method selected for production can greatly affect the speed with which the inoculum is produced. Speed of production is likely to be directly related to the cost of producing the inoculum and may also be related to the purity of the inoculum (Fravel and Lewis 1990). The production method also affects the type of propagule obtained. Production of specific propagule types is important since some propagules are more effective in biocontrol than others (Papavizas et al. 1984). The particular formulation can also affect the efficacy, shelf life, and ease of use of the product, as well as its cost (Fravel and Lewis 1990).

More information is also needed on the delivery systems. As discussed earlier in this chapter, delivery systems which take advantage of pathogen accessibility, characteristics of the biocontrol agent, and cropping system are needed to deliver the biocontrol agent to the right place at the right time. Delivery systems which exploit these aspects maximally, facilitate economically feasible delivery rates. Efficient delivery systems for *G. virens*, *P. gigantea*, *A. radiobacter*, *T. flavus*, *Trichoderma* spp., *Sporidesmium sclerotivorum*, *Pseudomonas* spp., and *E. cloacae* and have been developed and have been discussed previously in this chapter.

In some cases there may also be economic constraints to the widespread adoption of biocontrol. Knowledge of the economics of using biocontrol and of developing biocontrol is clearly necessary. The costs must be balanced by beneficial return to the growers. Although some studies have addressed these issues (Adams and Fravel 1990; Fravel and Lewis 1990), more in-depth information is needed. Regulatory constraints may also represent a cost to industry. It is not yet clear to what extent regulatory constraints will be a barrier to implementation of biocontrol.

10.3.2 Forecast for Biocontrol

Many of the same features of biocontrol agents which are viewed as disadvantages may also be seen as advantages. For example, the specificity of many biocontrol agents for a particular pathogen (e.g., *S. sclerotivorum* for *Sclerotinia* spp.) may be an advantage since such a specific agent is not likely to cause undesirable nontarget effects associated with the use of broad-spectrum chemical pesticides. The quantity and distribution of toxic compounds produced by biocontrol agents also should result in fewer nontarget effects than chemical pesticides since they are produced in very small quantities at the site of action as with gliotoxin (Lumsden et al. 1992). In addition, because biocontrol agents are living organisms, they have the potential to colonize the whole root from inoculum applied only to seed

(Ahmad and Baker 1987; Harman and Taylor 1990) and to provide disease control for more than one season (Adams and Fravel 1990).

Successful biocontrol requires an active population of the biocontrol agent in the right place at the right time. Thus, we need to consider interactions of individuals, populations, and communities over time and space to achieve our goal. Timing and placement of the biocontrol agent are often more critical than the population size of the biocontrol agent. For successful biocontrol, timing and placement must be based on epidemiological and ecological considerations. Understanding the principles of ecology underlying successful biocontrol systems can help in designing screening methods to search for new biocontrol agents. A good deal of thought must go into any screening method since the screening method dictates what will be found.

A feature to be considered in designing screening procedures is which mechanism(s) of biocontrol is most likely to work in a particular system. For example, with certain exceptions (Deacon 1991), parasitism is a slow process. Thus, if a strategy of protecting the infection court seems most appropriate for a particular production system, then screening should be designed to identify microbes providing biocontrol through competition, antibiosis, and/or induced resistance. Alternatively, if, after considering the total production system, destruction of the overwintering inoculum of the pathogen seems to be the most appropriate strategy, then a screening procedure selecting for parasites of the overwintering inoculum would be useful.

Marois and English (1989) indicate that mechanisms may also be related to biological diversity. As indicated previously, considerations about where and when the biocontrol agent is most likely to function are of paramount importance when designing screening procedures. In particular, an awareness of biological diversity in the system, pathogen vulnerability, and accessibility at different times/places is important to finding the "window of opportunity" for biocontrol. For example, in a system of high biological diversity where lowering diversity is not feasible, a specific parasite would be more likely to provide biocontrol than a general competitor. A general competitor would be a good choice for biocontrol in, for example, preventing pathogen reinvasion in a steamed soil.

To capitalize on features of biocontrol agents, plant pathogens, hosts, and various aspects of the environment for optimal biocontrol, we must first identify the essential features which drive the system as well as individual parameters which affect various aspects of the life histories of the organisms involved. Currently, however, our greatest successes have come from understanding a sufficient number of the components of the system adequately enough to exploit each component maximally, i.e., to encourage/stimulate microorganisms to do what they are already programmed to do. Total knowledge of the system is not necessary for successful biocontrol, although additional knowledge results in further refinement of the use of the biocontrol agent.

Many biocontrol agents and mycorrhizae are compatible with chemical pesticides and solarization (Fravel et al. 1985; Adams and Wong 1991; Afek et al. 1991; Tjamos et al. 1991). Biocontrol agents may also act synergistically to control soilborne diseases (Beale and Pitt 1990). Hence, these biocontrol agents can be integrated into current production systems. Environmental concerns about the use of chemical pesticides dictate a need for alternative control measures for soilborne plant pathogens which integrate reduced use of chemical pesticides with resistant plant varieties, cultural practices, and introduced biocontrol agents.

References

Adams PB (1986) Production of sclerotia of *Sclerotinia minor* on lettuce in the field and their distribution in soil after disking. Plant Dis 70:1043–1046

Adams PB, Fravel DR (1990) Economical control of Sclerotinia lettuce drop by *Sporidesmium sclerotivorum*. Phytopathology 80:1120–1124

Adams PB, Wong JA-L (1991) The effect of chemical pesticides on the infection of sclerotia of *Sclerotinia minor* by the biocontrol agent *Sporidesmium sclerotivorum*. Phytopathology 81:1340–1343

Afek U, Menge JA, Johnson ELV (1991) Interactions among mycorrhizae, soil solarization, metalaxyl, and plants in the field. Plant Dis 75:665–671

Ahmad JS, Baker R (1987) Rhizosphere competence of *Trichoderma harzianum*. Phytopathology 77:182–189

Alabouvette C (1986) Fusarium-wilt suppressive soils from the Châteaurenard region: review of a 10-year study. Agronomie 6:273–284

Alabouvette C (1990) Biological control of Fusarium wilt pathogens in suppressive soil. In: Hornby D (ed) Biological control of soil-borne plant pathogens. CAB International, Wallingford, pp 27–44

Baker KF, Cook RJ (1974) Biological control of plant pathogens. Freeman, San Francisco, 433 pp

Bakker AW, Schippers B (1987) Microbial cyanide production in the rhizosphere in relation to potato yield reduction and *Pseudomonas* spp.-mediated plant growth stimulation. Soil Biol Biochem 19:451–457

Beale RE, Pitt D (1990) Biological and integrated control of Fusarium basal rot of *Narcissus* using *Minimedusa polyspora* and other micro-organisms. Plant Pathol 39:477–488

Bliss DE (1951) The destruction of *Armillaria mellea* in citrus soils. Phytopathology 41:665–683

Bruehl GW (1975) Systems and mechanisms of residue possession by pioneer fungal colonists. In: Bruehl GW (ed) Biology and control of soil-borne plant pathogens. Am Phytopathol Soc, St Paul, MN, pp 77–83

Bull CT, Weller DM, Thomashow LS (1991) Relationship between root colonization and suppression of *Gaeumannomyces graminis* var. *tritici* by *Pseudomonas fluorescens* strain 2–79. Phytopathology 81:954–959

Burr TJ, Schroth MN, Suslow TV (1978) Increased potato yield by treatment of seedpieces with specific strains of *Pseudomonas fluorescens* and *P. putida*. Phytopathology 68:1377–1383

Chao WL, Nelson EB, Harman GE, Hoch H (1986) Colonization of the rhizosphere by biological control agents applied to seeds. Phytopathology 76:60–65

Deacon JW (1991) Significance of ecology in the development of biocontrol agents against soil-borne plant pathogens. Biocontr Sci Technol 1:5–20

Défago G, Haas D (1990) Pseudomonads as antagonists of soilborne plant pathogens: modes of action and genetic analysis. In: Bollag JM, Stotzky G (eds) Soil biochemistry, vol 6. Marcel Dekker, New York, 565 pp

Elad Y, Katan J, Chet I (1980) Physical, biological, and chemical control integrated for soilborne diseases in potatoes. Phytopathology 70:418–422

Fahima T, Henis Y (1990) Interactions between pathogen, host and biocontrol agent: multiplication of *Trichoderma hamatum* and *Talaromyces flavus* on roots of diseases and healthy hosts. In: Hornby D (ed) Biological control of soil-borne plant pathogens. CAB International, Wallingford, pp 165–180

Fravel DR, Lewis JA (1990) Production, formulation and delivery of beneficial microbes for biocontrol of plant pathogens. In: Chasin DG, Bode LE (eds) Pesticide formulation and application systems, vol 11, ASTM STP 1112. Available from American Society for Testing Materials, Philadelphia

Fravel DR, Marois JJ, Dunn MT, Papavizas GC (1985) Compatibility of *Talaromyces flavus* with potato seedpiece fungicides. Soil Biol Biochem 17:163–166

Fravel DR, Adams PB, Potts W (1992) Use of disease progress curves to study the effects of the biocontrol agent *Sporidesmium sclerotivorum* on lettuce drop. Biocontr Sci Technol 2:339–349

Fravel DR, Roberts DP (1991) In situ evidence for the role of glucose oxidase in the biocontrol of Verticillium wilt by *Talaromyces flavus*. Biocontr Sci Technol 1:91–99

Freeman S, Katan J (1988) Weakening effect on propagules of *Fusarium* by sublethal heating. Phytopathology 78:1656–1661

Garrett SD (1970) Pathogenic root-infecting fungi. Cambridge Univ Press, Cambridge, 294 pp

Harman GE, Taylor AG (1990) Development of an effective seed treatment system. In: Hornby D (ed) Biological control of soil-borne plant pathogens. CAB International, Wallingford, pp 416–426

Henis Y, Papavizas GC (1983) Factors affecting germinability and susceptibility to attack of sclerotia of *Sclerotium rolfsii* by *Trichoderma harzianum* in field soil. Phytopathology 73:1469–1474

Howell CR (1987) Relevance of mycoparasitism in the biological control of *Rhizoctonia solani* by *Gliocladium virens*. Phytopathology 77:992–994

Katan J (1981) Solar heating (solarization) of soil for control of soilborne pests. Annu Rev Phytopathol 19:211–236

Kerr A (1980) Biological control of crown gall through production of Agrocin 84. Plant Dis 64:25–30

Kim KK, Fravel DR, Papavizas GC (1990) Glucose oxidase as the antifungal principle of talaron from *Talaromyces flavus*. Can J Microbiol 36:760–764

Kloepper JW, Schroth MN (1978) Plant growth-promoting rhizobacteria on radishes. In: Station de Pathologie Végétale et Phytobactériologie (ed) Proc 4th Int Conf Plant Pathogenic Bact, vol 2. Gibert-Clarey, Tours, Angers, France, pp 879–882

Ko WH (1971) Biological control of seedling root rot of papaya caused by *Phytophthora palmivora*. Phytopathology 61:780–782

Lemanceau P, Alabouvette C (1991) Biological control of Fusarium diseases by fluorescent *Pseudomonas* and non-pathogenic *Fusarium*. Crop Prot 10:279–286

Leong J (1986) Siderophores: their biochemistry and possible role in the biocontrol of plant pathogens. Annu Rev Phytopathol 24:187–209

Lewis JA, Papavizas GC (1991) Biocontrol of cotton damping-off caused by *Rhizoctonia solani* in the field with formulations of *Trichoderma* spp. and *Gliocladium virens*. Crop Prot 10:396–402

Lifshitz R, Tabachnik M, Katan J, Chet I (1983) The effect of sublethal heating on sclerotia of *Sclerotium rolfsii*. Can J Microbiol 29:1607–1610

Locke JC, Marois JJ, Papavizas GC (1985) Biological control of Fusarium wilt of greenhouse grown chrysanthemums. Plant Dis 69:167–169

Lockwood JL (1986) Soilborne plant pathogens: concepts and connections. Phytopathology 76:20–27

Loper JE (1988) Role of fluorescent siderophore production in biological control of *Pythium ultimum* by a *Pseudomonas fluorescens* strain. Phytopathology 78: 166–172

Lumsden RD, Locke JC (1989) Biological control of damping-off caused by *Pythium ultimum* and *Rhizoctonia solani* with *Gliocladium virens* in soilless mix. Phytopathology 79:361–366

Lumsden RD, Locke JC, Adkins ST, Ridout CJ (1992) Isolation and localization of the antibiotic gliotoxin produced by *Gliocladium virens* from alginate prill in soil and soilless mix. Phytopathology 82:230–235

Lumsden RD, Locke JC, Walter JF (1991) Approval of *Gliocladium virens* by the U.S. EPA for biological control of *Pythium* and *Rhizoctonia* damping-off. Petria 1:138

Marois JJ, English JT (1989) Integration of biological control of plant pathogens into IPM programs. In: Proc Natl Integrated Pest Management Symp/Worksh, 25–28 April 1989. National IPM Coordinating Committee, Las Vegas, NV, pp 33–40

Marois JJ, Johnston SA, Dunn MT, Papavizas GC (1982) Biological control of Verticillium wilt of eggplant in the field. Plant Dis 66:1166–1168

Marois JJ, Mitchell DJ (1981) Effects of fumigation and fungal antagonists on the relationships of inoculum density to infection incidence and disease severity in Fusarium crown rot of tomato. Phytopathology 71:167–170

Marx DH (1972) Ectomycorrhizae as biological deterrents to pathogenic root infections. Annu Rev Phytopathol 10:429–454

Munnecke DE, Wilbur W, Darley EF (1976) Effect of heating or drying on *Armillaria mellea* or *Trichoderma viride* and the relation to survival of *A. mellea* in soil. Phytopathology 66:1363–1368

Nelson EB (1991) Exudate molecules initiating fungal responses to seeds and roots. In: Keister DL, Cregan PB (eds) Beltsville symposia in agricultural research 14. The rhizosphere and plant growth. Kluwer Academic, Dordrecht, pp 197–209

Nelson EB (1991) Current limits to biological control of fungal phytopathogens. In: Arora DK (ed) Handbook of applied mycology, vol 1. Soil and plants. Marcel Dekker, New York, pp 327–357

Nelson EB, Chao WL, Norton JM, Nash GT, Harman GE (1986) Attachment of *Enterobacter cloacae* to hyphae of *Pythium ultimum*: possible role in the biological control of Pythium preemergence damping-off. Phytopathology 76:327–335

Papavizas GC (1985) *Trichoderma* and *Gliocladium*: biology, ecology, and potential for biocontrol. Annu Rev Phytopathol 23:23–54

Papavizas GC, Dunn MT, Beagle-Ristaino J (1984) Liquid fermentation technology for experimental production of biocontrol fungi. Phytopathology 74:1171–1175

Park CS, Paulitz TC, Baker R (1988) Biocontrol of Fusarium wilt of cucumber resulting from interactions between *Pseudomonas putida* and non-pathogenic isolates of *Fusarium oxysporum*. Phytopathology 78:190–194

Paulitz TC (1990) The stimulation of *Pythium ultimum* by seed volatiles and the interaction of *Pseudomonas putida*. Phytopathology 80:994–995 (Abstr)

Paulitz TC, Loper JE (1991) Lack of a role for fluorescent siderophore production in the biological control of *Pythium* damping-off of cucumber by a strain of *Pseudomonas putida*. Phytopathology 81:930–935

Roberts DP (1991) Role of carbohydrates in spermosphere proliferation by *Enterobacter cloacae* EcCT501. Phytopathology 81:1215 (Abstr)

Rishbeth J (1975) Stump inoculation: a biological control of *Fomes annosus*. In: Bruehl GW (ed) Biology and control of soil-borne pathogens. Am Phytopathol Soc, St Paul, MN, pp 158–162
Rouxel F, Alabouvette C, Louvet J (1979) Recherches sur la résistance des sols aux maladies. IV. Mise en évidence due rôle des *Fusarium* autochtones dans la résistance d'un sol à la fusariose vasculaire du melon. Ann Phytopathol 11:199–207
Scher FM, Castagno JR (1986) Biocontrol: a view from industry. Can J Plant Pathol 8:222–224
Schippers B, Bakker AW, Bakker PAHM (1987a) Interactions of deleterious and beneficial rhizosphere microorganisms and the effect of cropping practices. Annu Rev Phytopathol 25:339–358
Schippers B, Lugtenberg B, Weisbeek PJ (1987b) Plant growth control by fluorescent pseudomonads. In: Chet I (ed) Innovative approaches to plant disease control. John Wiley & Sons, New York, pp 19–39
Schneider RW (ed) (1982) Suppressive soils and plant disease. Am Phytopathol Soc, St Paul, MN, 88 pp
Schroth MN, Hancock JC (1981) Selected topics in biological control. Annu Rev Microbiol 35:453–476
Sivan A, Chet I (1989) The possible role of competition between *Trichoderma harzianum* and *Fusarium oxysporum* on rhizosphere colonization. Phytopathology 79:198–203
Sivan A, Harman GE (1991) Improved competence in a protoplast fusion progeny of *Trichoderma harzianum*. J Gen Microbiol 137:23–29
Smith AM (1972) Nutrient leakage promotes biological control of dried sclerotia of *Sclerotium rolfsii* Sacc. Soil Biol Biochem 4:125–129
Suslow TV, Schroth MN (1982) Role of deleterious rhizobacteria as minor pathogens in reducing crop growth. Phytopathology 72:111–115
Sztejnberg A, Freeman S, Chet I, Katan J (1987) Control of *Rosellinia necatrix* in soil and in apple orchard by solarization and *Trichoderma harzianum*. Plant Dis 71:365–369
Thomashow LS, Weller DM (1988) Role of a phenazine antibiotic from *Pseudomonas fluorescens* in biological control of *Gaeumannomyces graminis* var. *tritici*. J Bacteriol 170:3499–3508
Thomashow LS, Weller DM, Bonsall RF, Pherson LS III (1990) Production of the antibiotic phenazine-1-carboxylic acid by fluorescent *Pseudomonas* species in the rhizosphere of wheat. Appl Envion Microbiol 56:908–912
Tjamos EC, Biris DA, Paplomatas EJ (1991) Recovery of olive trees with Verticillium wilt after individual application of soil solarization in established olive orchards. Plant Dis 75:557–562
Tjamos EC, Paplomatas EJ (1988) Long-term effect of soil solarization in controlling Verticillium wilt of artichokes in Greece. Plant Pathol 37:507–515
van der Plank JE (1963) Plant diseases: epidemics and control. Academic Press, New York
van Peer R, Niemann GJ, Schippers B (1991) Induced resistance and phytoalexin accumulation in biological control of Fusarium wilt of carnation by *Pseudomonas* sp. strain WCS417r. Phytopathology 81:728–734
Voisard C, Keel C, Haas D, Défago G (1989) Cyanide production by *Pseudomonas fluorescens* helps suppress black root rot of tobacco under gnotobiotic conditions. EMBO J 8:351–358
Waksman SA (1952) Soil microbiology. John Wiley & Sons, New York, 356 pp
Weller DM, Cook RJ (1983) Suppression of take-all of wheat by seed treatments with fluorescent pseudomonads. Phytopathology 73:463–469
Weller DM, Thomashow LS (1990) Antibiotics: evidence for their production and sites where they are produced. In: Baker RR, Dunn PE (eds) New directions in

biological control: alternatives for suppressing agricultural pests and diseases. Alan R Liss, New York, pp 703–711

Wei G, Kloepper JW, Tuzun S (1991) Induction of systemic resistance of cucumber to *Colletotrichum orbiculare* by selected strains of plant-growth promoting rhizobacteria. Phytopathology 81:1508–1512

Windels CE, Kommedahl T (1982) Rhizosphere effects of pea seed treatment with *Penicillium oxalicum*. Phytopathology 72:190–194

Yuen GY, Schroth MN, McCain AH (1985) Reduction in Fusarium wilt of carnation with suppressive soils and antagonistic bacteria. Plant Dis 69:1071–1075

11 Cultural Management

D.R. Sumner

11.1 Approaches to Cultural Management

11.1.1 Quarantine

Quarantine of plant material and seed has limited value in controlling soilborne pathogens. Most soils are already infested naturally with the most common pathogens, or man-derived quarantine measures are ineffective. Land that was not naturally infested in the prehistoric era probably was infested by contaminated seed and plant material before man became familiar with the germ theory. There are a few quarantines in effect that may be helpful in reducing the spread of pathogens into noninfested areas. In some countries, nurseries are prohibited from selling crucifers unless they are shown to be free of club root induced by *Plasmodiophora brassicae* (Gram 1960). The quarantine to prevent spread of the golden nematode (*Globodera rostochiensis*) from Long Island, New York, to other potato growing areas was instituted in 1944, and has apparently been successful (Brodie and Mai 1989). Using seed free of *Fusarium solani* f. sp. *cucurbitae* eliminates root and stem rot caused by the pathogen in squash because the pathogen does not survive in soil (Sumner 1976). Quarantine regulation concerning plant material, soil, and seed may help prevent the spread of other soilborne pathogens, but there has been little research to determine whether or not such quarantines have effectively reduced the spread of soilborne pathogens.

11.1.2 Soil Disinfestation

Early attempts to disinfest soil included formaldehyde and later soil fumigation, as described in Chapter 9 (this Vol.). Disinfestation with steam has been used commonly in greenhouse and nursery operations for almost 100 years (Johnson 1946; Baker 1962). Large, mobile boilers have been developed to steam heat up to several acres for high-value crops such as flowers and ornamentals. Aerated steam (60–75 °C) is used to treat soil for nursery crops and in fields to reduce the deleterious side effects of heating soil to

C.L. Campbell/D.M. Benson (Eds.)
Epidemiology and Management of Root Diseases

100 °C with pure steam (Baker 1962). Dry heat provided by electricity in boxes or ovens has been used on a limited scale. Flame sterilization in mobile, continuous soil pasteurizers is another effective alternative (Mitchell and Shaw 1975). Autoclaves, electric ovens, and microwave ovens are other alternatives for disinfestation of soil for home or small commercial operations. Populations of *R. solani*, *Pythium* spp., and *Fusarium* spp. were reduced below detectable levels when 0.5 to 4 kg of field soils in plastic bags were treated for a few minutes in a common household microwave oven (Ferriss 1984). There were no apparent deleterious affects on physical properties of the soils, or soil nutrients, and plant growth was good in treated soil. Thus, it may be practical for consumers to treat small amounts of soil or potting mixtures in microwave ovens before planting or transplanting.

Soil disinfestation for large-scale field, vegetable, and perennial crop systems is essentially unavailable because of costs. Soil fumigation, covered in Chapter 9, is the only method of soil disinfestation that has been used to any extent in large fields. Soil solarization, a relatively new practice, has potential and will be covered later in Section 11.4.

11.1.3 Tillage

The ability of farmers to use a variety of tillage implements, especially large, heavy machinery, was limited until the development of farm tractors early in this century. As larger tractors with increased horsepower and maneuverability became more common, moldboard and one-way disk plows capable of inverting the soil to great depths (30–90 cm) and subsoilers that fractured hardpans as deep as 90 cm were developed. In the 1920s, bulb growers in The Netherlands were aware that turning soil by hand or with machinery 75–90 cm deep, coupled with cow manure, aided in controlling soilborne pathogens (Drayton 1929). Even with bullocks and horses, plowing was the conventional method of preparing land worldwide in most cropping systems during the last century. With increased mechanization, it was possible for farmers to use clean tillage to prepare smooth seed beds rapidly over large acreages, and that became the norm after World War II.

In the 1950s, detailed research studies in peanut fields in the southeastern USA demonstrated the advantage of deep plowing and clean tillage to control *Sclerotium rolfsii*, the cause of southern stem rot (white mold). Burying organic matter and the inoculum of *S. rolfsii* deprives the pathogen of a food source to initiate colonization of the host at the soil line (Boyle 1956). This practice has been an accepted standard method of seed bed preparation for peanut culture in the peanut belt of the eastern and southeastern USA for more than 25 years. Also, plowing buries inoculum of *Rhizoctonia solani* (primarily colonized plant debris) and deprives the fungus of oxygen necessary for survival (Sumner et al. 1981). In Canada, moldboard plowing of wheat stubble from the previous year, in the spring before seeding wheat, was beneficial in reducing seedling disease caused by *Bipolaris sorokiniana* from 1951–1958. However, yields were usually not increased by plowing (Ledingham et al. 1960). In addition to burying propagules of certain pathogens, plowing also loosens the soil, improves aeration, and allows greater root development. Crop losses from other soilborne pathogens can be reduced by fracturing compacted soil (hardpans) 30–45 cm deep with large, heavy chisels. Root rot caused by *Fusarium solani* f. sp. *phaseoli* was less damaging to beans and peas in Washington following in-row subsoiling in sandy loam soil. The beneficial effect resulted from unusual rooting depth and root generation which reduced plant stress (Burke et al. 1972; Kraft and Wilkins 1989). In Georgia, increased root growth was observed with in-row subsoiling in cotton (Hussey and Roncadori 1977) in soil infested with *R. solani*, *Pythium* spp., *Fusarium* spp., and the nematodes *Hoplolaimus columbus* and *Helicotylenchus* spp.; and in-row subsoiling increased yield in soybean in soil infested with the nematodes *H. columbus* and *Meloidogyne incognita* (Minton and Parker 1975).

In perennial crops, tillage can be detrimental to established roots. In Georgia, disk-harrowing of peach orchards to control weeds damaged feeder roots and increased root injury by *Pythium* spp. and nematodes (Hendrix and Powell 1970). Controlling weeds by mowing or other practices that do not cause mechanical injury to roots is preferable in cultivated perennial crops.

11.1.4 Crop Rotations

The rotation of crops to encourage better plant health was probably the first method of disease control used by man. "Sick soil" was frequently referred to in the literature in the late 1800s and early 1900s before the etiology of root diseases and the relationship of root health to plant health and yield were known (Bolley 1901). Pammel (1888) suggested that rotations of cotton with small grains might be helpful in reducing cotton root rot induced by *Phymatotrichum omnivorum*. By rotating unlike crops, or crops of different susceptibility, losses to pathogens with relatively narrow host ranges can be reduced (Curl 1963).

Numerous publications from 1920 to 1960 began to enumerate the benefits of crop rotation in controlling diseases of specific crops. Some examples of long-range investigations are the control of root diseases in flue-cured tobacco in North Carolina including blackshank, caused by *Phytophthora nicotianae* var. *parasitica*; bacterial wilt, caused by *Pseudomonas solanacearum*; Fusarium wilt, caused by *F. oxysporum* f. sp. *nicotianae*; root knot, caused by *Meloidogyne incognita* (Clayton et al. 1944); cotton root rot caused by *P. omnivorum* in Texas (Ratcliffe 1934; Lyle et al. 1948); take-all on wheat caused by *Gaeumannomyces graminis* in New York (Kirby 1922) and Kansas (Sewell and Melchers 1924); common root rot caused by *Bipo-*

laris sorokiniana and *F. culmorum* in Canada (Broadfoot 1934; Ledingham 1961); corn root rot caused by *Fusarium* spp. and other fungi in Illinois (Koehler and Holbert 1930); pea root rot caused by *F. solani* f. sp. *pisi*, *Aphanomyces euteiches*, *P. ultimum*, *R. solani*, and *Ascochyta pinodes* in New York (Reinking 1942); tuber rot of potato, caused by *Actinomyces scabies* and *R. solani*, and Fusarium wilts, caused by *F. oxysporum* (Goss and Afanasiev 1938); and crown rot of sugar beet caused by *R. solani* (Schuster and Harris 1960) in Nebraska.

Allowing land to lie fallow to store moisture for crop production is a common parctice in the Great Plains of the USA and Canada. Fallowing as part of a crop rotation may reduce populations of soilborne pathogens in wheat (Russell 1930), but fallowing may have increased populations of *Pythium* spp. in Kansas before planting alfalfa (Grandfield et al. 1935). Also, Ledingham (1961) found that fallow maintained populations of *Cochliobolus sativus* for 2 to 3 years because of fungistatic properties that prevented conidial germination. In Brazil, fallowing 3 to 4 years effectively reduced losses from *B. sorokiniana* in wheat to only 9% compared with a 23% yield loss with only 1 or 2 years of fallow (Diehl et al. 1982). Populations of *R. solani* can be reduced greatly, but not eliminated, by a few months of fallow in Georgia (Bell and Sumner 1987). In Costa Rica, in soils infested with *Pseudomonas solanacearum*, disking and fallowing for 9 months successfully reduced the incidence of bacterial wilt in banana (Sequeira 1962).

In modern agriculture, crop rotation has become an accepted method of disease control to prevent yield decline in several crops. In China, consecutive crops susceptible to similar root pathogens are avoided in vegetable communes (Williams 1979). In potato, a rotation of one crop of potatoes every 2 to 4 years is common. Scholte (1989) in The Netherlands showed that a 4-year rotation with wheat-sugar beet-oats reduced disease from several soilborne pathogens including *Streptomyces* spp., *R. solani*, *Verticillium dahliae*, and *Meloidogyne* spp. The rotation increased yield 27% compared with continuous potato. In the USA, rotations of even 1 year with a small grain or grass crop reduce injury from *R. solani* (Specht and Leach 1987). In contrast, Rhizoctonia crown and root rot of sugar beet was reduced more by a rotation with fallow or sunflower than with wheat or sorghum on the southern high plains of Texas (Rush and Winter 1990).

Recent research supports investigations earlier in this century, and it is still unwise to use crops susceptible to similar pathogens in close rotations. In Virginia, crop losses to *Cylindrocladium crotalariae* can be reduced in both peanut and soybean by not having one of the crops preceded by the other (Krigsvold et al. 1977). In contrast, in Washington, the specialization of *F. solani* f. sp. *phaseoli* on bean and *F. solani* f. sp. *pisi* on English pea allows one crop to follow the other with little or no loss in yield compared with continuous cropping (Burke and Kraft 1974). Rotation of barley or wheat with crops nonsusceptible to common root rot is practiced on the

Canadian prairies (Ledingham 1961; Piening and Orr 1988). In Georgia, root diseases are usually increased in crops following legumes, and the avoidance of two legume crops in succession in irrigated, multiple-cropping systems or intensive cropping systems reduces root diseases caused by *R. solani* and other soilborne pathogens (Sumner et al. 1981, 1983; Sumner 1987). Several isolates of *R. solani* in several anastomosis groups and isolates of binucleate *Rhizoctonia* spp. colonize peanut shells and seeds, and may survive more than 1 year in soil (D.K. Bell unpubl.).

Sugar beet, corn, or spring wheat in rotation decreased basal rot of onion caused by *F. oxysporium* f. sp. *cepae* in Japan, and isolates of the pathogen from onions in rotation were less virulent than the isolates from onion in continuous culture (Higashida et al. 1982). Verticillium wilt is reduced by crop rotation in peppermint (Green 1967), but rotations have had mixed results with Verticillium wilt in cotton (Huisman and Ashworth 1976). Huisman and Ashworth (1976) suggest that initial levels of inoculum, soil type, and physical and chemical properties of the soil may be very influential in determining the success of crop rotations in controlling Verticillium wilt. If initial inoculum densities of microsclerotia are low, short crop rotations might be feasible to reduce wilt severity, but with high inoculum densities, initially, rotations between susceptible crops will be so long that farmers will not consider crop rotations for disease control as useful.

11.1.5 Planting Date and Plant Spacing

In many cultural systems, growers have very little leeway in adjusting planting date, but in some crops root disease intensity can be reduced by a change in date of planting. Fusarium root rot was more severe in dry beans in Washington with late-May than with mid-June plantings (Burke 1964). Yields were increased when planting was delayed until average soil temperatures at planting depth (5–6 cm) were 21 °C.

Root rot of corn caused by *R. solani* AG-2 type 2 was less severe in late spring plantings with soil temperatures of 13–30 °C than in early spring plantings with soil temperatures of 5–18 °C (Sumner and Dowler 1983). Also, a herbicide treatment of pendimethalin for weed control increased root disease severity in the early plantings but not the later plantings.

Numerous seed rots and seedling diseases of corn, cotton, cucurbits, beans, soybeans, and other crops can be reduced by delaying planting until soils are warmer in the spring in temperate climates. With cool season crops, such as crucifers, garden beets, and spinach, waiting until soil temperatures are cool in the fall or planting in late winter or early spring will increase plant stand. Delayed planting of small grains in the fall in the southern USA may reduce or eliminate reproduction of root-knot nematodes because of low soil temperatures (Johnson and Motsinger 1990).

Wider row spacing decreased root disease severity in bean, but narrow row spacing still increased yield of dry beans (Burke 1965; Silbernagel and Mills 1990). The incidence of Phymatotrichum root rot of cotton was later in the season with 1 or 5 than with 9 or 13 plants/m within the row. The proportion of diseased plants at the end of the season was less with 1 plant/m than with higher populations, but the relationship of plant spacing and disease incidence to cotton yield was not reported. With increased plant spacing. there is less spread of the pathogen from plant to plant, but strands of *P. omnivorum* may extend 1 m or more from a dead plant in field soil (Koch et al. 1987).

The severity of Sclerotinia wilt in sunflower caused by *S. sclerotiorum* was not affected by plant spacing in North Dakota. Populations varied from 37 000 to 74 000 plants/ha, and seed yields were similar at all plant spacings (Nelson et al. 1989). In contrast, Sclerotinia wilt was reduced and seed yield increased in Canada when plant populations were 26 000 to 49 000/ha, and plant spacings within the row were 36 cm or more compared with 15 or 25 cm (Hoes and Huang 1985). As plant spacings increase, more of the primary infection comes from soilborne inoculum than from plant-to-plant spread through infected roots.

11.2 Strategies for Use of Cultural Management

11.2.1 Tillage

Strategies for reducing disease losses to soilborne pathogens with tillage will vary with the cropping system and other cultural practices. In cropping systems including peanut in the southeastern USA, the control of southern stem rot is of primary importance and peanut is the most profitable crop in the rotation. Therefore, deep tillage with a moldboard plow usually precedes peanut to bury organic matter and propagules of *Sclerotium rolfsii*. Also, care is taken not to throw soil and debris next to the base of the plants and not to cover lateral branches with soil during cultivation (Boyle 1956; Garren 1961). Deep tillage is practiced with snapbean and other vegetable legumes preceded by crops that are susceptible to *Rhizoctonia solani* AG-4 (Sumner et al. 1986).

In soybean and cotton in Georgia, subsoiling under the row is used to fracture hardpans and achieve lower losses in soil infested with root knot or lance nematodes (Minton and Parker 1975; Hussey and Roncadori 1977). Subsoiling is expensive because large tractors are required to pull equipment capable of fracturing soil 40–45 cm deep. However, increased yield, resulting from the benefits of increased root growth in infested soil, and in sometimes allowing roots to penetrate into deeper soil not infested with root pathogenic fungi and nematodes, may justify the extra expense. Subsoiling is not always beneficial, as shown by research with vegetables in intensive cropping systems in Georgia. With fall snapbean or lima bean following corn in a rotation including soybean and squash or peanut, subsoiling under the row increased root disease severity and lowered yield compared with moldboard plowing (Sumner et al. 1986). Apparently *R. solani* and other soilborne pathogens that colonized soybean and peanut residues remained in the surface soil after subsoiling.

In Canadian dryland wheat, minimal tillage or no-till has commonly replaced conventional tillage with plows in the last 30 years, and the practice has not increased common root rot (Conner et al. 1987). Plowing decreased root rot in wheat following wheat in the seedling stages compared with surface tillage, but differences disappeared as the season progressed, and yields were not different (Ledingham et al. 1960). Similar results have been noted in Texas, where populations of *B. sorokiniana* in soil and the severity of common root rot were less with no-till than with conventional tillage (Mathieson et al. 1990). In contrast, Cephalosporium stripe of winter wheat (caused by *C. gramineum*) is more prevalent with conservation tillage or no tillage than with plowing or burning wheat residues (Latin et al. 1982; Bockus et al. 1983). Because the pathogen survives in wheat residues, crop rotation to nonhosts for 2–3 years, or host resistance, will be necessary to both prevent soil erosion and sustain wheat yields at an economic level.

Ecofallow is practiced in the Great Plains of the USA to control weeds, conserve soil moisture, and reduce soilborne pathogens. The system uses conservation tillage with a 3-year rotation of winter wheat-sorghum or corn-fallow or a 2-year rotation of winter wheat-fallow. In sorghum, stalk rot and lodging caused by *F. moniliforme* was reduced in ecofallow compared with conventional tillage (Doupnik and Boosalis 1980; Sumner et al. 1981).

11.2.2 Soil Fertility

There have been numerous, thorough reviews of the interaction of plant nutrition with root diseases (Huber and Watson 1974; Engelhard 1989; Huber 1990). Frequently, management of soil fertility can be used effectively to reduce losses to root diseases in different plant production systems. In wheat, fertilization with phosphorus has almost eliminated economic losses to Pythium root rot in the central wheat growing areas of the USA. Phosphorus promotes root development and helps seedlings to escape disease (Huber 1990). Earlier research in South Dakota indicated adequate nitrogen fertilization increased yields of barley and wheat infected with *P. arrhenomanes* (Bruehl 1953).

Common root of wheat and barley is reduced by chloride fertilization (Shefelbine et al. 1986; Timm et al. 1986). In Alberta, Canada, nitrate and phosphorus reduced common root rot in barley sown into wheat stubble or fallow and increased yield in barley sown into wheat stubble (Piening et al.

1983). Take-all of winter and spring wheat in the USA is reduced and yield increased by increasing nitrogen fertilization with the ammonium form of nitrogen (Taylor et al. 1983; Huber 1990). The rhizosphere pH is lowered by the ammonium form of nitrogen and increased by the nitrate form of nitrogen. The take-all pathogen does not grow in soil below pH 5, and the reduction in disease is apparently related to the inhibition of growth of the pathogen in the rhizosphere, even though the bulk soil pH is unchanged (Smiley and Cook 1973). In Western Australia take-all was reduced and yield increased with ammonium sulfate, but nitrogen fertilization did not influence Rhizoctonia root rot (MacNish 1988). Chloride fertilization increased yield of plants infested with take-all in Oregon in late-seeded wheat in acid soil (pH 5.5–6.2) that was fertilized adequately with nitrogen, phosphorus, and potassium (Taylor et al. 1983). In additional research in Oregon, chloride reduced the severity of take-all and increased grain yield of wheat on unlimed soil but not on limed soil. Chloride is a nitrification inhibitor, and may reduce take-all severity by inhibiting infection by the pathogen in moderately acid soil (Christensen and Brett 1985). In manganese-deficient soil, fertilization with manganese sulfate reduces take-all severity and increased yield in Australia (Wilhelm et al. 1988). Huber (1989) has found that populations of manganese-oxidizing bacteria are higher in the rhizosphere of wheat plants severely infected with take-all compared with plants with mild symptoms of take-all. It is known that manganese interacts with nitrogen metabolism, and manganese is more available at an acid pH (<5.5) than at an alkaline pH. Co-injection of manganese with ammonia stabilized with nitrapyrin (a nitrification inhibitor), increased yield, and decreased the severity of take-all (Huber 1989). In Washington, the minor nutrients zinc, copper, iron, and manganese were beneficial in reducing take-all (Reis et al. 1982).

The form of nitrogen used can be critical in controlling soilborne pathogens in corn. Inhibiting nitrification with anhydrous ammonia containing nitrapyrin reduced root and stalk rot of corn and increased yield in Indiana (Warren et al. 1975). However, in loamy sand and sand soil in Georgia, the nitrogen form had less influence on corn stalk rot than an adequate balance of potassium and nitrogen fertilization (Bullock et al. 1990; Sumner unpubl.). Adequate potassium nutrition has been beneficial in reducing Fusarium and Verticillium wilt of cotton. However, Fusarium wilt-resistant varieties are still needed to obtain acceptable yields (Bell 1989). Also, correcting potassium deficiency in pistachio reduced infection of thrifty trees by *V. dahliae* (Ashworth et al. 1985).

Calcium nutrition is important in reducing pod rot in peanut in the USA induced by *Pythium* spp., *R. solani*, and *Fusarium* spp. (Garren 1964; Hallock and Garren 1968). Applications of gypsum to maintain adequate calcium in peanut pods reduced pod rot and increased the percentage of sound, mature kernels and peanut yield in Georgia. Currently, the application of high rates of gypsum at flowering time is recommended to reduce the pod rot complex in many peanut producing states (Csinos and Bell 1989). Gypsum or gypsum plus dolomite mixed with soil naturally infested with *Pythium* spp. increased the number of healthy vegetables and ornamental plants in greenhouse cultures (Gill 1972). Calcium reduced damping-off of cucumber induced by *Pythium splendens* in fields in Hawaii (Kao and Ko 1986). An application of calcium nitrate at transplanting reduced the number of tomato plants killed by *S. rolfsii* in South Carolina (Sitterly 1962). The mechanism of control of *Pythium* and other diseases by calcium is unknown. Some of the theories included reduced polygalacturonase activity by the pathogens, reduced pathogen populations, suppression of growth and spore germination, and increased host resistance (Kao and Ko 1989).

Fusarium wilt of tomato, cucumber, watermelon, gladiolus, chrysanthemum, aster, carnation, and other crops is reduced by fertilization with nitrate nitrogen, lime, and low phosphorus (Jones et al. 1989). Fusarium yellows of celery was ameliorated with application of calcium nitrate and potassium chloride (Schneider 1985). Clubrot of crucifers has historically been controlled with lime. The effectiveness of control was increased by the use of small lime particles mixed thoroughly into soil with calcium nitrate (Dobson et al. 1983). In tobacco, Fusarium wilt decreased and leaf yield increased as pH increased after application of limestone (Stephenson et al. 1987).

In greenhouse operations, adjusting methods of fertilization may be beneficial in reducing disease and increasing quality in ornamentals. In soil infested with *P. splendens*, the use of high rates of a complete fertilizer increased root and plant quality in *Peperonia obtusifolia* (Chase and Poole 1984). High rates of fertilizer at transplanting increased control of root rot caused by *Phoma chrysanthemicola* in chrysanthemum (Menzies and Colhoun 1976). In contrast, Pythium root rot and increased plant mortality in poinsettia were associated with high fertilization (Moorman 1986).

11.2.3 Soil Water

Irrigation is used frequently to increase crop production in arid and semiarid areas of the world where surface or underground water supplies are available. The management of soil water can reduce root disease severity in several crops. Detailed research has been done by several researchers on the influence of irrigation management on root diseases caused by *Phytophthora* spp. In California, Phytophthora root rot of tomato caused by *P. parasitica* was more severe if fields were furrow-irrigated for 4 to 8 h every 14 days or with alternating 4- to 8-h and 24-h irrigations every 14 days compared with 4- to 8-h irrigation every 28 days (Ristaino et al. 1988). In safflower, moisture stress prior to irrigation increased root rot caused by *P. drechsleri* (Zimmer and Urie 1967). Root rot induced by *P. megasperma* f. sp. *medicaginis* was increased in irrigated alfalfa, and the fewest diseased plants occurred in nonirrigated plots (Wilkinson and Millar 1982). In citrus and avocado, less frequent irrigations that allowed the soil to dry between irrigations reduced disease severity caused by *Phytophthora* spp. (Zentmeyer and Richards 1952; Feld et al. 1990). Trickle irrigation reduced Phytophthora root rot in citrus and avocado (Palti and Shoham 1983). Also, less frequent, heavy irrigations reduced pod rot in peanut caused by *Pythium* spp. in Israel compared with frequent, light irrigation (Frank 1967).

In Russia, frequent irrigation increased the incidence of *Verticillium dahliae* in cotton (Babaev et al. 1979). In contrast, scab of potato in Denmark was reduced by frequent, light irrigations, but incidence differed with soil type (Ostergaard and Nielsen 1979). Pink snowmold, caused by *F. nivale*, was reduced in bentgrass by afternoon compared with morning watering in California (Madison et al. 1960). In India, bulb rot in onion caused by *F. solani* increased if onion was irrigated after leaf fall, and infection was greater with flood (15 cm) than with light (5 cm) irrigation (Singh and Bhatnagar 1981). Frequent, light irrigations reduced Verticillium wilt of potato in Idaho, and wilt (early dying) was less with sprinkler than with gravity (furrow) irrigation (Guthrie 1960). Sudden wilt in melon in California was increased by a water deficit stress followed by flooding (Gubler and Grogan 1982). Trickle irrigation can reduce susceptibility of many crops to *Macrophomina phaseolina* that increases with heat and drought stress (Palti and Shoham 1983). Levels of common root rot of spring barley in Montana were similar in irrigated and dryland fields, but with irrigation plants compensated for stand reduction and yield was increased (Grey et al. 1991).

Stalk rot of corn, caused by several soilborne pathogens, can be influenced by irrigation management. In California, a mild early season water stress increased colonization of roots by *F. moniliforme* and led to more stalk rot as the crop matured (Schneider and Pendery 1983).

Irrigating frequently to keep the soil moisture near or above field capacity (matric suction <30 kPa) increased grain yield and stalk rot in both silty clay loam and sandy soils in Nebraska compared with corn grown with less frequent irrigations or nonirrigated (Sumner 1968). In Georgia, rainfall and irrigation analyzed over years and soil type accounted for 30% of the variation in stalk rot, but irrigation did not influence root disease severity at pollination (Sumner and Hook 1985). Stalk rot increased with decreasing rainfall during the vegetative period and with increased rainfall during the reproductive period (Sumner and Hook 1985). Different fungi cause root and stalk rot of corn in different soils, and that may account for some of the variable effects of irrigation on stalk rot in different areas.

Water from farm ponds, sumps, and catch basins that collect waste irrigation water or surface drainage from surrounding fields may be a source of inoculum of root rot pathogens. *Phytophthora* spp. were common in waste irrigation water collected in sumps for recycling to citrus orchards in Arizona (Thomson and Allen 1974). Farm ponds receiving runoff water are a source of root pathogenic fungi including *Pythium* spp., *Phytophthora* spp., *Fusarium* spp., and *Rhizoctonia* spp. (Gill 1970; Shokes and McCarter 1979). Plant beds irrigated with canal water in Washington became heavily infested with a variety of plant parasitic nematodes, while those receiving well water did not (Faulkner and Bolander 1970).

Excessive irrigation or poor soil drainage after heavy rains that cause soil saturation for long periods can deplete oxygen supply and increase root decay from *Phytophthora* spp. in citrus, cranberry, and azalea (Stolzy et al. 1965; Benson et al. 1982; Caruso 1989); tuber rot from unknown causes in potato (Easton and Nagle 1977), and Fusarium root rot in bean (Miller and Burke 1977). Frosheiser and Barnes (1973) noted that alfalfa could tolerate wet soils in the absence of plant pathogens. Improving drainage in cranberry bogs and planting azalea on raised beds decreased losses to *Phytophthora* spp. (Benson et al. 1982; Caruso 1989).

In contrast, the deleterious effects of excess moisture can be used to advantage in some areas by flooding soils to control soilborne pathogens between crops. Kelman and Cook (1977) observed that flooding fields for paddy rice coupled with the heavy use of organic fertilizers may be key factors in the general absence of soilborne pathogens in China. In Honduras, flood irrigation for periods up to 18 months eradicated *F. oxysporium* f. sp. *cubense*, the causal agent of Panama disease, from soils in banana production (Stover et al. 1953). Flooding of land for 2–6 months in the Everglades of Florida reduced galling and root injury by *M. incognita* (Thames and Stoner 1953). A single rotation with paddy rice controlled Verticillium wilt of cotton in California for 2–3 years and increased yields compared with continuous cotton (Pullman and DeVay 1981).

11.2.4 Soil Temperature

Less research has been done on the effects of soil temperatures on root rot than on the effects of soil water, but there are several examples of reductions in root disease severity by manipulating soil temperatures. In container-grown azaleas, Phytophthora root rot was greater in the shade where temperature was lower (and moisture was greater) than on plants in the sun (Benson 1986). Raising soil temperatures a few degrees with soil-heating cables during cold periods in the winter reduced the severity of snowmold of turfgrass caused by an unidentified basidiomycete in Alberta (Lebeau 1966). The lower, more constant soil temperatures found with ecofallow in Nebraska may contribute to reduced stalk rot in corn (Doupnik and Boosalis 1980). Lowering soil temperature (from an average of 30–35 °C to 25 °C in the geocarposphere around the developing peanut pod), with or without irrigation, reduced the incidence of *Aspergillus flavus* and virtually eliminated aflatoxin contamination in peanut in Georgia (Hill et al. 1983). Cylindrocladium black rot in peanut, caused by *C. crotalariae* is sensitive to temperature increases above 25 °C. Delayed planting, and planting rows on raised beds rather than on flat ground reduced disease severity and increased pod yield for peanut in North Carolina (Sidebottom and Beute 1989).

11.2.5 Integrated Pest Management

Frequently, it is necessary to combine several aspects of cultural management with host resistance and chemical control methods to achieve satisfactory economic control of root diseases. A few examples of integrated root disease management will be given. In cotton, delaying planting and planting on raised beds increases soil temperature, and planting only as deep as necessary to have adequate moisture will reduce seedling diseases (Minton and Garber 1983). In addition, tillage to destroy and bury plant debris and crop rotation with grass or grain crops help to reduce initial inoculum of many soilborne pathogens. In potato, integrating the management of soil fertility, seed treatments, crop rotation and green manuring crops, irrigation timing, and cultivar resistance are all required to reduce scab (Keinath and Loria 1989).

In bean, combinations of subsoiling to reduce soil compaction, delayed planting, narrow row spacing, and resistant cultivars reduced Fusarium root rot and increased yield of dry beans in Washington (Burke 1964; Silbernagel and Mills 1990). In the eastern USA, deep tillage with a moldboard plow, and the use of seed or soil fungicides, and crop rotations were beneficial in controlling *R. solani* and *Pythium* spp. in snap bean (Lewis et al. 1983; Sumner et al. 1986). Wilt and root rot of pea in India caused by *F. solani* and *F. oxysporum* were lessened by late planting, irrigation timing, and phosphate fertilization (Maheshwari et al. 1981). In Canada, the pea root-rot disease complex was best managed with raised seed beds (to reduce soil compaction), crop rotation, fall tillage with a chisel or plow, and selection of herbicides that did not predispose plants to root rot (Tu 1987). To control *S. rolfsii* in carrots for processing, deep plowing and fertilization with calcium fertilizer and ammonium carbonate gave economical control (Punja et al. 1986). In Israel, integrating onion into a crop rotation directly ahead of peanut greatly decreased disease caused by *S. rolfsii* (Zeidan et al. 1986).

11.3 Limitations

11.3.1 Climate

Although many different root diseases in different crops have been reduced by manipulating soil water and temperature, climatic conditions may nullify cultural practices designed to reduce root disease intensity. Occasional periods of intense rainfall may occur in dry climates, precluding adjustments in soil moisture by irrigation. In Georgia, symptoms of decline in spinach were associated with wet soil, prolonged cloudiness, and abrupt temperature changes that interacted with plant nutrition and soilborne pathogenic fungi and nematodes (Sumner et al. 1976).

Many plants germinate and emerge into healthy seedlings and juvenile plants when temperatures are most favorable for the growth of the host and are least favorable for the soilborne pathogens (Leach 1947). For instance, seedling diseases may be more severe in cucurbits, peanut, corn, beans, soybeans, and cotton in cool, wet soils, whereas cool season crops such as crucifers may be more susceptible to soilborne pathogens at high soil temperatures. Thus, sudden, unexpected climatic changes may increase pre- and postemergence damping-off even when recommended planting dates are followed.

11.3.2 Physical Properties of Soils

Physical properties of soils are much more difficult to change than chemical properties, and frequently soil compaction, elasticity, drainage, porosity, and composition limit the use of cultural practices in root disease control. Phymatotrichum root rot of cotton is limited to alkaline, calcareous, montmorillonite clay soils with extensive capacity for expanding and shrinking in Texas, New Mexico, Arizona, and Mexico, and is of no consequence in cotton in other areas of the world (Lyda 1978). Pythium root rot in holly was influenced by the proportion of sand in container-grown media (Gugino et al. 1973), but in landscape areas, growers sometimes can do very little to change the physical properties of the soil when planting small shrubs and grass, and virtually nothing when planting large shrubs and trees.

The advantages of moldboard plowing and in-row subsoiling to reduce soil compaction and improve soil drainage and aeration were described under tillage practices (Sect. 11.1.3). However, those tillage practices require large tractors and large amounts of energy, and the effects of the deep tillage usually only last through one or two crops.

11.3.3 Production Systems

The constraints of climate, land availability, and economics limit the use of cultural controls in many areas. Some crops desirable in rotations for disease control may not be profitable, or adapted to the environment. In northern temperate climates, the choice of crops is very limited, whereas in subtropical and tropical regions numerous crops are available for designing rotations to minimize root disease severity. However, various cropping systems may be impractical for a number of reasons. Purchasing equipment to grow a variety of crops may not be feasible economically. Planting and harvesting equipment for corn and grain crops are different than those for vegetables, sugar beets, cotton, or peanuts. Hay and livestock operations may not be compatible with field and vegetable production systems. Perennial crops require entirely different management than annuals, and there may be more restraints on alternative cultural practices.

Labor availability and market potential will vary in different parts of the world. Water may not be available for irrigation, or it may not be possible to control excess water in high rainfall areas. Government programs may influence the use of tillage practices to control soil erosion and the profit potential of different crops. Animal manures and different organic agricultural wastes may be available at a reasonable cost in some areas but not in others. The use of herbicides and other pesticides may preclude what crops can be used subsequently in a rotation.

11.3.4 Residue Management

The beneficial effects of specific tillage practices in relation to controlling soilborne pathogens were discussed previously. This section will deal primarily with the detrimental effects of plant residues and phytotoxins from decomposing plant residues. Cochrane (1949) and Janes (1955) observed poor growth and yield of vegetables following sweet corn and vetch, and suggested that decomposing organic matter in cool soils produced toxins. Later, it was shown that phytotoxins are produced by decomposition of plant residues (McCalla and Haskins 1964) and, frequently, the formation of phytotoxins was associate with anaerobic conditions in heavy, poorly aerated, cool, wet soils (Patrick et al. 1964).

The tillage of residues from green crops of barley, rye, wheat, broad bean, Sudan grass, and broccoli into soil before planting spinach and lettuce apparently interacted with soilborne pathogens and increased root diseases in California (Patrick et al. 1963). Multiple cropping with two to four crops per year may produce large amounts of plant debris that is difficult to incorporate if crops are planted 1–2 weeks after harvest (Sumner et al. 1981). Conservation tillage systems may increase allelopathy and the production of phytotoxins from surface plant debris (Martin et al. 1990). The potential detrimental effects of the interaction of phytotoxins with soilborne pathogens have been investigated in many crops. Nevertheless, more research is needed under a variety of climatic conditions to determine the limitations of conservation tillage and multiple cropping systems.

11.3.5 Survival of Inoculum

Cultural practices have limited use in controlling some soilborne pathogens because inoculum levels are not reduced by tillage practices and crop rotation. Wilts caused by *F. oxysporum* and root rots caused by *F. solani* require long rotations to reduce disease severity in susceptible hosts. It was thought previously that the fungi survived as chlamydospores, but it is possible that the pathogens may be able to reproduce on nonhost crops and weeds (nonpathogenic invasion of living roots, Menzies 1963).

Other pathogenic fungi considered as normal inhabitants in many soils are *Streptomyces* spp., *Macrophomina phaseolina*, *Pythium* spp., and *Phytophthora* spp. (Menzies 1963). Sclerotia of *Phymatotrichum omnivorum* were found in soil down to 150 cm deep after 11 years of continuous cropping to nonhost crops of corn, oats, or sorghum (Rogers 1942). More recent research with *P. omnivorum* shows that sorghum roots are resistant to the pathogen, but that sclerotia are formed in soil planted with sorghum (Rush et al. 1984). Ratoon cropping of sorghum that allowed roots to stay alive longer into the winter, increased Phymototrichum root rot in a subsequent crop of cotton compared to disk plowing or killing the sorghum with the herbicide glyphosate (Rush and Gerik 1989). Conidia of *Cochliobolus sativus* survived at least 3 years after wheat in a rotation and caused common root rot in a succeeding wheat crop (Ledingtham 1961). Microsclerotia of *Verticillium* spp. survive for many years in soil, and cultural practices alone are of questionable value in reducing disease in susceptible hosts (Huisman and Ashworth 1976).

11.4 Novel Approaches

11.4.1 Equipment

Modification of planting practices may promote crop growth and minimize losses to soilborne pathogens. Direct drilling of winter wheat into stubble or pasture reduced the severity of take-all in England, and the technique may inhibit the spread of the fungus in nondisturbed soil (Brooks and Dawson 1968). Equipment designed to mix anhydrous ammonia, nitrification inhibitors, and other soil additives and to apply them simultaneously may promote growth and reduce certain soilborne pathogens in specific areas of fields where nitrogen loss occurs (Huber et al. 1981). Nitrification inhibitors have been injected successfully into swine manure with anhydrous ammonia, and the equipment may be useful with other animal manure (Miller et al. 1986). Fluid drilling promotes earlier, more even emergence and growth of some high-value crops, and it may be possible to add biocontrols or fungicides directly to the gels (Sumner et al. 1988). Fluid drilling may help make conservation tillage a more acceptable alternative for vegetable growers.

11.4.2 Organic Amendments

Composts are commonly used in China (Kelman and Cook 1977) and in home gardens in the USA. Recently, composed hardwood bark has been used successfully to reduce soilborne pathogens in container-grown plants (Hoitink and Fahy 1986). Amendments with crucifer residues in fields reduced Aphanomyces root rot in pea in New Zealand (Chan and Close 1987) and the USA (Muehlchen et al. 1990), and amendments with chicken manure reduced wilt caused by *F. oxysporum* in tomato in Japan (Homma et al. 1979). In Taiwan, field amendments with a mixture of bagasse, rice husks, oyster shell powder, urea, calcium superphosphate, and mineral ash reduced the severity of radish yellows, cabbage yellows, and Fusarium wilt in watermelon. Profitability was increased in all crops (Sun and Huang 1985). A mixture of milled pine bark, fertilizers, and glycerine reduced damping-off of pine seedlings caused by *R. solani*, *P. aphanidermatum*, and *F. moniliforme* var. *subglutinans* in field plots in Georgia (Huang and Kuhlman 1991). Hardwood bark suppressed several species of plant parasitic nematodes in container-grown plants in a greenhouse (Malek and Gartner 1975).

Composted sewage sludge or municipal refuse may be a satisfactory soil amendment to reduce root infection by soilborne nematodes and fungi in greenhouse and field crops (Hunt et al. 1973; Lumsden et al. 1983) where it is available, but more research is needed on the efficacy of sludges. There may be problems in variability among batches, treatment facilities, and geographical areas. Stabilizing swine manure with nitrification inhibitors increased yield and reduced stalk rot in corn (Sutton et al. 1990). Such techniques may have application in reducing soilborne pathogens in other crops.

11.4.3 Soil Solarization

Covering soil with transparent polyethylene for prolonged periods increases soil temperatures in warm climates with extended periods of sunshine. Irrigation has frequently been beneficial because wet soil is more beneficial for increasing soil conduction of heat and destruction of soilborne pathogens than dry soil (Pullman et al. 1981). Solarization has been shown to control pink root of onion and other soilborne pathogens in Israel (Katan et al. 1980) and to reduce Verticillium wilt of cotton in California (Pullman et al. 1981) and Fusarium wilt of cotton in Israel (Katan et al. 1983). *Verticillium dahliae* was controlled in established pistachio nut groves to a depth of 120 cm with a cover of clear polyethylene mulch, even through soil temperatures were not raised enough below 30 cm depth to explain the death of the pathogen. Verticillium wilt was reduced from 6.3% of the trees in nonmulched areas to 1.5–1.7% in mulched areas (Ashworth and Gaona 1982).

Mixing cabbage residues into soil before solarization was more effective than solarization alone in control of cabbage yellows in California (Ramirex-Villapudua and Munnecke 1987). Covering the soil with plastic greatly aided in control of the pathogen because fungitoxic gases from the decomposing cabbage residues were trapped in the soil (Ramirez-Villapudua and Munnecke 1988). Mycorrhizal fungi survived soil solarization and colonized cotton roots (Pullman et al. 1981). The survival of mycorrhizae and beneficial soil antagonists such as *Trichoderma* spp., other fungi, and bacteria could be very important in promoting root growth and preventing reinvasion of soil by plant pathogens. In Egypt, soil solarization decreased soil salinity, and increased control of soilborne pathogens and yield of successive crops of broad beans, onions, tomatoes, or clover for two or three seasons in furrow-irrigated soils (Abdel-Rahim et al. 1988). Thus, solarization could be much more economical if the costs could be spread over several crops.

Machinery has been developed to easily and rapidly install transparent plastic in fields, and methods of removing and reusing plastic are being developed. In time, soil solarization for a few weeks may become the method of choice for controlling soilborne pathogens on numerous crops in climates with long periods of hot weather and bright sunshine. Unfortunately, in many temperate climates and areas with less sunshine, soil solarization may not be effective in controlling disease unless it is combined with soil amendments or soil fungicides or fumigants.

11.5 Conclusions

Cultural practices were the first methods used to control soilborne pathogens, and they were practiced long before mankind had heard of the germ theory. Crop rotations are used widely around the world to reduce the severity of root diseases, foliar pathogens, weeds, and insects. However, because of economic considerations, climate, and limitations within production systems, monocropping has become common with wheat, corn, peanut, some vegetables, and other crops in some areas. The extent to which crop rotations are used in the future will be based not only on economic considerations, but on environmental regulations and availability of pest control alternatives.

Because of concerns about soil erosion, moldboard plowing may not be an option for growers much longer. Some countries already have regulations on the amount of plant residues that must be left on the soil surface to reduce erosion. Because conservation tillage methods may be mandated in the next decade, crop rotation may become even more important in root disease management than it was early in this century.

It is obvious from the examples given in this chapter that some cultural methods have broad application, whereas some methods used in one location to ameliorate root diseases may not necessarily work in another. Management of planting date, plant spacing, soil water and fertility, and plant residues will continue to be important methods of reducing root diseases in some production systems. In the future, there will be increased concern about pollution of soil water, and cultural practices may be coupled with biological additives to replace many of the soil chemicals used today. Equipment is available to inject or incorporate liquid and solid wastes from animal production and municipal sewage systems into soil. These cultural practices, with or without biological additives, may be used increasingly to augment control of soilborne pathogens in fields surrounding the source of the wastes. Soil solarization, coupled with biological additives, may become very common in warm, dry climates with increased restrictions on the use of soil chemicals.

References

Abdel-Rahim MF, Satour MM, Mickail KY, El-Eraki SA, Grinstein A, Chen Y, Katan J (1988) Effectiveness of soil solarization in furrow-irrigated Egyptian soils. Plant Dis 72:143–146

Ashworth LJ Jr, Gaona SA (1982) Evaluation of clear polyethylene mulch for controlling Verticillium wilt in established pistachio nut groves. Phytopathology 72:243–246

Ashworth LJ Jr, Gaona SA, Surber E (1985) Verticillium wilt of pistachio: the influence of potassium nutrition on susceptibility to infection of *Verticillium dahliae*. Phytopathology 75:1091–1093

Babaev FA, Mukhtarova LS, Kembel ME (1979) Wilt resistance of cotton and metabolism characteristics under different irrigation conditions. Vesnik Sel'skokhozy aist vennol Nauki Azer baidzhana No 1, pp 72–79 (Rev Plant Pathol 60:547) (in Russian)

Baker KF (1962) Principles of heat treatment of soil and planting materials. J Aust Inst Agric Sci 28:118–126

Bell AA (1989) Role of nutrition in diseases of cotton. In: Engelhard AW (ed) Soilborne plant pathogens: management of diseases with macro- and microelements. APS Press, St Paul, pp 167–204

Bell DK, Sumner DR (1987) Survival of *Rhizoctonia solani* and other soilborne basidiomycetes in fallow soil. Plant Dis 71:911–915

Benson DM (1986) Relationship of soil temperature and moisture to development of Phytophthora root rot of azalea. J Environ Hort 4:112–115

Benson DM, Shew HD, Jones RK (1982) Effects of raised and ground level beds and pine bark on survival of azalea and population dynamics of *Phytophthora cinnamoni*. Can J Plant Pathol 4:278–280

Bockus WW, O'Connor JP, Raymond PJ (1983) Effect of residue management method on incidence of Cephalosporium stripe under continuous winter wheat production. Plant Dis 67:1323–1324

Bolley HL (1901) Flax wilt and flax sick soil. North Dak Agric Exp Stn Bull 50, pp 25–60

Boyle LW (1956) Fundamental concepts in the development of control measures for southern blight and root rot of peanuts. Plant Dis Rep 40:661–665

Broadfoot WC (1934) Studies on foot and root rot of wheat. III. Effect of crop rotation and cultural practice on the development of foot rot of wheat. Can J Res 10C:95–114

Brodie BB, Mai WF (1989) Control of the golden nematode in the United States. Annu Rev Phytopathol 27:443–461

Brooks DH, Dawson MG (1968) Influence of direct drilling of winter wheat on incidence of take-all and eyespot. Ann Appl Biol 61:57–64

Bruehl GW (1953) Pythium root rot of barley and wheat. USDA Tech Bull 1084, 24 pp

Bullock DG, Gascho GJ, Sumner DR (1990) Grain yield, stalk rot, and mineral concentration of fertigated corn as influence by NPK. J Plant Nutr 13:915–937

Burke DW (1964) Time of planting in relation to disease incidence and yields of beans in central Washington. Plant Dis Rep 48:789–793

Burke DW (1965) Plant spacing and Fusarium root rot of beans. Phytopathology 55:757–759

Burke DW, Kraft JM (1974) Responses of beans and peas to root pathogens accumulated during monoculture of each crop species. Phytopathology 64:546–549

Burke DW, Miller DE, Hommes LD, Barker AW (1972) Counteracting bean root rot by loosening the soil. Phytopathology 62:306–309

Caruso FL (1989) Phytophthora root rot of cultivated cranberry in Massachusetts. Acta Hortic 241:307–311

Chan MKY, Close RC (1987) Aphanomyces root rot of peas. 3. Control by the use of cruciferous amendments. N Z J Agric Res 30:225–233

Chase AR, Poole RT (1984) Investigation into the roles of fertilizer level and irrigation frequency on growth, quality, and severity of Pythium root rot of *Peperomia obtusifolia*. J Am Soc Hortic Sci 109:619–622

Christensen NW, Brett M (1985) Chloride and liming effects on soil nitrogen form and take-all of wheat. Agron J 77:157–163

Clayton EE, Gaines JG, Smith TE, Shaw KJ, Graham TW (1944) Control of flue-cured tobacco root diseases by crop rotation. US Dep Agric Farmers Bull 1952, 12 pp

Cochrane VW (1949) Crop residues as causative agents of root rots of vegetables. Conn Agric Exp Stn (New Haven) Bull 526, 34 pp

Conner RL, Lindwall CW, Atkinson TG (1987) Influence of minimum tillage on severity of common root rot in wheat. Can J Plant Pathol 9:56–58

Csinos AS, Bell DK (1989) Pathology and nutrition in the peanut pod rot complex. In: Engelhard AW (ed) Soilborne plant pathogens: management of diseases with macro- and microelements. APS Press, St Paul, pp 124–136

Curl EA (1963) Control of plant diseases by crop rotation. Bot Rev 29:413–479

Diehl JA, Tinline RD, Kochlann RA, Shipton PJ, Rovira AD (1982) The effect of fallow periods on common root rot of wheat in Rio Grand do Sul, Brazil. Phytopathology 72:1297–1301

Dobson RL, Garbielson RL, Baker AS, Bennett L (1983) Effects of lime particle size and distribution and fertilizer formulation on clubroot disease caused by *Plasmodiophora brassicae*. Plant Dis 67:50–52

Doupnik B Jr, Boosalis MG (1980) Ecofallow – a reduced tillage system – and plant diseases. Plant Dis 64:31–35

Drayton FL (1929) Bulb growing in Holland and its relation to disease control. Sci Agric 9:494–509

Easton GD, Nagle ME (1977) Tuber rot – the result of ponding of irrigation water. Plant Dis Rep 61:1064–1066

Engelhard AW (ed) (1989) Soilborne plant pathogens: management of diseases with macro- and microelements. APS Press, St Paul, 217 pp

Faulkner LR, Bolander WK (1970) Agriculturally polluted irrigation water as a source of plant parasitic nematode infestation. J Nematol 2:368–374

Feld SJ, Menge JA, Stolzy LH (1990) Influence of drip and furrow irrigation on Phytophthora root rot of citrus under field and greenhouse conditions. Plant Dis 74:21–27

Ferriss RS (1984) Effects of microwave oven treatment on microorganisms in soil. Phytopathology 74:121–126

Frank ZR (1967) Effect of irrigation procedure on Pythium rot of groundnut pods. Plant Dis Rep 51:414–416

Frosheiser FI, Barnes DK (1973) Field and greenhouse selection for Phytophthora root rot resistance in alfalfa. Crop Sci 13:735–738

Garren KH (1961) Control of *Sclerotium rolfsii* through cultural practices. Phytopathology 51:120–124

Garren KH (1964) Landplaster and soil rot of peanut pods in Virginia. Plant Dis Rep 48:349–352

Gill DL (1970) Pathogenic *Pythium* from irrigation ponds. Plant Dis Rep 54:1077

Gill DL (1972) Effect of gypsum and dolomite on Pythium diseases of seedlings. J Am Soc Hortic Sci 97:467–471

Goss RW, Afanasiev MM (1938) Influence of rotations under irrigation on potato scab, Rhizoctonia and Fusarium wilt. Nebr Agric Exp Stn Bull 317, 18 pp

Gram E (1960) Quarantines. In: Horsfall JG, Dimond AE (eds) Plant pathology, an advanced treatise, vol 3. The diseased population epidemics and control. Academic Press, New York, pp 313–356

Grandfield CO, Lefebure CL, Metzger WH (1935) Relation between fallowing and the damping-off of alfalfa seedlings. J Am Soc Agron 27:800–806

Green RJ (1967) Control of Verticillium wilt of peppermint by crop rotation sequences. Plant Dis Rep 51:449–453

Grey WE, Engel RE, Mathre DE (1991) Reaction of spring barley to common root rot under several moisture regimes: effect on yield components, plant stand, and disease severity. Can J Plant Sci 71:461–472

Gubler WD, Grogan RG (1982) Factors affecting sudden wilt of melon in California. Phytopathology 72:986 (Abstr)

Gugino JL, Pokorny FA, Hendrix FF Jr (1973) Population dynamics of *Pythium irregulare* Buis. in container-plant production as influenced by physical structure of media. Plant Soil 39:591–602

Guthrie JW (1960) Early dying (Verticillium wilt) of potatoes in Idaho. Idaho Agric Exp Stn Res Bull 45, 24 pp

Hallock DL, Garren KH (1968) Pod breakdown, yield, and grade of Virginia type peanuts as affected by Ca, Mg, and K sulfates. Agron J 60:253–257

Hendrix FF Jr, Powell WM (1970) Control of root pathogens in peach decline sites. Phytopathology 60:16–19

Higashida S, Ohsaki I, Narita Y (1982) Effects of crop rotation on onion yields and its microbial factors. Bull Hokkaido Pref Agr Exp Stn No 48, pp 1–9

Hill RA, Blankenship PD, Cole RJ, Sanders TH (1983) Effects of soil moisture and temperature on preharvest invasion of peanuts by the *Aspergillus flavus* group and subsequent aflatoxin development. Appl Environ Microbiol 45:628–633

Hoes JA, Huang HC (1985) Effect of between-row and within-row spacing on development of Sclerotinia wilt and yield of sunflower. Can J Plant Pathol 7:98–102

Hoitink HAJ, Fahy PC (1986) Basis for the control of soilborne plant pathogens with composts. Annu Rev Phytopathol 24:93–114

Homma Y, Kubo C, Ishii M, Ohata K (1979) Effect of organic amendments to soil on suppression of disease severity of tomato Fusarium wilt. Bull Shikoku Agric Exp Stn 34, pp 89–101

Huang JW, Kuhlman EG (1991) Formulation of a soil amendment to control damping-off of slash pine seedlings. Phytopathology 81:163–170
Huber DM (1989) The role of nutrition in the take-all disease of wheat and other small grains. In: Engelhard AW (ed) Soilborne plant pathogens: management of diseases with macro- and microelements. APS Press, St Paul, MN, pp 46–74
Huber DM (1991) The use of fertilizers and organic amendments in the control of plant disease. In: Pimental D (ed) Handbook of pest management in agriculture. CRC Press, Boca Raton, pp 405–494
Huber DM, Karamesines PD, Warren HL, Nelson DW (1981) Equipment to mix chemical additives with anhydrous ammonia during application to soil. Agron J 73:1046–1048
Huber DM, Watson RD (1974) Nitrogen form and plant disease. Annu Rev Phytopathol 12:139–159
Huisman OC, Ashworth LJ Jr (1976) Influence of crop rotation on survival of *Verticillium albo-atrum* in soils. Phytopathology 66:978–981
Hunt PG, Hortense CC, Smart GC Jr (1973) Responses of plant parasitic and saprophytic nematode populations to composted municipal refuse. J Environ Qual 2:264–266
Hussey RS, Roncadori RW (1977) Vertical distribution of soil microorganisms following subsoiling in a cotton management system. Phytopathology 67:783–786
Janes BE (1955) Vegetable rotation studies in Connecticut. III. Effect of sweet corn and vetch on the growth of several crops which follow. Am Soc Hortic Sci Proc 65:324–330
Johnson AW, Motsinger RE (1990) Effects of planting date, small grain crop destruction, fallow, and soil temperature on the management of *Meloidogyne incognita*. J Nematol 22:348–355
Johnson J (1946) Soil-steaming for disease control. Soil Sci 61:83–91
Jones JP, Engelhard AW, Woltz SS (1989) Management of Fusarium wilt of vegetables and ornamentals by macro- and microelement nutrition. In: Engelhard AW (ed) Soilborne plant pathogens: management of diseases with macro- and microelements. APS Press, St Paul, pp 18–32
Kao CW, Ko WH (1986) The role of calcium and microorganisms in suppression of cucumber damping-off caused by *Pythium splendens* in a Hawaiian soil. Phytopathology 76:221–225
Katan J, Fishler G, Grinstein A (1983) Short- and long-term effects of soil solarization and crop sequence on Fusarium wilt and yield of cotton in Israel. Phytopathology 73:1215–1219
Katan J, Rotem I, Finkel Y, Daniel J (1980) Solar heating of the soil for the control of pink root and other soilborne diseases of vegetables. Phytoparasitica 8:39–50
Keinath AP, Loria R (1989) Management of common scab of potato with plant nutrients. In: Engelhard AW (ed) Soilborne plant pathogens: management of diseases with macro- and microelements. APS Press, St Paul, pp 152–166
Kelman A, Cook RJ (1977) Plant pathology in the People's Republic of China. Annu Rev Phytopathol 15:409–429
Kirby RS (1922) The take-all disease of cereals and grasses. Phytopathology 12:66–88
Koch DO, Jeger MJ, Gerik TJ, Kenerley CM (1987) Effects of plant density on progress of Phymatotrichum root rot in cotton. Phytopathology 77:1657–1662
Koehler B, Holbert JR (1930) Corn diseases in Illinois, their extent, nature, and control. Ill Agric Exp Stn Bull 354, 164 pp
Kraft JM, Wilkins DE (1989) The effects of pathogen numbers and tillage on root disease severity, root length, and seed yields in green peas. Plant Dis 73:884–887

Krigsvold DT, Garren KH, Griffin GJ (1977) Importance of peanut field cultivation and soybean cropping in the spread of *Clyindrocladium crotalariae* within and among peanut fields. Plant Dis Rep 61:495–499
Latin RX, Harder RW, Wiese MV (1982) Incidence of Cephalosporium stripe as influenced by winter wheat management practices. Plant Dis 66:229–230
Leach LD (1947) Growth rates of host and pathogen as factors determining the severity of pre-emergence damping-off. J Agric Res 75:161–179
Lebeau JB (1966) Pathology of winter injured grasses and legumes in Western Canada. Crop Sci 6:23–25
Ledingham RJ (1961) Crop rotations and common root rot in wheat. Can J Plant Sci 41:479–486
Ledingham RJ, Sallans BJ, Wenhardt A (1960) Influence of cultural practices on incidence of common root rot in wheat in Saskatchewan. Can J Plant Sci 40: 310–316
Lewis JA, Lumsden RD, Papavizas GC, Kantzes JG (1983) Integrated control of snap bean diseases caused by *Pythium* spp. and *Rhizoctonia solani*. Plant Dis 67:1241–1244
Lumsden RD, Lewis JA, Millner PD (1983) Effect of composted sewage sludge on several soilborne pathogens and diseases. Phytopathology 73:1543–1548
Lyda SD (1978) Ecology of *Phymatotrichum omnivorum*. Annu Rev Phytopathol 16:193–209
Lyle EW, Dunlap AA, Hill HO, Hargrove BD (1948) Control of cotton root rot by sweetclover in rotation. Texas Agric Exp Stn Bull 699, 21 pp
MacNish GC (1988) Changes in take-all (*Gaeumannomyces graminis* var. *tritici*), Rhizoctonia root rot (*Rhizoctonia solani*) and soil pH in continuous wheat with annual application of nitrogenous fertilizer in Western Australia. Aust J Exp Agric 28:333–341
Madison JH, Petersen LJ, Hodges TK (1960) Pink snowmold on bentgrass as affected by irrigation and fertilizer. Agron J 52:591–592
Maheshwari SK, Gupta JS, Jhooty JS (1981) Effect of various cultural practices on the incidence of the wilt and root rot of pea. Ind J Agric Res 15:149–151
Malek RB, Gartner JB (1975) Hardwood bark as a soil amendment for suppression of plant parasitic nematodes on container grown plants. Hortic Sci 10:33–35
Martin VL, McCoy EL, Dick WA (1990) Allelopathy of crop residues influences corn seed germination and early growth. Agron J 82:555–560
Mathieson JT, Rush CM, Bordovsky D, Clark LE, Jones OR (1990) Effects of tillage on common root rot of wheat in Texas. Plant Dis 74:1006–1008
McCalla TM, Haskins FA (1964) Phytotoxic substances from soil microorganisms and crop residues. Bacteriol Rev 28:181–207
Menzies JD (1963) Survival of microbial plant pathogens in soil. Bot Rev 29:79–122
Menzies SA, Colhoun J (1976) Control of Phoma root rot of chrysanthemum by the use of fertilizers. Trans Br Mycol Soc 67:455–462
Miller DE, Burke DW (1977) Effect of temporary excessive wetting on soil aeration and Fusarium root rot of beans. Plant Dis Rep 61:175–179
Miller ER, Jones DD, Huber DM, Sutton AL (1986) Technique and equipment for nitrification inhibitor addition to swine manure. Appl Eng Agric 2:186–189, 192
Minton EB, Garber RH (1983) Controlling the seedling disease complex of cotton. Plant Dis 67:115–118
Minton NA, Parker MB (1975) Interaction of four soybean cultivars with subsoiling and a nematicide. J Nematol 7:60–64
Mitchell DJ, Shaw LN (1975) Eradication of plant pathogenic fungi in soil and nursery potting mixtures with a mobile continuous soil pasteurizer. Plant Soil 42:591–600

Moorman GW (1986) Increased plant mortality caused by Pythium root rot of poinsettia associated with high fertilization rates. Plant Dis 70:160–162

Muehlchen AM, Rand RE, Parker JL (1990) Evaluation of crucifer green manures for controlling Aphanomyces root rot of peas. Plant Dis 74:651–654

Nelson BD, Hertsgaard DM, Holley RC (1989) Disease progress of Sclerotinia wilt of sunflower at varying plant populations, inoculum densities, and environments. Phytopathology 79:1358–1363

Ostergaard SP, Nielsen S (1979) Control of potato scab (*Streptomyces scabies*) by irrigation. Tidsskr Planteavl 83:201–204 (Rev Plant Pathol 59:33)

Palti J, Shoham H (1983) Trickle irrigation and crop disease management. Plant Dis 67:703–705

Pammel LH (1888) Root rot of cotton or "cotton blight". Texas Agric Exp Stn Annu Rep 1:50–65

Patrick ZA, Toussoun TA, Synder WC (1963) Phytotoxic substances in arable soils associated with decomposition of plant residues. Phytopathology 53:152–161

Patrick ZA, Toussoun TA, Koch LW (1964) Effect of crop-residue decomposition products on plant roots. Annu Rev Phytopathol 2:267–292

Piening LJ, Orr D (1988) Effects of crop rotation on common root rot of barley. Can J Plant Pathol 10:61–65

Piening LJ, Walker DR, Dagenais M (1983) Effect of fertilizer or root rot of barley on stubble and fallow land. Can J Plant Pathol 5:136–139

Pullman GS, DeVay JE (1981) Effect of soil flooding and paddy rice culture on the survival of *Verticillium dahliae* and incidence of Verticillium wilt in cotton. Phytopathology 71:1285–1289

Pullman GS, DeVay JE, Garber RH, Weinhold AR (1981) Soil solarization: effects on Verticillium wilt of cotton and soilborne populations of *Verticillium dahliae*, *Pythium* spp., *Rhizoctonia solani*, and *Thielaviopsis basicola*. Phytopathology 71:954–959

Punja ZK, Carter JD, Campbell GM, Rossell EL (1986) Effects of calcium and nitrogen fertilizers, fungicides, and tillage practices on incidence of *Sclerotium rolfsii* on processing carrots. Plant Dis 70:819–824

Ramirex-Villapudua J, Munnecke DE (1987) Control of cabbage yellows (*Fusarium oxysporium* f. sp. *conglutinans*) by solar heating of field soils amended with dry cabbage residues. Plant Dis 71:217–221

Ramirex-Villapudua J, Munnecke DE (1988) Effect of solar heating and soil amendments of cruciferous residues on *Fusarium oxysporum* f. sp. *conglutinans* and other organisms. Phytopathology 78:289–295

Ratcliffe GT (1934) Cotton root rot as affected by crop rotation and tillage at San Antonio, Texas. US Dep Agric Tech Bull 436, 31 pp

Reinking DA (1942) Distribution and relative importance of various fungi associated with pea root rot in commerical pea-growing areas in New York. NY Agric Exp Stn Tech Bull 264, 43 pp

Reis EM, Cook RJ, McNeal BL (1982) Effect of mineral nutrition on take-all of wheat. Phytopathology 72:224–229

Ristaino JB, Duniway JM, Marois JJ (1988) Influence of frequency and duration of furrow irrigation on the development of Phytophthora root rot and yield in processing tomato. Phytopathology 78:1701–1706

Rogers CH (1942) Cotton root rot studies with special reference to sclerotia, cover crops, rotations, tillage, seeding rates, soil fungicides, and effects on seed quality. Tex Agric Exp Stn Bull 614, 45 pp

Rush CM, Gerik TJ (1989) Relationship between post harvest management of grain sorghum and Phymatotrichum root rot in the subsequent cotton crop. Plant Dis 73:304–305

Rush CM, Gerik TJ, Lyda SD (1984) Interactions between *Phymatotrichum omnivorum* and *Sorghum bicolor*. Plant Dis 68:500–501
Rush CM, Winter SR (1990) Influence of previous crops on Rhizoctonia root and crown rot of sugar beet. Plant Dis 74:421–425
Russell RC (1930) Field studies of take-all in Saskatchewan. Sci Agric 10:654–668
Schneider RW (1985) Suppression of Fusarium yellows of celery with potassium, chloride, and nitrate. Phytopathology 75:40–48
Schneider RW, Pendery WE (1983) Stalk rot of corn: mechanism of predisposition by an early season water stress. Phytopathology 73:863–871
Scholte K (1989) Effects of crop rotation on the incidence of soil-borne pathogens and the consequences for potato production. PhD Thesis, Wageningen, 143 pp
Schuster ML, Harris L (1960) Incidence of Rhizoctonia crown rot of sugar beets in irrigated crop rotation. J Am Soc Sugar Beet Technol 11:128–136
Sequeira L (1962) Control of bacterial wilt of bananas by crop rotation and fallowing. Trop Agric Lond 39:211–217
Sewell MC, Melchers LE (1924) The effect of rotation and tillage on root-rot of wheat in Kansas, 1920–1924. J Am Soc Agron 16:768–771
Shefelbine PA, Mathre DE, Carlson G (1986) Effects of chloride fertilizer and systemic fungicide seed treatments on common root rot of barley. Plant Dis 70:639–642
Shokes FM, McCarter SM (1979) Occurrence, dissemination, and survival of plant pathogens in surface irrigation ponds in southern Georgia. Phytopathology 69: 510–516
Sidebottom JR, Beute MK (1989) Control of Cylindrocladium black rot of peanut with cultural practices that modify soil temperature. Plant Dis 73:672–676
Silbernagel MJ, Mills LJ (1990) Genetic and cultural control of Fusarium root rot in bush snap beans. Plant Dis 74:61–66
Singh JP, Bhatnagar DK (1981) Note on the irrigation of onion as a factor predisposing its bulbs to infection by *Fusarium solani* (Martius) Appel and Wollenweber in storage. Indian J Agric Sci 51:686–687
Sitterly WR (1962) Calcium nitrate for field control of tomato southern blight in South Carolina. Plant Dis Rep 46:492–494
Smiley RW, Cook RJ (1973) Relationship between take-all of wheat and rhizosphre pH in soils fertilized with ammonium vs nitrate nitrogen. Phytopathology 63: 882–890
Specht LP, Leach SS (1987) Effects of crop rotation on Rhizoctonia disease of white potato. Plant Dis 71:433–437
Stephenson MG, Parker MB, Gaines TP, Csinos AS (1987) Manganese and soil pH effects on yield and quality of flue-cured tobacco. Tob Sci 31:109–113
Stolzy LH, Letey J, Klotz LJ, Labanauskas CK (1965) Water and aeration as factors in root decay of *Citrus sinensis*. Phytopathology 55:270–275
Stover RH, Thorton NC, Dunlap VC (1953) Flood fallowing for eradication of *Fusarium oxysporum* f. *cubense*. I. Effect of flooding on fungus flora of clay loam soils in the Ulua Valley, Honduras. Soil Sci 76:225–238
Sumner DR (1968) The effect of soil moisture on corn stalk rot. Phytopathology 58:761–765
Sumner DR (1976) Etiology and control of root rot of summer squash. Plant Dis Rep 60:923–927
Sumner DR (1987) Root disease in crops following legumes in conservation tillage systems. In: Power JF (ed) Proc Natl Conf on the Role of legumes in conservation tillage systems. Athens, GA, April 27–29. Soil Cons Soc Am, Ankeny, IA, pp 74–75
Sumner DR, Doupnik B Jr, Boosalis MG (1981) Effects of reduced tillage and multiple cropping on plant diseases. Annu Rev Phytopathol 19:167–187

Sumner DR, Dowler CC (1983) Herbicide, planting date, and root disease interactions in corn. Plant Dis 67:513–517
Sumner DR, Dowler CC, Johnson AW, Glaze NC, Phatak SC, Chalfant RB, Epperson JE (1983) Root diseases of cucumber in irrigated, multiple-cropping system with pest management. Plant Dis 67:1071–1075
Sumner DR, Hook JE (1985) Irrigation management and root and stalk rot of corn. Plant Dis 69:239–243
Sumner DR, Ghate SR, Phatak SC (1988) Seedling diseases of vegetables in conservation tillage with soil fungicides and fluid drilling. Plant Dis 72:317–320
Sumner DR, Kays SJ, Johnson AW (1976) Etiology and control of root diseases of spinach. Phytopathology 66:1267–1273
Sumner DR, Smittle DA, Threadgill ED, Johnson AW, Chalfant RB (1986) Interactions of tillage and soil fertility with root diseases in snap bean and lima bean in irrigated multiple-cropping systems. Plant Dis 70:730–735
Sun DK, Huang JW (1985) Formulated soil amendment for controlling Fusarium wilt and other soilborne diseases. Plant Dis 69:917–920
Sutton AL, Huber DM, Jones DD, Kelly DT (1990) Use of nitrification inhibitors with summer applications of swine manure. Appl Eng Agric 6:296–300
Taylor RG, Jackson TL, Powelson RL, Christensen NW (1983) Chloride, nitrogen form, lime, and planting date effects on take-all root rot of winter wheat. Plant Dis 67:1116–1120
Thames WH Jr, Stoner WN (1953) A preliminary trial of lowland culture rice in rotation with vegetable crops as a means of reducing root-knot nematode infestations in the everglades. Plant Dis Rep 39:187–192
Thomson SV, Allen RM (1974) Occurrence of *Phytophthora* species and other potential plant pathogens in recycled irrigation water. Plant Dis Rep 58:945–949
Timm CA, Goos RJ, Johnson BE, Sobolick FJ, Stack RW (1986) Effect of potassium fertilizers on malting barley infected with common root rot. Agron J 78: 197–200
Tu JC (1987) Integrated control of the pea root rot disease complex in Ontario. Plant Dis 71:9–13
Warren HM, Huber DM, Nelson DW, Mann OW (1975) Stalk rot incidence and yield of corn as affected by inhibiting nitrification of fall-applied ammonium. Agron J 67:655–660
Wilhelm NS, Graham RD, Rovira AD (1988) Application of different sources of manganese sulfate decreases take-all (*Gaeumannomyces graminis* var. *tritici*) of wheat grown in a manganese deficient soil. Aust J Agric Res 39:1–10
Wilkinson HT, Millar RL (1982) Effects of soil temperature and moisture on activity of *Phytophthora megasperma* f. sp. *medicaginis* and alfalfa root rot in the field. Phytopathology 72:790–793
Williams PH (1979) Vegetable crop protection in the People's Republic of China. Annu Rev Phytopathol 17:311–324
Zeidan O, Elad Y, Hadar Y, Chet I (1986) Integrating onion in crop rotation to control *Sclerotium rolfsii*. Plant Dis 70:426–428
Zentmeyer GA, Richards SJ (1952) Pathogenicity of *Phytophthora cinnamomi* to avocado trees and the effect of irrigation on disease development. Phytopathology 42:35–37
Zimmer DE, Urie AL (1967) Influence of irrigation and soil infestation with strains of *Phytophthora drechsleri* on root rot resistance of safflower. Phytopathology 57:1056–1059

Subject Index

Springer-Verlag and the Environment

MIX
Papier aus verantwortungsvollen Quellen
Paper from responsible sources
FSC® C105338

If you have any concerns about our products,
you can contact us on
ProductSafety@springernature.com

In case Publisher is established outside the EU,
the EU authorized representative is:
Springer Nature Customer Service Center GmbH
Europaplatz 3, 69115 Heidelberg, Germany

Printed by Libri Plureos GmbH
in Hamburg, Germany